Dynamic Neuroscience

Zhe Chen • Sridevi V. Sarma
Editors

Dynamic Neuroscience

Statistics, Modeling, and Control

Springer

Editors
Zhe Chen
School of Medicine
New York University
New York, NY, USA

Sridevi V. Sarma
Institute for Computational Medicine
Johns Hopkins University
Baltimore, MD, USA

ISBN 978-3-030-10139-8 ISBN 978-3-319-71976-4 (eBook)
https://doi.org/10.1007/978-3-319-71976-4

Softcover re-print of the Hardcover 1st edition 2018

Printed on acid-free paper

This Springer imprint is published by Springer Nature
The registered company is Springer International Publishing AG
The registered company address is: Gewerbestrasse 11, 6330 Cham, Switzerland

Dedicated to the 60th birthday of
Professor Emery Neal Brown

Preface

A pre-preface note: This book is dedicated to Professor Emery N. Brown in honor of his great scientific contributions and his 60th birthday (in parallel with a symposium held in MIT on September 14–15, 2017). Noticing diverse scientific themes of Emery's work, the editors have invited Professor Rob Kass and Professor Ralph Lydic to write their personal reflections on Emery's scientific contributions in the fields of neurostatistics and anesthesiology, respectively.

Reflections from Rob Kass

It is a great pleasure to begin this volume by offering a short perspective on Emery N. Brown. There is much to admire in Emery's research, and also much to learn from his unique path to scientific leadership. As Emery's students and colleagues know, but some readers of this book may not, Emery combined his M.D., including a fellowship and board certification in anesthesiology, with a Ph.D. in statistics, and his professional life reflects this unusual training: he is in the operating room one day every week, and runs his human anesthesiology studies, at Massachusetts General Hospital (MGH), yet he teaches classes in statistics and has his main office in the Department of Brain and Cognitive Sciences at MIT. Together with his many collaborators, Emery has produced important new methods for analysis of neural data, advancing knowledge in a variety of scientific subdisciplines in the process, while his research on the mechanisms by which anesthesia produces altered states of arousal is putting the subject on a firm foundation. Emery is the only person to hold chaired professorships at both Harvard Medical School and MIT.

One clue in trying to understand Emery's trajectory comes from his undergraduate major in applied mathematics, which gave him two things: knowledge of elementary methods in dynamical systems and an appreciation for the importance of statistics. Emery brought these strands together in his PhD thesis work on time series analysis for circadian rhythms data. In addition, Emery recognized the potential for exploiting the connection between dynamical system models represented by differential equations, in continuous time, and certain time series models—including, especially, state-space models—that can be considered discrete analogues, where differential equations are replaced by difference equations. It is hard to overstate the importance of Emery's fundamental contribution, even though it will someday seem obvious: dynamic neural phenomena should be described using dynamic methods, which includes both mathematical models to aid understanding and statistical models to guide data analysis. Emery has made this big picture contribution, which I see as his enduring technical legacy, not through a single discovery, but through a steady stream of cases that, taken together, form a compelling whole. In retrospect, it is not hard to see the arc from Emery's scientific starting point in circadian rhythms, which he identified as a fertile subject in which differential equations and statistical inference could be brought together to move the science forward, to his numerous creative applications of state-space modeling in the analysis of neural data, to his recent discoveries, enabled by biophysical and statistical modeling, that are establishing a mechanistic neuroscience of anesthesia.

Beyond the seemingly prescient choices he made in his education, Emery has had a rare ability to adopt good strategies for achieving worthy goals. At many key points in his career, he has had the patience and confidence to turn away from short-term gain and has, instead, invested his time and energy in learning what he needed to know, and creating the environment necessary to achieve results he recognized as truly important. This has required not only foresight and dogged persistence but also a kind of dexterity in sidestepping the inevitable pitfalls. Emery himself emphasized this during an informal talk he gave many years ago at Carnegie Mellon University, to a group of underrepresented minority students: he spoke of some unpleasant childhood experiences, which began while public schools in his home state of Florida were segregated, and then continued when, after desegregation, he faced instances of blatant racial discrimination. During his remarks Emery stressed that while it is natural to be angered at such injustices, it is more productive to find ways around the obstacles they created; he added that everyone faces obstacles of various kinds, but the most successful among us are those who stay mindful of their goals and stay focused on finding paths to achievement. Those words have stayed with me as I have witnessed Emery's uncanny mastery of circumventing obstructions, which come in so many different forms. There is a limit to which each of us can be like our heroes, but we would all be wise to take the advice Emery offered to those students, and, at least in this respect, try to emulate his success. Finally, I'd like to add that, from many, many conversations, I know how deeply Emery cares about training at all levels. On the one hand, he has put great energy into his courses at MIT, and into specialized short courses around the world. On the other hand, he demonstrates sincere concern for all his many students and postdocs. This book is a great testimony to what Emery has done for his trainees, as they review some of their recent scientific advances, thereby giving back to him something of lasting value, in the best intellectual spirit, and with continuing affection.

Reflections from Ralph Lydic

It is my great pleasure to highlight selected aspects of Emery Brown's many contributions to anesthesia research. The editorial directives were to provide a brief, personal perspective. Of primary importance is the fact that the future of every discipline depends on the acquisition of new knowledge via recruitment and retention of investigators. This Festschrift, organized by Emery's collaborators and former students, effectively conveys the admiration and affection with which Emery is regarded. Emery is uniquely successful at providing novel and clinically relevant scientific discoveries while promoting his colleagues. Emery's ability to create an extended scientific family powerfully supplants the dehumanizing view of faculty and staff as a "human resource." Emery became an Assistant Professor of anesthesiology in 1993 and during the ensuing 24 years he has mentored more than 200 undergraduates, graduate students, and fellows. Mentees include 28 anesthesiology residents and faculty, some of whom remain actively engaged in research. Emery's role as a mentor has generated an incalculably positive "return on investment" for academic anesthesiology.

I view Emery's anesthesia-related research as a line segment originating decades ago from his successful mathematical modeling of sleep and circadian rhythms. Several of Emery's early papers are regarded as classics today. I first met Emery in the mid-1980s when he and I were associated with different Harvard training programs. It was kindness of Robert (Bob) McCarley that Emery and I had our first scientific exchange. A decade earlier Bob, using Lotka-Volterra type equations, had developed the first cellularly based, mathematical model of the mammalian sleep cycle. About the same time, Emery was independently developing dynamical systems models and nonlinear filtering techniques to characterize the oscillatory properties of biological rhythms. I was recording the discharge of dorsal raphe neurons across the sleep/wake cycle. We were keen to combine our respective approaches. In what may be an example of excessive synaptic delay, our collaborative plans finally matured in 2010 when Emery invited Nicholas Schiff and me to coauthor a review for the *New England Journal of Medicine*. Relative to the present volume, that review presents a structurally based story model concerning the neural networks that generate states of anesthesia, sleep, and coma.

Anesthesia and sleep are distinctly different states of consciousness identified by constellations of physiological and behavioral traits. Emery's early mathematical modeling of sleep is thematically related to his current anesthesiology research. The present prologue was written 117 years after the first demonstration of ether anesthesia. In contrast, modern humans have existed for about 160,000 years. The ratio of these two numbers illustrates the small percentage of time that humans have been able to reliably manage surgical pain. At the beginning of the current millennium, the 6 June 2000 issue of the *New England Journal of Medicine* listed anesthesia as "one of the most important medical developments of the past thousand years." At present, there are approximately 60,000 cases of general anesthesia each day in the United States. When delivered by expert caregivers, anesthesia is remarkably safe. In no case, however, is it understood how anesthetics eliminate waking consciousness. Understanding the mechanisms of anesthetic action will be a significant scientific advance. At the level of clinical care, such an understanding will help develop countermeasures that diminish or eliminate unwanted anesthetic side effects such as nausea and vomiting, as well as inadequate pain relief currently experienced by about half of postsurgical patients. All of the desired effects produced by general anesthetics, sedatives, and opiates are products of the nervous system. Thus, anesthesiology can be viewed as a branch of clinical neuroscience. In addition, Emery's anesthesia research has the potential to advance neurology, sleep disorders medicine, pain medicine, and biological psychiatry. An enhanced understanding of the neuronal mechanisms of anesthesia also is likely to contribute to consciousness studies.

Directly relevant to the present volume is Emery's creative application of computational biology to problems relevant for anesthesiology. One challenge for such a research program is complexity. Anesthetic effects on any dependent measure vary by trait and by scale. Sources of variability include the class and amount of drug administered, brain regions acted upon, sex, age, species, and time, to name a few. All of these features also are subject to individual allelic variability. Obviously, a further complexity is that traits and scales vary as a function of disease states.

Rapidly advancing technologies will help address many of the forgoing complexities. For example, the 2014 Intel microprocessor containing more than 35 million transistors per square mm will soon be surpassed by a smaller chip that triples the number of transistors per square mm. Computational neuroscience can build large data matrices that are ideal for novel analytic approaches, such as Jim Grey's "fourth science paradigm" using "big data." Supercomputer development also holds exciting promise for advancing computational neuroscience. The 26 June 2016 issue of *Science* reported that 167 of the World's top 500 supercomputers are in China and have a total capacity of 211 Pflop/s compared to 165 supercomputers in the United States with a cumulative capacity of 173 Pflop/s. I am confident that Emery's research, discoveries by his trainees and colleagues, and books such as this one will continue to advance computational neuroscience that enhances anesthesia.

Pittsburgh, PA, USA — Robert E. Kass
Knoxville, TN, USA — Ralph Lydic
July 2017

Acknowledgements

We are deeply indebted to Professor Emery Brown, without whom this book would be impossible. Most chapter contributors are former graduate students or postdoctoral fellows with strong academic careers and have greatly benefited from their training with Emery. It goes without saying that all of us are forever grateful for Emery's mentorship.

We are also grateful to the editors and staff members of Springer for supporting this book project. In particular, Marta Moldvai has been a constant contact with us during the development of this project, from the moment we submitted our proposal for this book. From the start of the project, Zoe Kennedy and Murugesan Tamilsevan have also been helping us keep everything in order and to deliver on time.

We are indebted to Professor Rob Kass and Professor Ralph Lydic, who have kindly accepted our invitation to write the preface of this book. Their personal reflections on their own relationships with Emery reflect Emery's broad knowledge and impact in multiple fields. In preparing the book proposal, Professor Kass has also offered some advice, including the title of this book. This book also benefited intellectually from a special symposium held at Massachusetts Institute of Technology (MIT) on September 14–15, 2017, for which we also thank Ms. Sheri Leone and Ms. Kara Pavone who offered tremendous help in logistics and communications. We would like to express our deep gratitude to the generous financial support on the symposium from the Burroughs Wellcome Fund, Massachusetts General Hospital (MGH), and MIT. Finally, we also like to thank our dedicated colleagues for valuable comments and proofreading effort.

New York, NY, USA — Zhe Chen
Baltimore, MD, USA — Sridevi V. Sarma

Contents

Contributors

Behtash Babadi University of Maryland, College Park, MD, USA

Zhe Chen New York University School of Medicine, New York, NY, USA

ShiNung Ching Washington University at St Louis, St Louis, MO, USA

Uri T. Eden Boston University, Boston, MA, USA

Amit Etkin Stanford University School of Medicine, Stanford, CA, USA

Rose T. Faghih University of Houston, Houston, TX, USA

Loren M. Frank University of California, San Francisco, San Francisco, CA, USA

Ralph Lydic University of Tennessee, Knoxville, TN, USA

Robert E. Kass Carnegie Mellon University, Pittsburgh, PA, USA

Corey Keller Stanford University School of Medicine, Stanford, CA, USA

Michelle McCarthy Boston University, Boston, MA, USA

Michael J. Prerau Massachusetts General Hospital, Harvard Medical School, Boston, MA, USA

Pierre Sacré Johns Hopkins University, Baltimore, MD, USA

Sridevi V. Sarma Johns Hopkins University, Baltimore, MD, USA

Maryam M. Shanechi University of Southern California, Los Angeles, CA, USA

Hideaki Shimazaki Kyoto University, Kyoto, Japan and Honda Research Institute Japan, Saitama, Japan

Long Tao Boston University, Boston, MA, USA

Sujith Vijayan School of Neuroscience, Virginia Tech, Blacksburg, VA, USA

Wei Wu Stanford University, CA, USA

Editors' Biosketch

Zhe Chen is Assistant Professor at the New York University School of Medicine, having previously worked at the RIKEN Brain Science Institute, Harvard Medical School, and Massachusetts Institute of Technology. He is a senior member of the IEEE and an editorial board member of *Neural Networks* (Elsevier) and *Journal of Neural Engineering* (IOP). Professor Chen has received a number of awards including the Early Career Award from the Mathematical Biosciences Institute and has had his work funded by the US National Science Foundation and the National Institutes of Health. He is the lead author of the book *Correlative Learning: A Basis for Brain and Adaptive Systems* (Johns & Wiley, 2007) and the editor of the book *Advanced State Space Methods for Neural and Clinical Data* (Cambridge University Press, 2015).

Sridevi V. Sarma is Associate Professor at the Johns Hopkins University, having previously worked at Massachusetts Institute of Technology and Harvard Medical School. She is an associate editor of *IEEE Transactions on Neural Systems and Rehabilitation Engineering*. Professor Sarma is a recipient of the GE faculty for the future scholarship, a L'Oreal For Women in Science fellow, the Burroughs Wellcome Fund Careers at the Scientific Interface Award, the Krishna Kumar New Investigator Award from the North American Neuromodulation Society (NANS), and the Presidential Early Career Award for Scientists and Engineers (PECASE).

Acronyms

ACA	Aligned Cluster Analysis
ACC	Anterior Cingulate Cortex
ACTH	Adrenocorticotropic Hormone
ADA	Adaptive Detrend Algorithm
ADJUST	Automatic artifact Detection based on the Joint Use of Spatial and Temporal features
AIC	Akaike Information Criterion
BARS	Bayesian Adaptive Regression Splines
BCI	Brain Computer Interface
BIC	Bayesian Information Criterion
BMI	Brain-Machine Interface
BOLD	Blood Oxygen Level Dependent
CA	Cornu Ammonis
CCA	Canonical Correlation Analysis
CIF	Conditional Intensity Function
CLDA	Closed-Loop Decoder Adaptation
CNS	Central Nervous System
CRH	Corticotropin Releasing Hormone
CS	Compressed Sensing
DBS	Deep Brain Stimulation
DC	Direct Current
DIC	Deviance Information Criterion
DNN	Deep Neural Network
DP	Dirichlet Process
ECoG	Electrocorticography
EDA	Electrodermal Activity
EEG	Electroencephalography
EKG	Electrocardiogram
EM	Expectation Maximization
EMD	Empirical Mode Decomposition
EMG	Electromyography

FASTER	Fully Automated Statistical Thresholding for EEG artifact Rejection
FCSS	Fast Compressible State-Space
FEF	Frontal Eye Field
FIS	Fixed-Interval Smoothing
FLDA	Fisher Linear Discriminant Analysis
fMRI	Functional Magnetic Resonance Imaging
GCV	Generalized Cross-Validation
GLM	Generalized Linear Model
GMFP	Global Mean-Field Power
GP	Gaussian Process
GPST	Geometric Singular Perturbation Theory
GS	Golden Section
HACA	Hierarchical Aligned Cluster Analysis
HC	Healthy Control
HDP	Hierarchical Dirichlet Process
HMM	Hidden Markov Model
HSMM	Hidden Semi-Markov Model
HPA	Hypothalamic-Pituitary-Adrenal
HRF	Hemodynamic Response Function
HTC	High-threshold Thalamo-cortical
IC	Independent Component
ICA	Independent Component Analysis
ICD	Intermittent Context-Dependence
IF	Integrate-and-Fire
IPS	Intraparietal Sulci
IRLS	Iterative Reweighted Least Squares
ISI	Interspike Interval
KDE	Kernel Density Estimation
KF	Kalman Filter
KKT	Karush-Kuhn-Tucker
LDS	Linear Dynamical Systems
LFP	Local Field Potential
LIF	Leaky Integrate-and-Fire
LIP	Lateral Intraparietal
LNP	Linear-Nonlinear-Poisson
LOC	Loss of Consciousness
LQG	Linear Quadratic Gaussian
LTD	Long-Term Depression
LTP	Long-Term Potentiation
LTS	Low-Threshold Spiking
MAP	Maximum a posteriori
MAR	Multivariate Autoregressive
MARA	Multiple Artifact Rejection Algorithm
MCMC	Markov Chain Monte Carlo
MEC	Medial Entorhinal Cortex

MEG	Magnetoencephalography
MI	Mutual Information
ML	Maximum Likelihood
MMSE	Minimum Mean-Squared Error
MODWT	Maximal Overlap Discrete Wavelet Transform
MP	Matching Pursuit
MSIT	Multi-Source Interference Task
MTL	Medial Temporal Lobe
NHP	Non-Human Primate
NMI	Normalized Mutual Information
OFC	Optimal Feedback Control
OFC	Orbitofrontal Cortex
OLE	Optimal Linear Estimator
PCA	Principal Component Analysis
PD	Parkinson's Disease
PLDS	Poisson Linear Dynamical Systems
PPF	Point Process Filter
PPM	Point Process Model
PTSD	Post-Traumatic Stress Disorder
PV	Population Vector
RANSAC	Random Consensus
RBM	Restricted Boltzmann Machine
RE	Reticular Nucleus
RLS	Recursive Least Squares
SCR	Skin Conductance Response
SEP	Somatosensory Evoked Potential
SMC	Sequential Monte Carlo
SP	Spectrotemporal Pursuit
SSM	State-Space Model
STRF	Spectrotemporal Receptive Field
TC	Thalamo-cortical
TEP	Transcranial Magnetic Stimulation Evoked Potential
TMS	Transcranial Magnetic Stimulation
TORS	Temporally Orthogonal Ripple Combinations
VB	Variational Bayes

Chapter 1
Introduction

Zhe Chen and Sridevi V. Sarma

1.1 Background

In today's modern age, an enormous amount of neural data have been recorded or collected (Stevenson and Kording 2011). It remains a great challenge to process and analyze this "big data." By nature, neural signals are stochastic (noisy) signals measured from dynamic processes in the brain at various spatiotemporal scales. The term "*dynamic*" is emphasized because the neural signals are generated from biophysical processes (i.e., neurons) that have memory. Neural signals are often modeled as non-stationary stochastic processes. Unlike other physical signals, neural signals are driven by complex behaviors of experimental subjects. In some cases, multi-modal neural data are simultaneously collected at different spatial and temporal scales. The development of efficient quantitative methods to characterize these recordings and extract information that reveals underlying neurophysiological mechanisms remains an active and important research field.

To date we have witnessed tremendous advances and growing interests in applying statistics, signal processing, control and modeling methods to neuroscience. Meanwhile, new applications encounter emerging problems and challenges. Therefore, it is important to recognize these challenges and frequently exchange innovative ideas among researchers at both computational and experimental ends as well as those at the interface. We will review some important research topics and progresses of applying quantitative methods for neuroscience data. The concept of *dynamics* is emphasized throughout the book.

Z. Chen (✉)
New York University School of Medicine, New York, NY, USA
e-mail: zhe.chen3@nyumc.org

S.V. Sarma
Johns Hopkins University, Baltimore, MD, USA
e-mail: sree@jhu.edu

Z. Chen, S.V. Sarma (eds.), *Dynamic Neuroscience*,
https://doi.org/10.1007/978-3-319-71976-4_1

1.2 Statistics and Signal Processing in Neuroscience

Analysis of neurophysiological or behavioral data from neuroscience investigations is a fundamental task in computational and statistical neuroscience (Brown et al. 2004; Kass et al. 2005). The task can be challenging when the following one or more scenarios are present: (i) The dimensionality of the data is scaled up from an order of tens to hundreds or even larger; (ii) The data are either super noisy with a very low signal-to-noise (SNR) ratio or large variability (across trials or time); (iii) The exact quantitative mapping between neural codes and the measured behavior is always partially unknown, given partial observations of behavioral measures and neural recordings.

The core of statistics is data science. The data are generated and collected from neuroscience experiments at neurophysiological, imaging, or behavioral levels. In their article "What is statistics" published in *American Statistician*, Brown and Kass defined two fundamental principles for statistical thinking (Brown and Kass 2009; Kass et al. 2014):

1. Statistical models of regularity and variability may be used to express knowledge and uncertainty about a signal in the presence of noise, via inductive reasoning.
2. Statistical methods may be analyzed to determine how well they are likely to perform.

The first principle focuses on the construction of statistical models, whereas the second principle focuses on the evaluation of statistical inference procedures. Depending on the model assumption, statistical models can be *parametric, semiparametric*, or *nonparametric*, with a trend in growing model complexity. It is important to stress the motto: "all models are wrong, but some are useful," in a sense that statistical models need to be adapted or modified according to the need, which may or may not fully reflect the truth of data generating process.

Based on these two principles, likelihood or Bayesian methods can be developed for neural data analysis (Pawitan 2001; Brown et al. 2003; Gelman et al. 2004; Robert 2007). Among Bayesian inference methods, there are techniques based on deterministic optimization (such as the Laplace method or variational method) or stochastic sampling (such as the sequential Monte Carlo or Markov chain Monte Carlo methods). Examples of statistical models and inference are shown in Table 1.1.

Signal processing is a discipline that encompasses the fundamental theory, applications, algorithms, and implementations of processing or transferring infor-

Table 1.1 Examples of statistical models and inference

	Likelihood inference	Bayesian inference
Parametric	Linear regression, GLM	Particle filter (Doucet et al. 2001)
Semiparametric	Finite mixture models	BARS (DiMatteo et al. 2001)
Nonparametric	KDE, Kernel regression	GP, Dirichlet process mixtures (Hjort et al. 2010)

mation contained in signals. Roughly speaking, we can classify fundamental tasks of statistical signal processing into three categories: *signal recovery* (examples of such include a wide range of inverse problems such as filtering, detection, denoising, and deconvolution), *representation and visualization* (examples of such include spectral analysis, subspace identification, compression), and *prediction and control* (the problem of control will be addressed separately in the next section). The goal of neural signal processing is to combine statistics, signal processing, and optimization methods to process neural data of diverse sources. Unlike traditional signal processing assumptions, neural signal processing often deals with neurophysiological signals with non-Gaussianity, non-stationarity, and heterogeneity (Chen 2017). Over the past decades, many statistical signal processing tools have been developed for various neuroscience applications. We will briefly review a few important applications in this area.

1.2.1 Neural Coding

From a neural coding perspective, there are *encoding* and *decoding* phases (Fig. 1.1). The goal of neural encoding is to elucidate the representation and transmission of information in the nervous system (Perkel and Bullock 1968), whereas the goal of decoding is to extract as much information about a stimulus as possible from

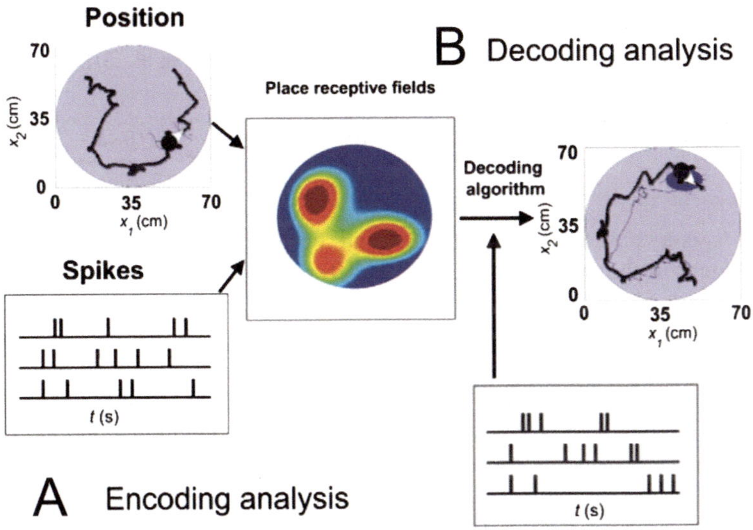

Fig. 1.1 Schematic diagram of neural coding and decoding analyses. (**a**) In the hippocampal encoding analysis, sorted spikes from three neurons are correlated with the measured animal's spatial position during run behavior to construct three place receptive fields. (**b**) In the hippocampal decoding analysis, sorted spikes are plugged in the designed decoding algorithm to reconstruct animal's spatial position in time. Adapted from (Brown et al. 2004)

neural signals. Depending on coding strategy, neural codes can be categorized as rate code, timing code, correlation code, or synchronous firing. At a single neuron level, the neuronal tuning information is characterized by its receptive field (RF), which defines how its spiking activity changes in response to a stimulus (see Fig. 1.2c for an example). For instance, in visual systems, the neuronal tuning function is characterized by a 2D spatiotemporal RF; whereas in auditory systems, it is characterized as a spectrotemporal RF (STRF). In the case of auditory neurons, the classic STRF model assumes a linear relationship between the time-dependent neuronal response $r(t_k)$ and the time-frequency spectrum of acoustic stimuli $s(k, \omega)$:

$$r(t_k) = r_0 + \sum_{\tau} \sum_{\omega} STRF(\tau, \omega) s(k - \tau, \omega) \tag{1.1}$$

where r_0 denotes the baseline firing rate, $s(k - \tau, \omega)$ denotes the stimulus energy at different tonotopic locations ω and different time delays $\tau > 0$. The strength and nature of the influences, whether being excitatory (positive) or suppressive (negative), is described by the STRF gain function $STRF(\tau, \omega)$. The traditional method for mapping the neuronal RF is *reverse correlation* (Ringach and Shapley 2004); see Fig. 1.2a for an illustration. In order to characterize the response nonlinearity and non-Gaussianity and to account for the spiking history, a so-called linear-nonlinear Poisson (LNP) model has been developed (Fig. 1.2b), which consists of a linear filter, followed by a pointwise static nonlinearity and a Poisson random number generator. The LNP model is essentially a generalized linear model (GLM). Specifically, to account for non-Poisson spiking (such as the refractory period and bursting), one can incorporate a post-spike history filter and model the instantaneous firing rate as (Truccolo et al. 2005; Calabrese et al. 2011)

$$\lambda(t_k) = f\Bigg(\theta_0 + \sum_{f} \sum_{\tau} \underbrace{STRF(\tau, \omega) s(k - \tau, \omega)}_{\text{stimulus effect}} + \sum_{l} \underbrace{h(l) n(k - l)}_{\text{spike-history effect}}\Bigg) \tag{1.2}$$

where θ_0 is a constant, $h(l)$ is a finite-length post-spike filter, $n(k - l)$ denotes the spike count in the previous l-th window before time index k, and $f(\cdot)$ is a static nonlinearity, which can be an exponential: $f(u) = \exp(u)$, or $f(u) = \log(1 + \exp(u))$, or a custom function (e.g., $f(u) = \exp(u)$ for $u \leq 0$ and $f(u) = 1 + u + \frac{u^2}{2}$ for $u > 0$).

Neural population decoding is aimed to exploit various coding strategies and extract information to reconstruct the sensory input or motor command. In the case of rate code, neurons are generally considered to communicate information by increasing or decreasing their firing rates. In the case of others, neuron population can use specific spatiotemporal patterns of spiking activities and silent intervals. The early representative work on population decoding include the "*population vector*" (Georgopoulos et al. 1986), which computes the sum of preferred directions of a population of neurons, weighted by respective spike counts, and the *optimal linear estimator* (Bialek et al. 1991), which is based on a static linear regression model

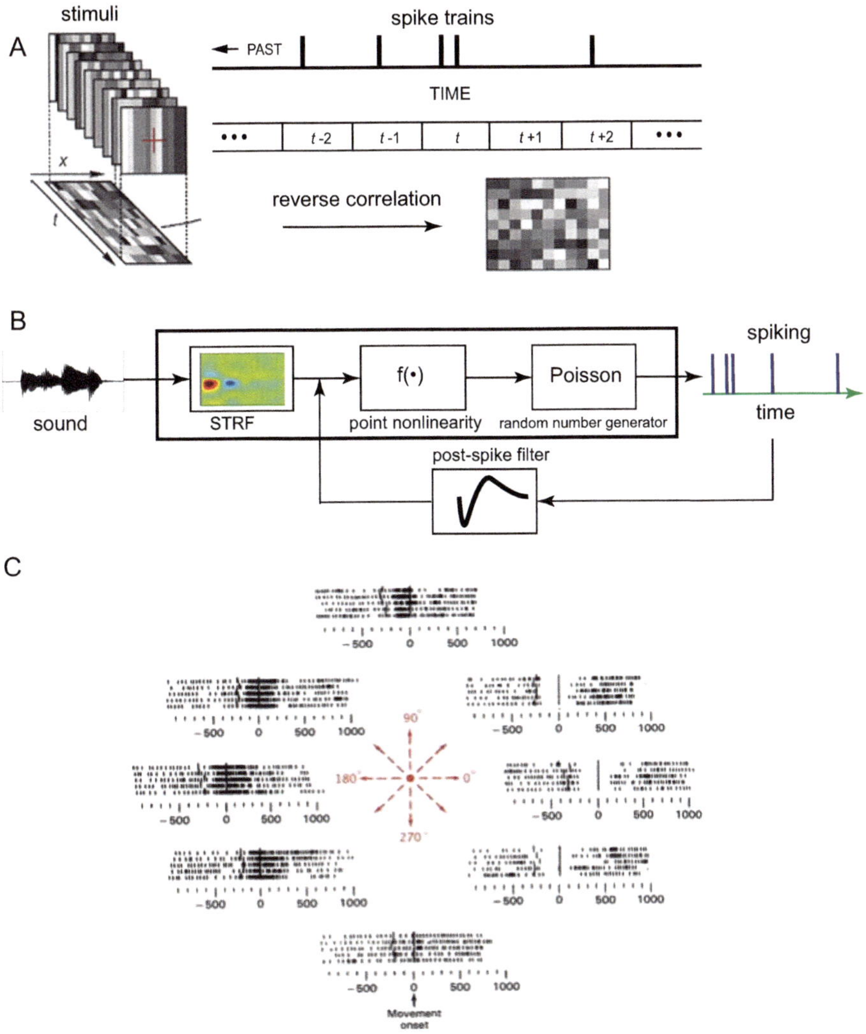

Fig. 1.2 Illustration of neural encoding. (**a**) Reverse correlation: the receptive field is estimated by regressed with spikes with temporally shifted stimuli using a linear Gaussian model. (**b**) A linear-nonlinear-Poisson (LNP) model for an auditory neuron. Each neuron has a stimulus filter or $STRF(\tau, \omega)$, and a post-spike filter $h(l)$ that captures dependencies on the neuron's own spiking history. Summed filter output passes through a static nonlinearity $f(\cdot)$ to produce the instantaneous spike rate. (**c**) Spike trains of a single motor cortical neuron while a monkey performed a reaching task in each of eight directions. Each of the eight spike rasters displays five repetitions of the reach. Time 0 indicates initiation of movement. In this example, this neuron has preferred (greater) firing activity when the movement was roughly in the leftward direction. Adapted from (Chen 2017) and (Brockwell et al. 2007) with permission (Copyright: IEEE)

between the kinematics and population firing rate. However, the strong limitation of this model is its failure to capture temporal dynamics of population codes. Later on, the linear state space model (i.e., Kalman filter) was subsequently proposed to decode motor population codes (Wu et al. 2006, 2009). The Kalman filter consists of two equations: the *state equation* and *measurement equation* (Kalman 1960). For instance, the state equation characterizes the dynamics of motor kinematics, and the measurement equations assume a Gaussian likelihood based on the population firing rate. Furthermore, the linear Gaussian model was extended to Poisson-GLM with latent variables (Lawhern et al. 2010). To date, most population coding analyses have used decoding algorithms based on spike count observations or rate codes (Rieke et al. 1997; Zhang et al. 1998; Zemel et al. 1998).

Another line of research in population decoding methods is built upon point processes (Brown 2005). The pioneering application of decoding analysis at a millisecond resolution was demonstrated in the rodent hippocampus (Brown et al. 1998), and was later extended to motor cortical areas (Truccolo et al. 2008; Shanechi et al. 2012). Since the point process model assumes a non-Gaussian likelihood, Gaussian approximation methods have been used to derive a recursive point process filter (Smith and Brown 2003; Eden et al. 2004; Barbieri et al. 2004). In addition, other numerical methods such as the particle filter or sequential Monte Carlo have been used for population decoding methods (Brockwell et al. 2004; Ergun et al. 2007). Notably, most neural encoding models for receptive fields are *parametric*. However, due to the complexity of statistical dependency between the input and responses, parametric models are limited in their representation power. As a result, a few *nonparametric* methods have been proposed (Truccolo and Donoghue 2007; Coleman and Sarma 2010; Agarwal et al. 2016). Another Bayesian nonparametric method is Gaussian process (GP), which has also been developed in population decoding methods (Huys et al. 2007).

To date, most decoding methods are based upon sorted spikes. However, spike sorting is a complex, time-consuming, and error-prone process, and it often discards many non-clustered spikes (Lewicki 1998). To overcome this limitations, several efforts have been dedicated to decoding unsorted ensemble spikes (Ventura 2008, 2009; Chen et al. 2012; Kloosterman et al. 2014). One idea is to treat the cell identity as a missing variable, and assume the temporal evolution of the stimulus is smooth, from which a maximum likelihood-based decoding method is derived using an expectation maximization (EM) algorithm (Ventura 2008). In another approach, information in covariates that modulate neuronal firing is exploited in addition to spike waveform information, and this can lead to improved spike sorting and decoding results (Ventura 2009). Yet another idea is to model the spike waveform features by a temporal marked point process (Kloosterman et al. 2014). In the rodent hippocampal example, we can directly map the high-dimensional features of unsorted hippocampal spikes (denoted by vector $\mathbf{a}$, which are treated as a proxy of the unit identity) to the animal's position (denoted by $\mathbf{x}$), from which we estimate the generalized rate function $\lambda(\mathbf{a}, \mathbf{x})$ using *nonparametric* or *semiparametric* density estimation methods. This idea was further extended by incorporating temporal priors (Deng et al. 2015).

Finally, from a practical viewpoint, neural decoding algorithms based on continuous neural signals, such as the local field potential (LFP) and electrocorticography (ECoG), are frequently used (Zhuang et al. 2009; Bansal et al. 2012; Gilja et al. 2012; Stavisky et al. 2015). The Kalman filter, factor analysis, and the hidden Markov model (HMM) are among the common tools for these algorithms.

1.2.2 *The Inverse Problems*

In neuroscience, it is often the case that neural sources are not easily within reach to record from. This motivates the development of solving the inverse problem for neuroscience applications. That is, we would like to infer neural activity at the source from measurable data collected in regions that are either spatially or functionally connected to the source. Nearly all inverse problems are ill-posed, therefore the solutions to inverse problems are non-unique. We briefly review three types of inverse problems in neuroscience: *source localization, deconvolution*, and *denoising or artifact rejection*.

The first type of inverse problem is the electroencephalography (EEG) or magnetoencephalography (MEG) source imaging or localization (see Fig. 1.3). EEG and MEG represent two noninvasive functional brain imaging methods, whose extracranial recordings measure electric potential differences and extremely weak magnetic fields generated by the electric activity of the neural cells, respectively (Wendel et al. 2009). The goal of source localization is to estimate the location and

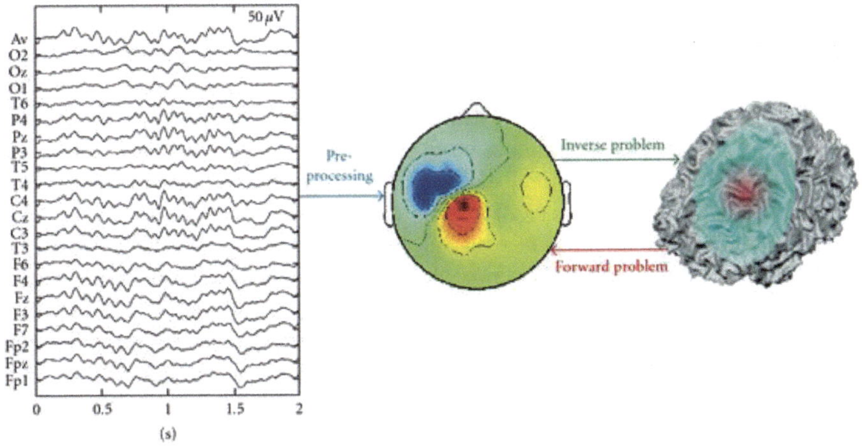

Fig. 1.3 Illustration of source imaging. Signal processing (e.g., denoising, artifact rejection) starts at the preprocessing step. The inverse problem attempts to locate the sources from recorded measurements, whereas the forward problem assumes a source definition in order to calculate a potential distribution map. From (Wendel et al. 2009)

strengths of the current sources that generate the multichannel EEG or MEG signals (Wu et al. 2016). Traditional signal processing methods are based on spatial filtering, such as LORETA (Pascual-Marqui 1999). However, it is important to exploit other constraints, such as temporal priors (Lamus et al. 2012) or sparsity priors (Babadi et al. 2014).

The second type of inverse problem is deconvolution of imaging signals. In human neuroscience experiments, functional magnetic resonance image (fMRI) is a powerful technology to measure macroscopic brain activity by detecting changes associated with blood flow. Cerebral blood flow is coupled with neuronal activation, therefore the hemodynamic dynamics is dependent on the underlying neural dynamics. Although the hemodynamic response is relatively slow, simultaneous fast fMRI-EEG recordings can reveal also detect oscillatory neural activity in human brains (Lewis et al. 2016). Since fMRI is an indirect measure that rests upon a mapping from neuronal activity $s(t)$ to the blood oxygen level dependent (BOLD) signal via hemodynamic effects, specified by a linear convolution operator (Gitelman et al. 2003)

$$y(t) = s(t) \otimes h(t) + n(t) \tag{1.3}$$

where $\otimes$ denotes convolution and $n(t)$ denotes the measurement. The goal of deconvolution (i.e., unfolding the convolution process) is to reconstruct the neural dynamics $s(t)$ via an estimated or measured hemodynamic response function (HRF) $h(t)$. In a more general setting, the HRF can be spatiotemporal (Aquino et al. 2014), and the interaction model can be nonlinear (Penny et al. 2005). Once the neural activity is reconstructed from multiple brain regions, one can further infer the directional interactions or functional connectivity between those areas.

Calcium ions generate versatile intracellular signals that control key functions in all types of neurons. Imaging calcium in neurons is particularly important because calcium signals exert their highly specific functions in well-defined cellular subcompartments (Grienberger and Konnerth 2012). Today, confocal and two-photon microscopy for calcium imaging has become a widely used tool to investigate large-scale neuronal activity in animal's brain. The powerful imaging tool has enabled us to detect spatiotemporal activation patterns of neural assemblies and to uncover neuronal population dynamics (see an illustration in Fig. 1.4). Similar to fMRI, the goal of deconvolution is to infer neuronal spike activity from calcium imaging trace (Vogelstein et al. 2009, 2010; Onativia et al. 2013). This process consists of a serial of signal processing operations: spatial filtering, denoising, deconvolution, and demixing. Developing fast and efficient algorithms for large-scale calcium imaging data has been an active research topic (Pnevmatikakis et al. 2006; Theis et al. 2016; Deneux et al. 2016; Rahmati et al. 2016; Friedrich et al. 2017).

This brings us to the last type of inverse problem, which is signal denoising. Denoising is often a preprocessing step for all neural data analyses, since nearly all neural measurements are corrupted by various sources of noise (e.g., electrical, mechanical, movement, etc.). Recently, harmonic regression has been introduced in neural signal processing for denoising calcium imaging data (Malik et al. 2011)

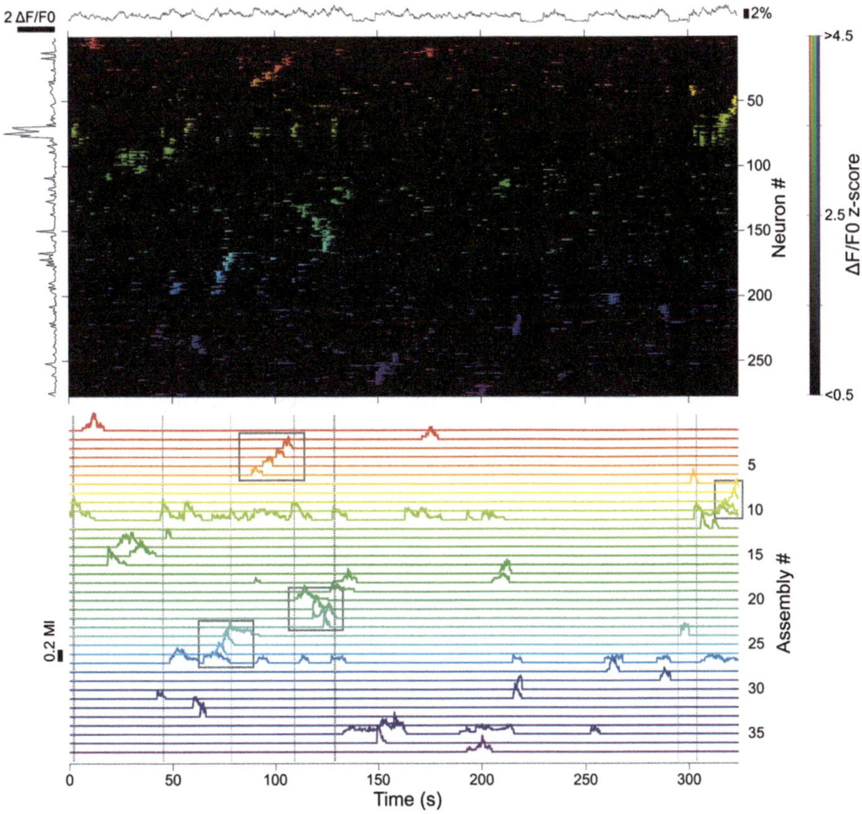

Fig. 1.4 Detection of neuronal assemblies from calcium imaging. *Top:* raster plot of the z-scored $\Delta F/F0$ of 277 neurons distributed in 37 assemblies, from the total imaged population of 1025 neurons. Neurons are sorted and color-coded according to the assembly to which they belong to (color bar on the right). Black trace on top, fluctuations of the number of active neurons in the total imaged population; Black trace on the left, average neuronal responses to a whisker-object contact. *Bottom:* activation dynamics of the detected assemblies, color-coded as in the raster plot. From (Romano et al. 2017)

and for EEG-fMRI artifact rejection (Krishnaswamy et al. 2016). In general, the observed signal $y(t)$ is modeled as the sum of two statistically independent processes $y(t) = s(t) + n(t)$: a signal process and a noise process. The idea of harmonic regression is to represent the signal or noise process as a harmonic series

$$s(t) = \mu_0 + \sum_{r=1}^{R} A_r \cos(\omega r t) + B_r \sin(\omega r t) \tag{1.4}$$

where order R denotes the number of harmonics, μ_0 is a constant, $[A_r, B_r]$ together define the amplitude and phase of the r-th harmonic, and ω specifies the fundamental frequency. The residual noise (or signal) process $n(t)$ will be modeled as a p-order autoregressive (AR) process. The unknown harmonic coefficients and AR coefficients can be estimated by maximum likelihood methods (Brown et al. 2004).

1.2.3 Analysis of Plasticity or Dynamics of Single Neurons and Populations

Neuroscience experiments often consist of multiple independent trials, and neuronal spike activity can exhibit a large variability between trials, for reasons due to learning, adaptation, remapping, or top-down attention. Consequently, receptive fields of neurons are dynamic, i.e., neuronal responses to relevant stimuli change with experience. Experience-dependent change and neural plasticity has been documented in a number of brain regions, and characterizing such neural plasticity is important since the plasticity may reveal neural mechanisms in learning.

Two pioneering studies that characterized dynamics from neuronal spiking data were performed by Brown et al. (2001) and Czanner et al. (2008). In one experiment, a rat continuously navigates along track and learns which directions lead to reward. To track the rat's hippocampal place fields during learning and navigation, they derived an instantaneous steepest-descent adaptive filter algorithm based on an instantaneous log-likelihood for point process observations, and used the adaptive point process filter (Brown et al. 2001). Their approach was motivated from the most popular adaptive filter algorithm (i.e., the least-mean square filter) in signal processing, and they demonstrated that decoding of the rat's position improved with an adaptive filter over a static filter. In a second study, Brown and colleagues developed a likelihood-based modeling approach for analyzing between-trial hippocampal neuronal dynamics while monkeys performed an associative learning task (Czanner et al. 2008).

It is worth noting that learning experiments, behavioral data generated can be analyzed using a state-space model with a discrete observation process (Chen et al. 2010; Chen 2015). For example, a typical learning experiment consists of a sequence of trials on which a subject executes a task correctly or incorrectly. In the behavioral learning analysis, the objective is to estimate the learning curve, i.e. the probability of a correct response as a function of trial number give all the entire sequence correct and incorrect responses in the experiment. This has been formulated as a dynamic system and the inference is given by a state-space smoothing algorithm (Wirth et al. 2003; Smith et al. 2004, 2005, 2007). This state space approach can also be extended to behavioral experiments with mixed observations (binary and continuous) (Prerau et al. 2009) or trivariate responses (Wong et al. 2014). In this book, Chap. 7 describes the state-space modeling approach in detail.

At the population level, the assessment of second and higher-order neuronal correlation is also an important task in neuroscience experiments, as cooperative activity between simultaneously recorded neurons is expected to organize dynamically during behavior and cognition. One important approach is to estimate time-varying spike interactions for multivariate point process observations by means of a state-space analysis (Shimazaki et al. 2012). Another approach is to identify the functional connectivity or (directional) Granger causality between neuronal assemblies within or across different brain areas. Several maximum likelihood and Bayesian methods have been developed in the literature for neural spike train data (Okatan et al. 2005; Stevenson et al. 2009; Chen et al. 2011; Kim et al. 2011) and EEG signals (Stokes and Purdon 2017). The functional connectivity is usually assumed stationary within a task or trials, but an extended solution to a non-stationary scenario has been considered (Zhou et al. 2016).

Finally, data smoothing and high-dimensional data visualization has become an increasingly important topic in neuroscience. Exploiting temporal dynamics structure and latent structure of neural data has proved useful in various applications (Yu et al. 2009; Cunningham and Yu 2014; Chen et al. 2012, 2014; Ba et al. 2014; Kobak et al. 2016).

1.3 Modeling and Control in Neuroscience

Control theory is a field that entails the analysis of dynamical systems and the synthesis of controllers that actuate these systems to meet specific objectives (e.g. tracking a signal, rejecting disturbances, stabilizing an unstable system). If the actuation is done in the absence of system response measurements, then the control system (dynamical system plus controller) is said to be *open loop*. If the controller generates an actuation based on response measurements, then the control system is said to be *closed-loop*. Consider the simple most widely used closed-loop control system: a thermostat. The objective of the thermostat is to regulate the temperature in a room at a desired set point. That is, the heat or air condition should be increased or decreased, based on room temperature measurements, to remain steady at the desired set point. Here, the dynamical system is the evolving room temperature, which is actuated by heat or air conditioning, and the thermostat is the sensor that measures the temperature, compares it to the set point, and adjusts the heat or air conditioner.

Control theory has emerged as an important field in neuroscience because it has become possible to more easily manipulate the chemical and electrical patterns in the brain (the dynamical system to be controlled) with drugs that cross the blood–brain barrier, electrical stimulation delivered through electrodes implanted into the brain, or via light delivered through optical fibers that excites genetically manipulated neurons. Traditionally, these actuating mechanisms are applied either in open-loop or in closed-loop at a very slow rate. For example, a drug may be given to a Parkinson's disease (PD) patient to suppress movement disorders including resting tremor, rigidity, and bradykinesia (slowness of movements). After a few days or weeks on the medication, the patient's responsiveness is measured. Depending on how well the drug suppresses the patient's symptoms, the dosage is either increased or decreased or the medication is changed altogether (slow closed-loop control).

More recently, deep brain stimulation (DBS) has been used in clinical practice to suppress pathological neural network dynamics and restore behavior. DBS works as an exogenous localized control input into the network (Benabid et al. 2009). It injects pulses of electrical current in well-defined anatomical sites, but its effects spread throughout the network (Benabid et al. 2009; Montgomery and Gale 2002, 2008; Perlmutter and Mink 2006). For example, in Parkinson's disease, a specific motor-related neuronal network exhibits pathological oscillations and synchronization that are hypothesized to cause movement disorders described above (Kühn et al. 2009; Gale et al. 2008, 2009; Sarma et al. 2012; Santaniello et al. 2012). DBS applied to one structure in the motor network can suppress these symptoms if the electrode is placed precisely and if the DBS signal parameters are set appropriately (Benabid et al. 2009; Kuncel et al. 2006; Lang and Lozano 1998). Therapeutic stimulation, however, operates in open-loop and is typically high in power which leads to several problems: frequent surgical battery replacements, adverse side effects, long-term tissue damage, and non-adaptation of stimulation parameters to patient's needs (Butson and McIntyre 2008; Tommasi et al. 2008; Wei and Grill 2009; Zahodne et al. 2009). Consequently, closed-loop designs have been proposed to overcome these drawbacks. It is worth noting that DBS is also used to suppress seizures in epilepsy patients (Sohal and Sun 2011; Colpan et al. 2007; Gluckman et al. 2001; Ehrens et al. 2015; Good et al. 2009) (see Fig. 1.5), suppress involuntary movements in dystonia, stop ticks in Tourette's syndrome, and to improve outcomes for clinically depressed patients (Ressler and Mayberg 2007; Wichmann and DeLong 2006).

An alternative to electrical stimulation that can more precisely target individual structures and neurons is optogenetic stimulation. Here, targeted neurons are injected with a virus that allows neurons to grow photoreceptors so that light at specific wavelengths can either activate or inhibit action potentials. Optogenetics has proven to be a valuable tool in neuroscience studies as networks in the brain can be targeted and controlled to study functionality. However, despite substantial advancements in these genetic tools, their use is largely restricted to perturbative paradigms wherein neurons in the targeted network are stimulated *en masse*.

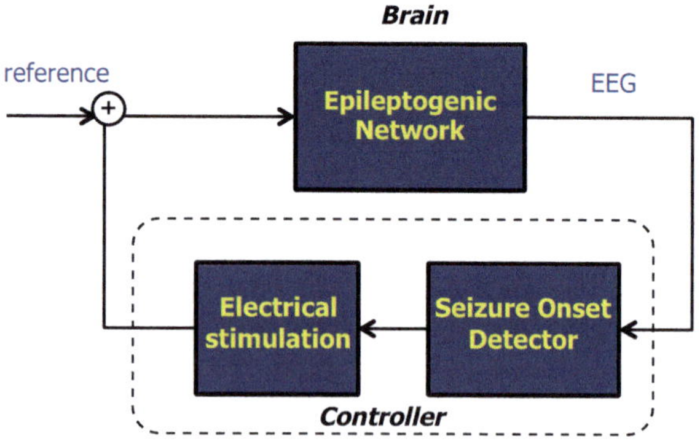

Fig. 1.5 Closed-loop control for seizure detection and electrical stimulation. In this schematic diagram, the closed-loop control continuously steers the epileptogenic neural network away from seizure genesis entirely using adaptive stimulation patterns that change with EEG measurements

Thus, there is an as yet unmet need for new methods that will allow for finer spatial and temporal control of neural activity at single-neuron and millisecond specificity (Grosenick et al. 2015; Ching and Ritt 2013). Ultimately, the goal is to engineer optogenetic inputs (light-based waveforms) that, despite impinging on many neurons simultaneously (i.e., a broadcast-type input), can control cells at an individual level. However, neural dynamics present several nontrivial analytical challenges that preclude the direct application of classical control theory to resolve this goal, thus necessitating new innovations in analysis such as spiking-based notions of controllability and reachability (Ching and Ritt 2013). In particular, the nonlinearity of the underlying network dynamics and large scale of the networks in question may require the use of statistical methods and model-free learning approaches (Nandi et al. 2017), together with established optimization theory in order to arrive at an effective and scalable solution for this type of neurostimulation problem.

Control theory has also emerged as critical when designing brain-machine interfaces (BMIs) that entail brain measurements interacting with external computers and devices. We will mention two applications here: (i) titration of anesthetic drugs to regulate a medically induced coma and (ii) actuation of prosthetics for amputees. A medically induced coma is a drug-induced state of profound brain inactivation and unconsciousness used to treat refractory intracranial hypertension and to manage treatment-resistant epilepsy. The state of coma is achieved by continually monitoring the patient's brain activity with an EEG and manually titrating the anesthetic infusion rate to maintain a specified level of burst suppression, an EEG marker of profound brain inactivation in which bursts of electrical activity alternate with periods of quiescence or suppression. The medical coma is often required for several days and is currently regulated by a team of nurses who monitor the EEG 24/7 (slow closed-loop control). A more rational approach would be to implement a BMI that monitors the EEG and adjusts the anesthetic infusion rate in real time to maintain the specified target level of burst suppression as demonstrated in (Shanechi et al. 2013).

Other, perhaps more popular BMI systems are those for control of movement, referred to as motor BMIs. Motor BMIs enable subjects to control external devices or even their own limbs by directly modulating their neural activity (Schwartz et al. 2006; Lebedev and Nicolelis 2006; Donoghue 2008; Thakor 2013; Shanechi 2017). To do so, BMIs record neural activity from motor cortical areas, use a mathematical algorithm, termed decoder, to estimate the subject's motor intent, use the decoded intent to actuate and control an external device or the native limb, and provide visual feedback of the generated movement to the subject (Fig. 1.6). Thus motor BMI systems can be viewed as closed-loop control systems. Both noninvasive (e.g., EEG) and invasive neural signal modalities have been used in motor BMIs. However, the highest levels of performance have been achieved using invasive modalities, in particular ensemble spiking activities.

A major component of the BMI system is the decoder. In the vast majority of BMI decoders, the input is taken as the binned spike counts. Early BMIs used linear filters such as the population vector, the optimal linear estimator (OLE), and the Wiener filter to process binned spike counts. Later work incorporated some modeling of

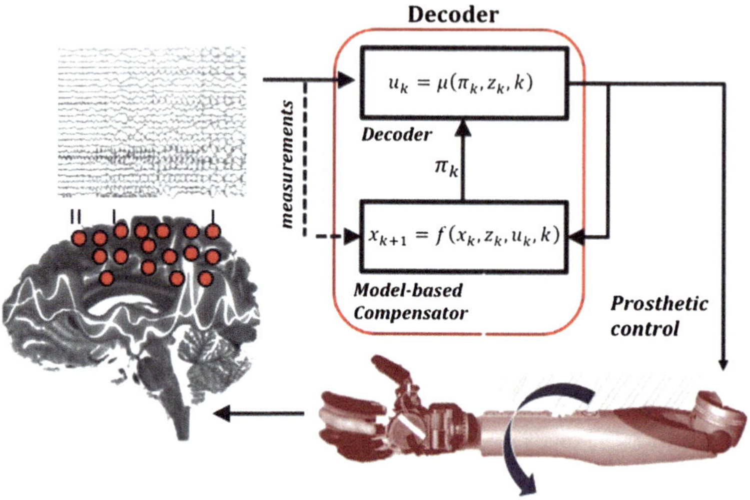

Fig. 1.6 Schematic diagram of motor BMI for prosthetic control. Neural signals are translated into actuation of prosthetic arm and hand via a decoder which can then send feedback signals to brain via vision and electrical stimulation of residual nerves.

movement by building linear state-space models and using a Kalman filter to process the counts. Until recently, these decoders were trained in open loop during a training session in which neural activity was recorded from subjects while they moved their own arms or imagined movements. However, recent work has shown that training the decoder during closed-loop BMI operation improves BMI performance (Taylor et al. 2002; Velliste et al. 2008; Gilja et al. 2012; Orsborn et al. 2012; Collinger et al. 2013; Hochberg et al. 2012; Shanechi et al. 2016, 2017). In addition to binned spike counts, BMI decoders can also use as input the binary time-series of spike events. These binary time-series can be modeled as point processes (Brown et al. 1998; Kass and Ventura 2001; Truccolo et al. 2005). Point process filters have been studied in offline or numerical simulation studies (Brown et al. 1998; Eden et al. 2004; Srinivasan et al. 2006; Shanechi et al. 2013). Recent algorithmic advances using optimal feedback-control modeling and adaptive point process filtering have led to closed-loop BMIs that use spikes directly at the millisecond timescale (Shanechi et al. 2013, 2016), resulting in improved BMI performance (Shanechi et al. 2017).

By providing an experimenter-defined control system, BMIs provide a new tool to study the brain's control mechanisms and the sensorimotor factors affecting them. For example, previous studies have explored the effect of sensorimotor delays on BMI performance (Willett et al. 2013). Recently, it has been shown that rapid sensorimotor control and feedback rates enabled by a point process filter significantly improved BMI performance (Shanechi et al. 2017). A closed-loop control framework could also help explain potential changes in neural representation in BMI control, for example, in response to perturbation or as a result of learning (Ganguly and Carmena 2009; Jarosiewicz et al. 2008; Chase et al. 2012).

Finally, control theory has been very useful in understanding how neuronal activity and behavior emerge from complex neural systems. In particular, feedback control models (e.g. state-space models) are used to understand how latent variables influence neural and behavioral measurements (our decision making work and learning studies and more) or to simply explain how and why control systems in the central nervous system operate the way they do.

As an example, secretion of cortisol and some hormones is stimulated by a well-known sequence of pulsatile events governed by a natural control system. Cortisol controls the body's metabolism and response to inflammation and stress. Cortisol is released in pulses from the adrenal glands in response to pulses of adrenocorticotropic hormone (ACTH) released from the anterior pituitary; in return, cortisol has a negative feedback effect on ACTH release (Faghih 2014). An important question in neuroendocrine data analysis involves determining the timing and amplitude of ACTH secretory events from concurrent time series of blood ACTH and cortisol levels. The solution to this problem has important implications for understanding normal and pathological neuroendocrine states. Simultaneous recording of ACTH and cortisol is not typical, and determining the number, timing, and amplitudes of pulsatile events from simultaneously recorded data is challenging because of several factors: (i) stimulator ACTH pulse activity, (ii) kinematics of ACTH and cortisol, (iii) the sampling interval, and (iv) the measurement error (Faghih et al. 2015). By taking advantage of the sparse nature of hormone pulses and adding more constraints for recovering hormone pulses, a solution to this can be achieved (Faghih et al. 2014). This solution is extendable to the analysis of pathological conditions related to cortisol as well as the analysis of concurrent measurements of other pairs of pulsatile endocrine hormones whose interactions are controlled through feedback loops (Faghih et al. 2015).

Considering that pulsatile cortisol release relays distinct signaling information to target cells, it is crucial to understand the physiology underlying pulsatile cortisol release. Understanding the underlying nature of the pulsatile release of cortisol via mathematical formalization can be beneficial to understanding the pathological neuroendocrine states and could lay the basis for a more rigorous physiologically based approach for administering cortisol therapeutically (Faghih et al. 2015). Traditional control-theoretic methods do not normally consider the intermittent control that is observed in pulsatile control of cortisol release. A plausible solution is to build a controller that minimizes the number of secretory events that result in cortisol secretion, as a way of minimizing the energy required for cortisol secretion, and maintains the blood cortisol levels within a specific circadian range while following the first-order dynamics underlying cortisol secretion (Faghih et al. 2015). This novel approach results in pulse control where the pulses and the obtained blood cortisol levels have rhythms that are in agreement with the known physiology of cortisol secretion (Faghih et al. 2015). The proposed formulation is a first step in developing intermittent controllers for curing cortisol deficiency. It is possible to personalize the medication and use an impulse controller to mimic the physiology of a healthy subject so that patients maintain hormonal levels (e.g. cortisol levels) that are similar to healthy subjects (Fig. 1.7). Furthermore, inspired by the pulse

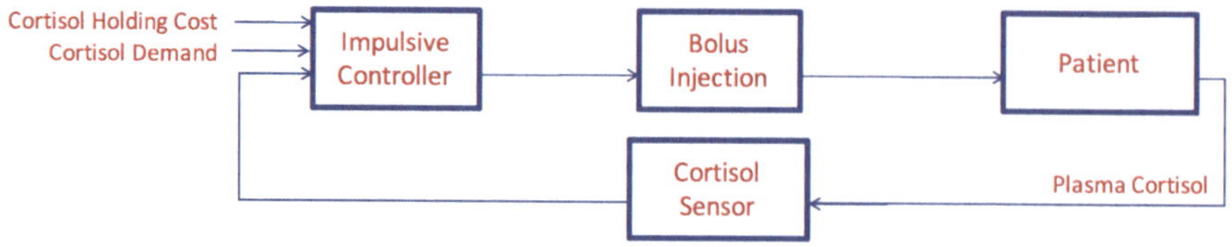

Fig. 1.7 Block diagram of the envisioned cortisol closed-loop controller. By considering the patient's time-varying cortisol demand and holding cost, and based on the circulatory cortisol levels measured by cortisol sensor, the impulsive controller determines the timing and amount of the next bolus of synthetic cortisol to be injected to the patient. Upon injection to the patient, circulatory cortisol first increases reaching the patient's holding cost, and then decays based on the patient's metabolism. Cortisol sensor measures the circulatory cortisol so that based on the patient's cortisol demand and holding cost, the controller can determine the timing and amount of the next synthetic cortisol injection (from Dr. Rose T. Faghih)

controller proposed in this research, in BMI design, it is possible to design pulse controllers instead of continuous controllers to improve the battery life of the brain implant (Faghih et al. 2015).

In the above introduction, we provide examples of where control theoretic tools are being applied in neuroscience. It is worth noting that the models used to analyze neural systems may differ. Some are phenomenological, where the timing between spike events or firing rates is directly modeled, while others are more biophysically based, wherein the mechanisms of the generation of spike events or population activity are modeled explicitly. Point process models are typically used to characterize timing between spike events as a function of task variables (e.g., stimuli, behavior, spiking history), while linear time invariant models are used to characterize continuous firing rates that also may be parameterized by task variables (D'Aleo et al. 2017; Shenoy et al. 2013).

Biophysical-based modeling can also be described at several scales (Schliebs and Kasabov 2014). On the microscopic level, the neuron model is described by the flow of ions throughout the channels of the membrane potential. The flow may depend on the presence or absence of various chemical messenger molecules. Examples of such include the Hodgkin-Huxley (H-H) model (Hodgkin and Huxley 1952) and the compartment models that describe separate segments of a neuron by a set of ionic equations. On the macroscopic level, the neuron is treated as a homogeneous unit, receiving and emitting spikes according to some defined internal dynamics. However, the principles of how a spike is generated and carried through the synapse, dendrite and soma are irrelevant. These models are often known as integrate-and-fire (IF) models.

The single neuron is the fundamental building block of neural networks in specific brain circuits. From spiking neuron models, we can further simulate a biologically realistic neural network (Li et al. 2017). To do so, we need to further specify the cell type (e.g., excitatory vs. inhibitory neurons, regular vs. bursting neurons), network connectivity and synaptic strength. Notably, although the H-H model can reproduce the biophysical mechanism more accurately, the simulation of the model is computational costly. To efficiently simulate a large-scale network of spiking neurons, it is therefore preferred to use a mathematically simpler neuron model, such as the leaky integrate-and-fire (LIF) neuron model (Knight 1972; Abbott 1999) or the Izhikevich model (Izhikevich 2006).

Both phenomenological and biophysically based models come with pros and cons. Biophysically based models are more realistic and describe mechanisms neuronal function, but are nonlinear and more difficult to analyze. Phenomenological models are easier to analyze but do not describe detailed mechanisms of neuronal and neural processes. Depending on the question of interest, one may be preferred over the other. In this book, each chapter uses a model appropriate for the study being performed.

1.4 Roadmap

This edited volume has two aims. First, it collects recent advances in statistics, signal processing, modeling and control methods in neuroscience. Second, it welcomes innovative or cross-disciplinary ideas along this line of research, and discusses important issues in neural data analysis (e.g., goodness-of-fit assessment, uncertainty evaluation, prior information, curse of dimensionality, model selection, etc.).

The contributors are solicited to cover representative research areas (signal processing, system identification, modeling and control) in important neuroscience and anesthesiology applications. All contributors have previously trained with Professor Emery Brown. The topics of this edited volume will include: state-space model, likelihood and Bayesian inference, variational and Monte Carlo methods, compressed sensing, deconvolution, system identification, EEG/MEG inverse problem, transcranial magnetic stimulation (TMS), statistical mechanics, neural decoding and BMIs. Research applications have covered a variety of species including rodents, ferrets, nonhuman primates, and human subjects.

This book will be relevant to a broad audience (electrical or biomedical engineers, statisticians, physicists, computer scientists, and neuroscientists), and it can be used as complementary teaching material for graduate students in related research fields. The book will also emphasize several important issues that will promote rigorous neural data analysis, such as data and software sharing, proper use of statistical assumption or statistical tests. The reader is assumed to have basic knowledge in probability, statistics, signal processing, and control theory.

The book is divided into two parts according to the technical content of chapters. Part I consists of five chapters (Chaps. 2 to 6). Chapter 2 by Eden and colleagues presents a state-space analysis paradigm to characterize complex and multi-scale neural observations. Chapter 3 by Chen discusses latent variable modeling of neural population dynamics. Chapter 4 by Prerau and Eden proposes a distribution-based approach to decode contexts using neurons with intermittent context-dependence. Chapter 5 by Babadi discusses signal processing methods that integrate sparsity and dynamics to identification and inverse problems in neuroscience. Chapter 6 by Wu and colleagues discusses artifact rejection methods for concurrent TMS-EEG data.

Part II consists of six chapters (Chaps. 7 to 12). Chapter 7 by Sarma and Sacré use dynamic models to characterize complex human behaviors and neural responses. Chapter 8 by Shanechi discusses inference algorithms for BMIs. Chapter 9 by Ching discusses modeling and controlling for neuronal inactivation. Chap. 10 by Faghig presents a sparse system identification approach for physiological signals. Chapter 11 by Shimazaki discusses a new neural engine hypothesis motivated from statistical mechanics and information theory. Finally, Chap. 12 by Vijayan and McCarthy discusses a mathematical modeling framework for inferring neuronal network mechanisms underlying anesthesia-induced brain oscillations.

1.5 Further Reading

In the end note, we want to stress that by no means this edited book attempts to cover the fast-growing field of computational and statistical neuroscience. Therefore, as complementary materials for the current collection, interested reader may find valuable resources from the following books:

- Chen, Z. (Ed.) (2015). *Advanced state space methods for neural and clinical data*. Cambridge: Cambridge University Press.
- Cohen, M. X. (2014). *Analyzing neural time series data: Theory and practice*. Cambridge: MIT Press.
- Grün, S. & Rotter, S. (Eds.) (2010). *Analysis of parallel spike trains*. New York: Springer.
- Hady, A. E. (Ed.) (2016). *Closed loop neuroscience*. Amsterdam: Elsevier.
- Kass, R. E., Eden, U. T. & Brown, E. N. (2014). *Analysis of neural data*. New York: Springer.
- Kramer, M. A. & Eden, U. T. (2016). *Case studies in neural data analysis: A guide for the practicing neuroscientist*. Cambridge: MIT Press.
- Oweiss, K. G. (Ed.) (2010). *Statistical signal processing for neuroscience and neurotechnology*. Cambridge: Academic.
- Ozaki T. (2012). *Time series modeling of neuroscience data*. Boca Raton: CRC Press.
- Rao, R. P. N. (2013). *Brain-computer interfacing: An introduction*. Cambridge: Cambridge University Press.
- Schiff, S. J. (2012). *Neural control engineering: The emerging intersection between control theory and neuroscience*. Cambridge: MIT Press.

References

Abbott, L. F. (1999). Lapicque's introduction of the integrate-and-fire model neuron (1907). *Brain Research Bulletin*, *50*, 303–304.

Agarwal, R., Chen, Z., Kloosterman, F., Wilson, M. A., & Sarma, S. V. (2016). A novel nonparametric approach for neural encoding and decoding models of multimodal receptive fields. *Neural Computation*, *28*, 1356–1387.

Aquino, K., Robinson, P., Schira, M., & Breakspear, M. (2014). Deconvolution of neural dynamics from fMRI data using a spatiotemporal hemodynamic response function. *Neuroimage, 94*, 203–215.

Ba, D., Babadi, B., Purdon, P. L., & Brown, E. N. (2014). Robust spectrotemporal decomposition by iteratively reweighed least squares. *Proceedings of National Academy of Sciences, USA, 111*(50), E5336–E5345.

Babadi, B., Obregon-Henao, G., Lamus, C., Hämäläinen, M. S., Brown, E. N., & Purdon, P. L. (2014). A subspace pursuit-based iterative greedy hierarchical solution to the neuromagnetic inverse problem. *Neuroimage, 87*, 427–443.

Bansal, A. K., Truccolo, W., Vargas-Irwin, C. E., & Donoghue, J. P. (2012). Decoding 3D reach and grasp from hybrid signals in motor and premotor cortices: Spikes, multiunit activity, and local field potentials. *Journal of Neurophysiology, 107*, 1337–1355.

Barbieri, R., Frank, L. M., Nguyen, D. P., Quirk, M. C., Solo, V., Wilson, M. A., & Brown, E. N. (2004). Dynamic analyses of information encoding in neural ensembles. *Neural Computation, 16*(2), 277–307.

Benabid, A. L., Chabardes, S., Mitrofanis, J., & Pollak, P. (2009). Deep brain stimulation of the subthalamic nucleus for the treatment of Parkinson's disease. *Lancet Neurology, 8*(1), 67–81.

Bialek, W., Rieke, F., de Ruyter van Steveninck, R. R., & Warland, D. (1991). Reading a neural code. *Science, 252*, 1854–1857.

Brockwell, A. E., Kass, R. E., & Schwartz, A. B. (2007). Statistical signal processing and the motor cortex. *Proceedings of the IEEE, 95*(5), 891–898.

Brockwell, A. E., Rojas, A. L., & Kass, R. E. (2004). Recursive Bayesian decoding of motor cortical signals by particle filtering. *Journal of Neurophysiology, 91*(4), 1899–1907.

Brown, E. N. (2005). The theory of point processes for neural systems. In C. Chow, B. Gutkin, D. Hansel, C. Meunier, & J. Dalibard (Eds.), *Methods and models in neurophysics* (pp. 691–726). Amsterdam: Elsevier.

Brown, E. N., Barbieri, R., Eden, U. T., & Frank, L. M. (2003). Likelihood methods for neural data analysis. In J. Feng (Ed.), *Computational neuroscience: A comprehensive approach* (pp. 253–286). Boca Raton: CRC Press.

Brown, E. N., Frank, L. M., Tang, D., Quirk, M. C., & Wilson, M. A. (1998). A statistical paradigm for neural spike train decoding applied to position prediction from ensemble firing patterns of rat hippocampal place cells. *Journal of Neuroscience, 18*, 7411–7425.

Brown, E. N., & Kass, R. E. (2009). What is statistics? *The American Statistician, 7*, 456–461.

Brown, E. N., Kass, R. E., & Mitra, P. P. (2004). Multiple neural spike train data analysis: State-of-the-art and future challenges. *Nature Neuroscience, 7*, 456–461.

Brown, E. N., Ngyuen, D. P., Frank, L. M., Wilson, M. A., & Solo, V. (2001). An analysis of neural receptive field plasticity by point process adaptive filtering. *Proceedings of National Academy of Sciences USA, 98*, 12261–12266.

Brown, E. N., Solo, V., Choe, Y., & Zhang, Z. (2004). Measuring period of human biological clock: Infill asymptotic analysis of harmonic regression parameter estimates. In *Methods in enzymology* (Vol. 383, pp. 382–405). Amsterdam: Elsevier.

Butson, C. R., & McIntyre, C. C. (2008). Current steering to control the volume of tissue activated during deep brain stimulation. *Brain Stimulation, 1*(1), 7–15.

Calabrese, A., Schumacher, J. W., Schneider, D. M., Paninski, L., & Woolley, S. M. N. (2011). A generalized linear model for estimating spectrotemporal receptive fields from responses to natural sounds. *PLoS One, 6*(1), e16104.

Chase, S. M., Kass, R. E., & Schwartz, A. B. (2012). Behavioral and neural correlates of visuomotor adaptation observed through a brain-computer interface in primary motor cortex. *Journal of Neurophysiology, 108*(2), 624–644.

Chen, Z. (Ed.) (2015). *Advanced state space methods in neural and clinical data*. Cambridge: Cambridge University Press.

Chen, Z. (2017). A primer on neural signal processing. *IEEE Circuits and Systems Magazine, 17*(1), 33–50.

Chen, Z., Barbieri, R., & Brown, E. N. (2010). State-space modeling of neural spike train and behavioral data. In K. Oweiss (Ed.), *Statistical signal processing for neuroscience and neurotechnology* (pp. 175–218). Amsterdam: Elsevier.
Chen, Z., Gomperts, S. N., Yamamoto, J., & Wilson, M. A. (2014). Neural representation of spatial topology in the rodent hippocampus. *Neural Computation*, *26*(1), 1–39.
Chen, Z., Kloosterman, F., Brown, E. N., & Wilson, M. A. (2012). Uncovering spatial topology represented by rat hippocampal population neuronal codes. *Journal of Computational Neuroscience*, *33*(2), 227–255.
Chen, Z., Kloosterman, F., Layton, S., & Wilson, M. A. (2012). Transductive neural decoding for unsorted neuronal spikes of rat hippocampus. In *Proceedings of IEEE Engineering in Medicine and Biology Conference* (pp. 1310–1313).
Chen, Z., Putrino, D. F., Ghosh, S., Barbieri, R., & Brown, E. N. (2011). Statistical inference for assessing functional connectivity of neuronal ensembles with sparse spiking data. *IEEE Transactions on Neural Systems and Rehabilitation Engineering*, *19*(2), 121–135.
Ching, S., & Ritt, J. T. (2013). Control strategies for underactuated neural ensembles driven by optogenetic stimulation. *Frontiers in Neural Circuits*, *7*, 54.
Coleman, T. P., & Sarma, S. S. (2010). A computationally efficient method for nonparametric modeling of neural spiking activity with point processes. *Neural Computation*, *22*(8), 2002–2030.
Collinger, J. L., Wodlinger, B., Downey, J. E., Wang, W., Tyler-Kabara, E. C., Weber, D. J., et al. (2013). High-performance neuroprosthetic control by an individual with tetraplegia. *The Lancet*, *381*, 557–564.
Colpan, M. E., Li, Y., Dwyer, J., & Mogul, D. J. (2007). Proportional feedback stimulation for seizure control in rats. *Epilepsia*, *48*(8), 594–603.
Cunningham, J. P., & Yu, B. M. (2014). Dimensionality reduction for large-scale neural recordings. *Nature Neuroscience*, *17*(11), 1500–1509.
Czanner, G., Eden, U. T., Wirth, S., Yanike, M., Suzuki, W. A., & Brown, E. N. (2008). Analysis of between-trial and within-trial neural spiking dynamics. *Journal of Neurophysiology*, *99*(5), 2672–2693.
D'Aleo, R., Rouse, A., Schieber, M., & Sarma, S. V. (2017). An input-output linear time invariant model captures neuronal firing responses to external and behavioral events. In *Proceedings of IEEE Engineering in Medicine and Biology Conference*.
Deneux, T., Kaszas, A., Szalay, G., Katona, G., Lakner, T., Grinvald, A., et al. (2016). Accurate spike estimation from noisy calcium signals for ultrafast three-dimensional imaging of large neuronal populations in vivo. *Nature Communications*, *7*, 12190.
Deng, X., Liu, D. F., Kay, K., Frank, L. M., & Eden, U. T. (2015). Clusterless decoding of position from multiunit activity using a marked point process filter. *Neural Computation*, *27*(7), 1438–1460.
DiMatteo, I., Genovese, C. R., & Kass, R. E. (2001). Bayesian curve fitting with free-knot splines. *Biometrika*, *88*, 1055–1071.
Donoghue, J. P. (2008). Bridging the brain to the world: A perspective on neural interface systems. *Neuron*, *60*(3), 511–521.
Doucet, A., de Freitas, N., & Gordon, N. (Eds.) (2001). *Sequential Monte Carlo methods in practice*. New York: Springer.
Eden, U. T., Frank, L. M., Barbieri, R., Solo, V., & Brown, E. N. (2004). Dynamic analysis of neural encoding by point process adaptive filtering. *Neural Computation*, *16*(5), 971–998.
Ehrens, D., Sritharan, D., & Sarma, S. (2015). Closed-loop control of a fragile network: Application to seizure-like dynamics of an epilepsy model. *Frontiers in Neuroscience*, *9*, 58.
Ergun, A., Barbieri, B., Eden, U. T., Wilson, M. A., & Brown, E. N. (2007). Construction of point process adaptive filter algorithms for neural systems using sequential monte carlo methods. *IEEE Transactions on Biomedical Engineering*, *54*(3), 419–428.
Faghih, R. T. (2014). System Identification of Cortisol Secretion: Characterizing Pulsatile Dynamics. Ph.D. thesis. Cambridge: Massachusetts Institute of Technology.

Faghih, R. T., Dahleh, M. A., Adler, G., Klerman, E., & Brown, E. N. (2014). Deconvolution of serum cortisol levels by using compressed sensing. *PLoS One*, *9*(1), e85204.

Faghih, R. T., Dahleh, M. A., Adler, G., Klerman, E., & Brown, E. N. (2015). Quantifying pituitary adrenal dynamics: Deconvolution of concurrent cortisol and adrenocorticotropic hormone data by compressed sensing. *IEEE Transactions on Biomedical Engineering*, *62*(10), 2379–2388.

Faghih, R. T., Dahleh, M. A., & Brown, E. N. (2015). Optimization formulation for characterization of pulsatile cortisol secretion. *Frontiers in Neuroscience*, *9*, 228.

Friedrich, J., Zhou, P., & Paninski, L. (2017). Fast online deconvolution of calcium imaging data. *PLoS Computational Biology*, *13*(3), e1005423.

Gale, J. T., Amirnovin, R., Williams, Z. M., Flaherty, A. W., & Eskandar, E. N. (2008). From symphony to cacophony: Pathophysiology of the human basal ganglia in Parkinson disease. *Neuroscience & Biobehavioral Review*, *32*(3), 378–387.

Gale, J. T., Shields, D. C., Jain, F. A., Amirnovin, R., & Eskandar, E. N. (2009). Subthalamic nucleus discharge patterns during movement in the normal monkey and Parkinsonian patient. *Brain Research*, *3*, 240–245.

Ganguly, K., & Carmena, J. M. (2009). Emergence of a stable cortical map for neuroprosthetic control. *PLoS Biology*, *7*(7), e1000153.

Gelman, A., Carlin, J. B., Stern, H. S., & Rubin, D. B. (2004). *Bayesian data analysis* (2nd ed.). London: Chapman & Hall/CRC Press.

Georgopoulos, A. P., Schwartz, A. B., & Kettner, R. E. (1986). Neuronal population coding of movement direction. *Science*, *233*, 1416–1419.

Gilja, V., Nuyujukian, P., Chestek, C. A., Cunningham, J. P., Yu, B. M., Fan, J. M., et al. (2012). A high-performance neural prosthesis enabled by control algorithm design. *Nature Neuroscience*, *15*, 1752–1757.

Gitelman, R., Penny, W., Ashburner, J., & Friston, K. (2003). Modeling regional and pyschophysiologic interactions in fMRI: The importance of hemodynamic deconvolution. *Neuroimage*, *19*, 200–207.

Gluckman, B. J., Nguyen, H., Weinstein, S. L., & Schiff, S. J. (2001). Adaptive electric field control of epileptic seizures. *Journal of Neuroscience*, *21*(2), 590–600.

Good, L. B., Sabesan, S., Marsh, S. T., Tsakalis, K., Treiman, D., & Iasemidis, L. (2009). Control of synchronization of brain dynamics leads to control of epileptic seizures in rodents. *International Journal of Neural Systems*, *19*(3), 173–196.

Grienberger, C., & Konnerth, A. (2012). Imaging calcium in neurons. *Neuron*, *73*(5), 862–885.

Grosenick, L., Marshel, J. H., & Deisseroth, K. (2015). Closed-loop and activity-guided optogenetic control. *Neuron*, *86*(1), 106–139.

Hjort, N. L., Holmes, C., Müller, P., & Walker, S. G. (Eds.) (2010). *Bayesian nonparametrics*. Cambridge: Cambridge University Press.

Hochberg, L. R., Bacher, D., Jarosiewicz, B., Masse, N. Y., Simeral, J. D., Vogel, J., et al. (2012). Reach and grasp by people with tetraplegia using a neurally controlled robotic arm. *Nature*, *485*, 372–375.

Hodgkin, A. L., & Huxley, A. F. (1952). A quantitative descrip-tion of membrane current and its application to conduction and excitation in nerve. *Journal of Physiology*, *117*(4), 500–544.

Huys, Q. J. M., Zemel, R. S., Natarajan, R., & Dayan, P. (2007). Fast population coding. *Neural Computation*, *19*, 404–441.

Izhikevich, E. M. (2006). *Dynamical systems in neuroscience: The geometry of excitability and bursting*. Cambridge: MIT Press.

Jarosiewicz, B., Chase, S. M., Fraser, G. W., Velliste, M., Kass, R. E., & Schwartz, A. B. (2008). Functional network reorganization during learning in a brain-computer interface paradigm. *Proceedings of the National Academy of Sciences USA*, *105*(49), 19486–19491.

Kalman, R. E. (1960). A new approach to linear filtering and prediction problems. *Transactions of the ASME–Journal of Basic Engineering*, *82*, 35–45.

Kass, R. E., Eden, U. T., & Brown, E. N. (2014). *Analysis of neural data*. New York: Springer.

Kass, R. E., & Ventura, V. (2001). A spike-train probability model. *Neural Computation*, *13*(8), 1713–1720.

Kass, R. E., Ventura, V., & Brown, E. N. (2005). Statistical issues in the analysis of neuronal data. *Journal of Neurophysiology, 94*, 8–25.

Kim, S., Putrino, D., Ghosh, S., & Brown, E. N. (2011). A granger causality measure for point process models of ensemble neural spiking activity. *PLoS Computational Biology, 7*(3), e1001110.

Kloosterman, F., Layton, S., Chen, Z., & Wilson, M. A. (2014). Bayesian decoding of unsorted spikes in the rat hippocampus. *Journal of Neurophysiology, 111*(1), 217–227.

Knight, B. W. (1972). Dynamics of encoding in a population of neurons. *Journal of General Physiology, 59*, 734–766.

Kobak, D., Brendel, W., Constantinidis, C., Feierstein, C. E., Kepecs, A., Mainen, Z. F., et al. (2016). Demixed principal component analysis of neural population data. *eLife, 5*, e10989.

Krishnaswamy, P., Bonmassar, G., Poulsen, C., Pierce, E. T., Purdon, P. L., & Brown, E. N. (2016). Reference-free removal of EEG-fMRI ballistocardiogram artifacts with harmonic regression. *NeuroImage, 128*, 398–412.

Kühn, A. A., Tsui, A., Aziz, T., Ray, N., Brücke, C., Kupsch, A., et al. (2009). Pathological synchronisation in the subthalamic nucleus of patients with parkinson's disease relates to both bradykinesia and rigidity. *Experimental Neurology, 215*, 380–387.

Kuncel, A. M., Cooper, S. E., Wolgamuth, B. R., Clyde, M. A., Snyder, S. A., Montgomery, E. B. J., et al. (2006). Clinical response to varying the stimulus parameters in deep brain stimulation for essential tremor. *Movement Disorder, 21*, 1920–1928.

Lamus, C., Hamalainen, M. S., Temereanca, S., Long, C. J., Brown, E. N., & Purdon, P. L. (2012). A spatiotemporal dynamic distributed solution to the MEG inverse problem. *NeuroImage, 63*(2), 894–909.

Lang, A. E., & Lozano, A. M. (1998). Parkinson's disease. First of two parts. *New England Journal of Medicine, 15*, 1044–1053.

Lawhern, V., Wu, W., Hatsopoulos, N. G., & Paninski, L. (2010). Population decoding of motor cortical activity using a generalized linear model with hidden states. *Journal of Neuroscience Methods, 189*, 267–280.

Lebedev, M. A., & Nicolelis, M. A. (2006). Brain-machine interfaces: Past, present and future. *Trends in Neurosciences, 29*(9), 536–546.

Lewicki, M. S. (1998). A review of methods for spike sorting: The detection and classification of neural action potentials. *Network, 9*(4), R53–R78.

Lewis, L. D., Setsompop, K., Rosen, B. R., & Polimeni, J. R. (2016). Fast fMRI can detect oscillatory neural activity in humans. *Proceedings of National Academy of Sciences, USA, 113*, E6679–E6685.

Li, X., Chen, Q., & Xue, F. (2017). Biological modelling of a computational spiking neural network with neuronal avalanches. *Philosophical Transactions of the Royal Society of London A: Mathematical, Physical and Engineering Sciences, 375*(2096), 1–16.

Malik, W. Q., Schummers, J., Sur, M., & Brown, E. N. (2011). Denoising two-photon calcium imaging data. *PLoS One, 6*(6), e20490.

Montgomery, E. B., & Gale, J. T. (2002). Deep brain stimulation for parkinsons disease: Disrupting the disruption. *Lancet Neurology, 1*, 225–231.

Montgomery, E. B., & Gale, J. T. (2008). Mechanisms of action of deep brain stimulation (DBS). *Neuroscience & Biobehavioral Review, 32*, 388–407.

Nandi, A., Kafashan, M., & Ching, S. (2017). Control analysis and design for statistical models of spiking networks. *IEEE Transactions on Control of Network Systems*, in press. https://doi.org/10.1109/TCNS.2017.2687824.

Okatan, M., Wilson, M., & Brown, E. (2005). Analyzing functional connectivity using a network likelihood model of ensemble neural spiking activity. *Neural Computation, 17*, 1927–1961.

Onativia, J., Schultz, S. R., & Dragotti, P. L. (2013). A finite rate of innovation algorithm for fast and accurate spike detection from two-photon calcium imaging. *Journal of Neural Engineering, 10*, 046017.

Orsborn, A. L., Dangi, S., Moorman, H. G., & Carmena, J. M. (2012). Closed-loop decoder adaptation on intermediate time-scales facilitates rapid bmi performance improvements independent of decoder initialization conditions. *IEEE Transactions on Neural Systems and Rehabilitation Engineering, 20*(4), 468–477.

Pascual-Marqui, R. D. (1999). Review of methods for solving the EEG inverse problem. *International Journal of Bioelectromagnetism, 1*(1), 75–86.
Pawitan, Y. (2001). *In all likelihood: Statistical modelling and inference using likelihood*. Gloucestershire: Clarendon Press.
Penny, W., Ghahramani, Z., & Friston, K. (2005). Bilinear dynamical systems. *Philosophical Transactions on Royal Society of London B, 360*, 983–993.
Perkel, D. H., & Bullock, T. H. (1968). *Neural coding: By Donald H. Perkel and Theodore Holmes Bullock*. Neurosciences Research Program (NRP).
Perlmutter, J. S., & Mink, J. W. (2006). Deep brain stimulation. *Annual Review in Neuroscience, 29*, 229–257.
Pnevmatikakis, E. A., Soudry, D., Gao, Y., Machado, T. A., Merel, J., Pfau, D., et al. (2016). Simultaneous denoising, deconvolution, and demixing of calcium imaging data. *Neuron, 89*(2), 285–299.
Prerau, M. J., Smith, A. C., Eden, U. T., Kubota, Y., Yanike, M., Suzuki, W., et al. (2009). Characterizing learning by simultaneous analysis of continuous and binary measures of performance. *Journal of Neurophysiology, 102*(5), 3060–3072.
Rahmati, V., Kirmse, K., Marković, D., Holthoff, K., & Kiebel, S. J. (2016). Inferring neuronal dynamics from calcium imaging data using biophysical models and Bayesian inference. *Nature Communications, 12*(3), e1004835.
Ressler, K. J., & Mayberg, H. (2007). Targeting abnormal neural circuits in mood and anxiety disorders: From the laboratory to the clinic. *Nature Neuroscience, 10*, 1116–1124.
Rieke, F., Warland, D., de Ruyter van Steveninck, R. R., & Bialek, W. (1997). *Spikes: Exploring the neural code*. Cambridge: MIT Press.
Ringach, D., & Shapley, R. (2004). Reverse correlation in neurophysiology. *Cognitive Science, 28*, 147–166.
Robert, C. P. (2007). *The Bayesian choice: From decision-theoretic foundations to computational implementation* (2nd ed.). New York: Springer.
Romano, S. A., Pérez-Schuster, V., Jouary, A., Boulanger-Weill, J., Candeo, A., Pietri, T., et al. (2017). An integrated calcium imaging processing toolbox for the analysis of neuronal population dynamics. *PLoS Computational Biology, 13*(6), e1005526.
Santaniello, S., Montgomery, E. B., Gale, J. T., & Sarma, S. V. (2012). Non-stationary discharge patterns in motor cortex under subthalamic nucleus deep brain stimulation: A review. *Frontiers in Integrative Neuroscience, 6*, 35.
Sarma, S. V., Cheng, M. L., Eden, U. T., Williams, Z., Brown, E. N., & Eskandar, E. N. (2012). The effects of cues on neurons in the basal ganglia in Parkinson's disease. *Frontiers in Integrative Neuroscience, 6*, 40.
Schliebs, S., & Kasabov, N. (2014). Computational modeling with spiking neural networks. In N. Kasabov (Ed.), *Springer handbook of bio-/neuroinformatics* (pp. 625–646). Berlin: Springer.
Schwartz, A. B., Cui, X. T., Weber, D. J., & Moran, D. W. (2006). Brain-controlled interfaces: Movement restoration with neural prosthetics. *Neuron, 52*(1), 205–220.
Shanechi, M. M. (2017). Brain-machine interface control algorithms. *IEEE Transactions on Neural Systems and Rehabilitation Engineering, 25*(10), 1725–1734.
Shanechi, M. M., Chemali, J. J., Liberman, M., Solt, K., & Brown, E. N. (2013). A brain-machine interface for control of medically-induced coma. *PLoS Computational Biology, 9*(10), e1003284.
Shanechi, M. M., Hu, R. C., Powers, M., Wornell, G. W., Brown, E. N., & Williams, Z. M. (2012). Neural population partitioning and a concurrent brain-machine interface for sequential motor function. *Nature Neuroscience, 15*(12), 1715–1722.
Shanechi, M. M., Orsborn, A. L., & Carmena, J. M. (2016). Robust brain-machine interface design using optimal feedback control modeling and adaptive point process filtering. *PLoS Computational Biology, 12*(4), e1004730.
Shanechi, M. M., Orsborn, A. L., Moorman, H. G., Gowda, S., Dangi, S., & Carmena, J. M. (2017). Rapid control and feedback rates enhance neuroprosthetic control. *Nature Communications, 8*, 13825.

Shanechi, M. M., Williams, Z. M., Wornell, G. W., Hu, R., Powers, M., & Brown, E. N. (2013). A real-time brain-machine interface combining motor target and trajectory intent using an optimal feedback control design. *PLoS One*, *8*(4), e59049.
Shanechi, M. M., Wornell, G. W., Williams, Z. M., & Brown, E. N. (2013). Feedback-controlled parallel point process filter for estimation of goal-directed movements from neural signals. *IEEE Transactions on Neural Systems and Rehabilitation Engineering*, *21*, 129–140.
Shenoy, K. V., Sahani, M., & Churchland, M. M. (2013). Cortical control of arm movements: A dynamical systems perspective. *Annual Review of Neuroscience*, *36*, 337–359.
Shimazaki, H., Amari, S., Brown, E. N., & Gruen, S. (2012). State-space analysis of time-varying higher-order spike correlation for multiple neural spike train data. *PLoS Computational Biology*, *8*(3), e1002385.
Smith, A. C., & Brown, E. N. (2003). Estimating a state-space model from point process observations. *Neural Computation*, *15*(5), 965–991.
Smith, A. C., Frank, L. M., Wirth, S., Yanike, M., Hu, D., Kubota, Y., et al. (2004). Dynamic analysis of learning in behavioral experiments. *Journal of Neuroscience*, *24*, 447–461.
Smith, A. C., Stefani, M. R., Moghaddam, B., & Brown, E. N. (2005). Analysis and design of behavioral experiments to characterize population learning. *Journal of Neurophysiology*, *93*, 1776–1792.
Smith, A. C., Wirth, S., Suzuki, W. A., & Brown, E. N. (2007). Bayesian analysis of interleaved learning and response bias in behavioral experiments. *Journal of Neurophysiology*, *97*, 2516–2524.
Sohal, V. S., & Sun, F. T. (2011). Responsive neurostimulation suppresses synchronized cortical rhythms in patients with epilepsy. *Neurosurgery Clinics of North America*, *22*(4), 481–488.
Srinivasan, L., Eden, U. T., Willsky, A. S., & Brown, E. N. (2006). A state-space analysis for reconstruction of goal-directed movements using neural signals. *Neural Computation*, *18*, 2465–2494.
Stavisky, S. D., Kao, J. C., Nuyujukian, P., Ryu, S. I., & Shenoy, K. V. (2015). A high performing brain-machine interface driven by low-frequency local field potentials alone and together with spikes. *Journal of Neural Engineering*, *12*, 036009.
Stevenson, I. H., & Kording, K. P. (2011). How advances in neural recording affect data analysis. *Nature Neuroscience*, *14*, 139–142.
Stevenson, I. H., London, B. M., Oby, E. R., Sachs, N. A., Reimer, J., Englitz, B., et al. (2009). Bayesian inference of functional connectivity and network structure from spikes. *IEEE Transactions on Neural Systems and Rehabilitation Engineering*, *17*, 203–213.
Stokes, P. A., & Purdon, P. L. (2017). A study of problems encountered in Granger causality analysis from a neuroscience perspective. *Proceedings of National Academy of Sciences, USA*, *114*(34), E7063–E7072.
Taylor, D. M., Tillery, S. I. H., & Schwartz, A. B. (2002). Direct cortical control of 3D neuroprosthetic devices. *Science*, *296*, 1829–1832.
Thakor, N. V. (2013). Translating the brain-machine interface. *Science Translational Medicine*, *5*, 210–217.
Theis, L., Berens, P., Froudarakis, E., Reimer, J., Rosón, M. R., Baden, T., et al. (2016). Benchmarking spike rate inference in population calcium imaging. *Neuron*, *90*(3), 471–482.
Tommasi, G., Lanotte, M., Albert, U., Zibetti, M., Castelli, L., Maina, G. et al. (2008). Transient acute depressive state induced by subthalamic region stimulation. *Journal of Neurological Sciences*, *273*, 135–138.
Truccolo, W., & Donoghue, J. P. (2007). Nonparametric modeling of neural point processes via stochastic gradient boosting regression. *Neural Computation*, *19*(3), 672–705.
Truccolo, W., Eden, U. T., Fellows, M. R., Donoghue, J. P., & Brown, E. N. (2005). A point process framework for relating neural spiking activity to spiking history, neural ensemble, and extrinsic covariate effects. *Journal of Neurophysiology*, *93*(2), 1074–1089.
Truccolo, W., Fiehs, G. M., Donoghue, J. P., & Hochberg, L. R. (2008). Primary motor cortex tuning to intended movement kinematics in humans with tetraplegia. *Journal of Neuroscience*, *28*(5), 1163–1178.

Velliste, M., Perel, S., Spalding, M. C., Whitford, A. S., & Schwartz, A. B. (2008). Cortical control of a prosthetic arm for self-feeding. *Nature*, *453*, 1098–1101.

Ventura, V. (2008). Spike train decoding without spike sorting. *Neural Computation*, *20*(4), 923–963.

Ventura, V. (2009). Traditional waveform based spike sorting yields biased rate code estimates. *Proceedings of National Academy of Science, USA*, *106*, 6921–6926.

Vogelstein, J., Packer, A., Machado, T. A., Sippy, T., Babadi, B., Yuste, R., & Paninski, L. (2010). Fast nonnegative deconvolution for spike train inference from population calcium imaging. *Journal of Neurophysiology*, *104*, 3691–3704.

Vogelstein, J., Watson, B., Packer, A., Yuste, R., Jedynak, B., & Paninski, L. (2009). Spike inference from calcium imaging using sequential Monte Carlo methods. *Biophysical Journal*, *97*(2), 636–655.

Wei, X. F., & Grill, W. M. (2009). Impedance characteristics of deep brain stimulation electrodes in vitro and in vivo. *Journal of Neural Engineering*, *6*, 046008.

Wendel, K., Väisämen, O., Malmivuo, J., Gencer, N. G., Vanrumste, B., Durka, P., et al. (2009). EEG/MEG source imaging: Methods, challenges, and open issues. *Computational Intelligence and Neuroscience*, *2009*, 656092.

Wichmann, T., & DeLong, M. (2006). Deep brain stimulation for neurologic and neuropsychiatric disorders. *Neuron*, *52*(1), 197–204.

Willett, F. R., Suminski, A. J., Fagg, A. H., & Hatsopoulos, N. G. (2013). Improving brain-ĂŞmachine interface performance by decoding intended future movements. *Journal of Neural Engineering*, *10*(2), 026011.

Wirth, S., Yanike, M., Frank, L. M., Smith, A. C., Brown, E. N., & Suzuki, W. A. (2003). Single neurons in the monkey hippocampus and learning of new associations. *Science*, *300*, 1578–1584.

Wong, K. F. K., Smith, A. C., Pierce, E. T., Harrell, P. G., Walsh, J. L., Salazar-Gomez, A. F., et al. (2014). Statistical modeling of behavioral dynamics during propofol-induced loss of consciousness. *Journal of Neuroscience Methods*, *227*, 65–74.

Wu, W., Gao, Y., Bienenstock, E., Donoghue, J. P., & Black, M. J. (2006). Bayesian population decoding of motor cortical activity using a Kalman filter. *Neural Computation*, *18*(1), 80–118.

Wu, W., Kulkarni, J. E., Hatsopoulos, N. G., & Paninski, L. (2009). Neural decoding of hand motion using a linear state-space model with hidden states. *IEEE Transactions on Neural Systems and Rehabilitation Engineering*, *17*, 370–378.

Wu, W., Nagarajan, S., & Chen, Z. (2016). Bayesian machine learning: EEG/MEG signal processing measurements. *IEEE Signal Processing Magazine*, *33*(1), 14–36.

Yu, B. M., Cunningham, J. P., Santhanam, G., Ryu, S. I., Shenoy, K. V., & Sahani, M. (2009). Gaussian-process factor analysis for low-dimensional single-trial analysis of neural population activity. *Journal of Neurophysiology*, *102*(1), 614–635.

Zahodne, L. B., Young, S., Darrow, L. K., Nisenzon, A., Fernandez, H. H., Okun, M. S., et al. (2009). Examination of the lille apathy rating scale in Parkinson disease. *Movement Disorder*, *24*(5), 677–683.

Zemel, R. S., Dayan, P., & Pouget, A. (1998). Probabilistic interpretation of population codes. *Neural Computation*, *10*(2), 403–430.

Zhang, K., Ginzburg, I., McNaughton, B. L., & Sejnowski, T. J. (1998). Interpreting neuronal population activity by reconstruction: Unified framework with application to hippocampal place cells. *Journal of Neurophysiology*, *79*(2), 1017–1044.

Zhou, B., Moorman, D., Behseta, S., Ombao, H., & Shahbaba, B. (2016). A dynamic bayesian model for characterizing cross-neuronal interactions during decision making. *Journal of American Statistical Association*, *111*, 459–471.

Zhuang, J., Truccolo, W., Vargas-Irwin, C., & Donoghue, J. P. (2009). Decoding 3-D reach and grasp kinematics from high-frequency local field potentials in primate primary motor cortex. *IEEE Transactions on Biomedical Engineering*, *57*(7), 1774–1784.

Part I
Statistics & Signal Processing

Chapter 2
Characterizing Complex, Multi-Scale Neural Phenomena Using State-Space Models

Uri T. Eden, Loren M. Frank, and Long Tao

2.1 Introduction

Understanding neural encoding requires describing the relationship between the input, or stimulus, presented to a neuron or neural circuit, and its output, or response. Early models in many neural systems often focused on the responses of individual neurons (Felleman and Kaas 1984; Chapin 1986; Girman et al. 1999) to simple stimuli (Kuffler 1953; Hubel and Wiesel 1962; Rodieck 1965; Jones and Palmer 1987), assuming stationarity and treating anatomically connected brain areas in isolation. In the past few decades there has been a massive expansion in our ability to record neural activity: we can now record from many more neurons, across multiple brain areas, and at multiple spatial and temporal scales. These technological advances have enabled neuroscientists to analyze more complex neural coding and communication properties of both stimulus and response. First, on the response side, recent models have relaxed or removed the assumption of stationarity, admitting models that capture response dynamics such as adaptation and plasticity (Rao and Ballard 1997; Brown et al. 2001; Frank et al. 2002; Eden et al. 2004). Furthermore, an increase in the number of neurons that can be simultaneously recorded has enabled modeling not only of receptive field properties of individual neurons but also modeling of the ways that neural populations coordinate to represent stimulus features or to extract particular types of information from their inputs (Paninski et al. 2004; Chapin 2004; Shanechi et al. 2012; Archer et al. 2014). Recording from multiple brain areas has likewise permitted study of coordination on a larger scale,

U.T. Eden (✉) · L. Tao
Boston University, Boston, MA, USA
e-mail: tzvi@bu.edu

L.M. Frank
University of California, San Francisco, San Francisco, CA, USA
e-mail: loren@phy.ucsf.edu

Z. Chen, S.V. Sarma (eds.), *Dynamic Neuroscience*,
https://doi.org/10.1007/978-3-319-71976-4_2

in this case between brain areas (Bullmore and Sporns 2009; Stephen et al. 2014). Second, on the stimulus side, recent analyses have built more complex models in which stimuli are dynamic and multidimensional. The resulting models are more applicable to real-world situations. Finally, while many earlier encoding analyses often focused on neural phenomena that could be described by spiking alone, more recent studies have sought to describe phenomena that involve electrophysiological signals at multiple spatial and temporal scales, incorporating the summed activity of many neurons in the form of the LFP, and analyzing coordination of both LFP and spikes across brain regions.

One illustrative example of this increase in the complexity of neural coding analyses is the study of place specific activity in the structures of the medial temporal lobe (MTL) of the rat. Early analyses in this domain examined the responses of individual neurons in hippocampus to a rat's location during simple spatial navigation tasks. Initially, these place fields were often modeled as static and unimodal (O'Keefe and Dostrovsky 1971; Muller et al. 1987). Subsequent analyses incorporated additional levels of complexity, for example by integrating information from spiking and LFP to identify phase precession of place specific spiking to the theta rhythm of the LFP (O'Keefe and Recce 1993; Skaggs and McNaughton 1996), by building dynamic models to capture place field plasticity during learning (Brown et al. 2001; Frank et al. 2002; Eden et al. 2004; Huang et al. 2009), by examining the coding properties of other signals related to spatial navigation such as velocity and head direction (Knierim et al. 1995), by examining more complicated spatial field structures such as grid cell patterns (Hafting et al. 2005), and by examining coordination and neural processing across multiple areas of the MTL, as in analyses of the entorhinal grid cell activity that gives rise to hippocampal place field structure (O'Keefe and Burgess 2005; Fuhs and Touretzky 2006; McNaughton et al. 2006; Solstad et al. 2006).

Two common themes underlie the evolution of experimental and analytical approaches in this field. First is the progression from simple, static, low-dimensional stimulus response relationships to complex, dynamic, high-dimensional representations. Understanding the mechanisms and effects of such phenomena requires the ability to integrate information from multiple sources across neural ensembles, brain regions, and scales. Second is the goal of identifying and estimating variables that are difficult to observe directly. These might include the information available from the entire active neuronal ensemble, from a particular brain region or an even more abstract notion like the current learning state of the animal. Estimating the value of these variables and understanding how they are transformed by neural systems is in fact a central goal of systems neuroscience, but experimental neuroscientists are often limited in the statistical and data analysis tools available to address directly the associated estimation problem.

The state-space paradigm, whose application to complex neural phenomena has been pioneered by Emery Brown and his colleagues, provides a natural statistical modeling approach for integrating information across multiple sources and scales, for discovering low dimensional representations of behavioral and cognitive states, and for expressing confidence about estimates and inferences (Brown et al. 1998;

Eden et al. 2004; Archer et al. 2014; Smith and Brown 2003; Chen et al. 2010). State-space methods have a long history in the engineering literature, where the observed signals are assumed to have helpful properties such as stationarity, linearity, and Gaussianity. Their application to neural coding analyses, where signals such as spike trains rarely have such properties, is more recent.

The fundamental idea of the state-space approach is to define two probability models. The first describes the evolution of an unobserved dynamic signal, called the state process. The second describes how this state affects the observed data. Using these two models, it is often possible to derive expressions to estimate the unobserved state as well as the parameters for both models, providing a clear path to estimate underlying but unobservable variables from brain activity.

One early application of this paradigm was to decode movement trajectories of a rat actively exploring its environment using spiking data from a population of place cells (Brown et al. 1998; Zhang et al. 1998; Barbieri et al. 2004). A place cell will increase its firing rate above baseline when the animal is in a particular location in space, known as the cell's place field. In this application of the state-space paradigm, the state process represents the movement of the rat in space, the observation model describes the place field(s) of each cell, and a point process filter is derived to decode the rat's movement trajectory at each instant. This constituted a test of the fidelity of the hippocampal place code, as the decoded spatial trajectory could be checked against the known location of the rat. Other early methodological derivations and applications included an expectation-maximization (EM) algorithm to estimate a dynamic cognitive learning state from binary (correct vs. incorrect) task performance data (Smith et al. 2004; Coleman et al. 2006), and receptive field models with dynamic parameters to track plasticity during learning in hippocampal place fields and elsewhere (Brown et al. 2001; Frank et al. 2002; Eden et al. 2004). In recent years, this paradigm has been adapted to address many of the more complicated classes of phenomena described above. Some recent applications include tracking dynamic spiking rhythms in the subthalamic nucleus of Parkinsonian patients performing reaching tasks (Deng et al. 2013) and fitting parameters of dynamical systems and conductance-based models of spiking neurons (Meng et al. 2011).

In this chapter, we will review the fundamental features of the state-space paradigm, discuss successful applications of the paradigm to various neural data analysis problems, and introduce a novel extension of these methods to better understand the phenomenon of hippocampal replay. We present the basic structure of state-space models that include point-process observations and develop the filter equations used for estimating dynamic signals from neural spiking. We then discuss two recently published applications of this paradigm as illustration of its power and versatility. Finally, we provide a new specification of this approach to address the problem of identifying hippocampal ripple-replay events in the rat. Replay is defined as the activation of a set of neurons that recapitulate patterns of activity associated with specific behaviors in the absence of the animal executing those behavior. Ripple replay is seen when sequences of hippocampal place cells are reactivated in patterns that are similar to those seen during active exploration, but typically on a much faster timescale and when animals are still (Carr et al. 2011; Buzsáki 2015). The goals of this new analysis are: first, to define a hippocampal replay event in terms of a specific non-local representation of position, the rhythms in the LFP, and the spiking patterns of a population of place cells; second, to compute, at each instant, the probability of a replay event occurring; and third, to decode the information content of each replay event. To address these challenges, we develop a new state-space model that includes one state variable that indicates whether or not a replay is occurring and another semi-latent state that is given by the rat's observed position during active exploration, but during replay events is considered an unobserved process. By assuming that the place cells fire similarly during exploratory movement and during replay, we are able to decode the replay state and the probability of replay occurring at each instant in time. We illustrate this new method on hippocampal data from a rat performing a spatial memory task.

The remainder of this chapter is organized as follows. In Sect. 2.2, we present the fundamental structure of the state-space paradigm for spike train observations, and derive a point process filter algorithm to estimate dynamic signals from spiking data. In Sect. 2.3, we briefly highlight a couple of recent applications of the state-space paradigm to different classes of problems. In Sect. 2.4, we discuss the hippocampal ripple-replay estimation problem and derive a new state-space model to address it, illustrating the result on a hippocampal dataset. In Sect. 2.5, we discuss some reasons for the success of this state-space approach and some future directions for these methods.

2.2 State-Space Models

The state-space paradigm is well suited to neural data analysis problems where an observed signal is influenced by some set of unknown or unobserved factors, which may themselves change in time. The unknown or unobserved signals are called latent states. For example, if we record extracellularly from a neuron, we might use a latent state to describe the dynamics of an unobserved membrane potential and of the membrane conductance of a particular set of ions. We can use such an approach to solve problems related to estimating the latent signals, fitting models between the latent and observed signals, and performing statistical tests about their relationship.

To construct a state-space model, we must define a pair of statistical models. The first model, called the state model, describes the probabilistic dynamics of the latent state process. The second model, called the observation model, describes how the latent state influences the probability distribution of the observation process at each time. In the example above of extracellularly observed spiking as a function of a latent state for the membrane potential, the state model might be defined by a stochastic set of Hodgkin-Huxley equations (Meng et al. 2011), and the observation model might define the probability of observing a spike in the next instant, given the current value of the membrane potential.

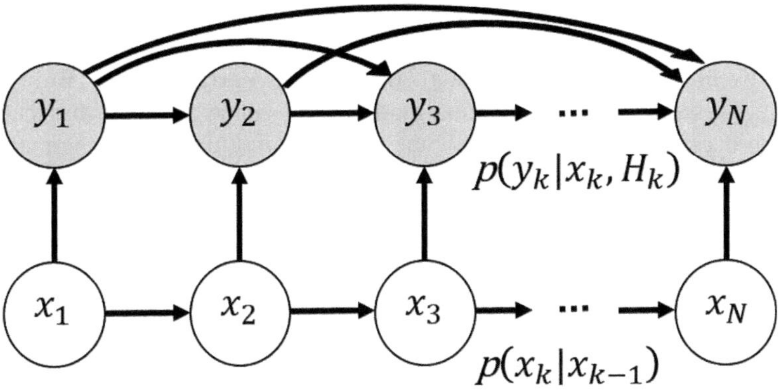

Fig. 2.1 The state-space paradigm. An unobserved state process x_k undergoes stochastic dynamics and influences an observation process y_k

To establish some mathematical notation, let x_k denote the state process and y_k denote the observation process at time t_k. For notational convenience, we will also define H_k to be the past history of the observation process prior to time t_k. While we could define these processes in either continuous or discrete time, here we will focus on a discrete time representation, in which case, $H_k = y_{1:k-1}$ represents the collection of discrete observations between time steps 1 and $k - 1$.

Using this notation, we can define the state model as a conditional probability distribution $p(x_k|x_{1:k-1})$, where $x_{1:k-1}$ represents the history of the state variable between time steps 1 and $k - 1$. We typically further assume that the state is a Markov process, which means that given the value of the state at any time, its future values are independent of any of its past values. Mathematically, this means that $p(x_k|x_{1:k-1}) = p(x_k|x_{k-1})$. We will make this common assumption here, but it is easy to extend the methods for states with longer history dependence structure.

We can similarly define the observation model using a conditional probability distribution $p(y_k|x_k, H_k)$. Here we have assumed that the observation process depends only on the value of the state at the current time, although it can still depend on past values of the observation process. The influence of each of the state and observation processes on each other is shown as a graphical model in Fig. 2.1. From this illustration, it is clear that the state variable at each time step influences both the observation at that time and the state at the next time step.

Here, we are particularly concerned with observations processes that include neural spiking data. In that case, the observation process y_k is equal to, or has as a component, the number of spikes fired in a sequence of small time intervals. We will write $y_k = \Delta N_k$, where ΔN_k is the number of spikes that occur between times t_{k-1} and t_k, which is called the spike increment process. We define a neural spiking model by writing an expression for the conditional intensity of firing, $\lambda(t|H_t)$, which defines the instantaneous probability of seeing a spike around time t, $\lambda(t|H_t) = \lim_{\Delta t \to 0} \Pr(\text{spike in } [t, t + \Delta t)|H_t)/\Delta t$ (Daley and

Vere-Jones 2003; Brown et al. 2003), as a function of the state process. Once we have an expression for this conditional intensity, the observation distribution is given by $p(\Delta N_k|x_k, H_k) \approx \exp\{\log(\lambda(t_k|H_k))\Delta N_k - \lambda(t_k|H_k)\Delta t_k\}$ (Brown et al. 2003). These state and observation models fully define the joint distribution of these processes. They are therefore the basic building blocks for computing any probabilities associated with these states.

For example, a common problem for state-space models is estimating the probability distribution of x_k given all of the observations up to and including time t_k; that is, determining the conditional probability density $p(x_k|y_k, H_k)$. This is called the filtering problem, and when the observations are spikes, the solution to this problem is called a point process filter. For the state-space model of Hodgkin-Huxley spiking described above, this would mean estimating the membrane potential and ionic currents at each time based on all of the spiking up to that time.

We can solve the point process filter by applying Bayes' rule to the desired conditional probability density, $p(x_k|\Delta N_k, H_k)$, called the filter density, to switch the current state and observation terms, x_k and ΔN_k. The filter density can then be expressed as

$$p(x_k|\Delta N_k, H_k) \propto p(\Delta N_k|x_k, H_k)p(x_k|H_k) \tag{2.1}$$

The first term on the right-hand side of Eq. (2.1) is the observation distribution. The second term, $p(x_k|H_k)$, called one-step prediction density, defines the conditional probability of the state at time t_k given all of the observation up to, but not including, the most recent. This one-step prediction density can be computed using the *Chapman-Kolmogorov equation*

$$p(x_k|H_k) = \int p(x_k|x_{k-1})p(x_{k-1}|\Delta N_{k-1}, H_{k-1})dx_{k-1} \tag{2.2}$$

where $p(x_{k-1}|\Delta N_{k-1}, H_{k-1})$ in the integrand is the filter density from previous time t_{k-1}. Plugging (2.2) into (2.1), we get

$$p(x_k|\Delta N_k, H_k) \propto p(\Delta N_k|x_k, H_k) \int p(x_k|x_{k-1})p(x_{k-1}|\Delta N_{k-1}, H_{k-1})dx_{k-1} \tag{2.3}$$

Equation (2.3) provides an iterative formula to calculate the filter density at each time step from the density at the previous time using the state and observation models.

Typically, the integral in Eq. (2.3) does not have an analytical solution and we need to solve it numerically or find a suitable approximation. When x_k is a scaler or is low dimensional, simple numerical methods such as Riemann sums might be sufficiently accurate and computationally efficient to compute the filter density. If x_k is high dimensional, alternative methods such as Gaussian approximate solutions (Brown et al. 1998; Eden et al. 2004; Smith and Brown 2003), and sequential Monte

Carlo methods (Doucet et al. 2001; Ergun et al. 2007) have been used successfully to solve point process filter problems.

2.3 Applications of the State-Space Paradigm

In the previous section, we introduced the fundamental structure of the state-space paradigm and derived the point process filter algorithm for spike train observations. Specific instantiations of this approach have been successfully applied to a number of different neural coding problems, including spike train filtering and smoothing, stimulus decoding, estimating spatially varying firing rates, and reconstructing goal-directed hand movement, among many others. Here we will highlight two recent applications of the state-space paradigm to neural spiking data for two very different classes of problems.

2.3.1 Decoding Movement Trajectories from a Place Cell Population in Hippocampus

Huang et al. used a state-space approach to decode the movement trajectories and future turn decisions of a rat navigating through a maze from ensemble spiking from hippocampal place cells (Huang et al. 2009). A rat was trained to navigate up the stem of a T-shaped maze and alternate between left and right turns, before returning to the base of the stem through one of two return arms (see Fig. 2.2b). This decoding problem presented a number of statistical challenges: the state and observation models needed to be designed to capture information about position and future turn direction simultaneously, each neuron potentially had multimodal place field structure, the track itself has a topological structure that made computing the integrals for the point process filter challenging and led to multimodal filter distributions.

Careful selection of the state and observation models allowed for each of these challenges to be addressed. The state was defined using a linearization of the track, where negative values denoted trajectories that included a left turn and positive values denoted trajectories that included a right turn at the top of the stem. This means that any position on the stem is coded using two values, one positive and one negative. When the point process filter decodes a position on the stem, the sign of the decoded value is also used to predict the future turn decision. Note that this state process is not completely observed at every time—if we observe the rat on the stem at time t, we don't know whether this corresponds to a positive or negative state value until later when the rat makes a turn. However, the state-space framework is designed to capture states that may not be fully observable.

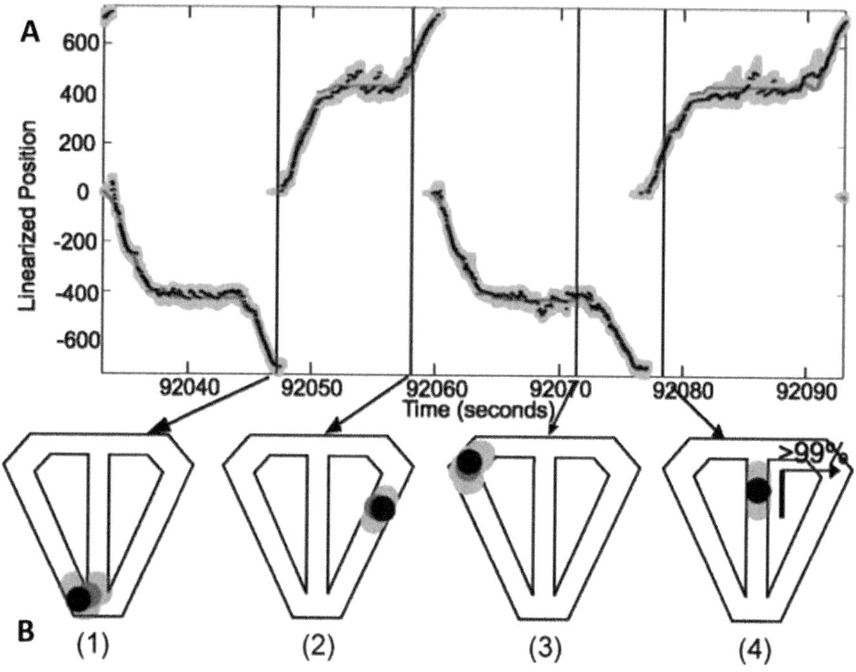

Fig. 2.2 Decoding movement trajectories using a point process filter. (**a**) The plot of the linearized position versus time for 1 min of the experiment. The continuous dark gray path represents the actual position of the animal, and the discontinuous black points represent the predicted position of the animal, with light gray being 95% confidence bounds for the estimate. The estimated positions in black almost overlap the true positions in dark gray. (**b**) Actual and predicted position with 95% confidence bounds of the animal, mapped back to the original T-maze. Adapted from Huang et al. (2009)

An observation model was selected by setting the conditional intensity of spiking to be a spline based function of this linearized state variable. This both allowed the place fields to have peaks at multiple positions along the track, and to have different rates at the same position on the stem for periods preceding right vs left turns. The shape of this spline-based place field model was fit for each of 47 neurons individually during a first encoding period along with an empirical model of the rat's movement. Then, in a second decoding phase, a point process filter was used to estimate the distribution of the rat's position and its next turn decision at each time point. The decoding result shown in Fig. 2.2 demonstrates the capability of the state-space paradigm to characterize the complicated dependence relations between the spiking of the hippocampal population and the rat's movements.

2.3.2 Estimating Biophysical Neural Models from Spikes

In contrast to the first example that used a state-space model to capture the receptive field structure of a population of neurons in response to external, behavioral signals, Meng et al. used the state-space paradigm to estimate a dynamical model for the internal ionic currents and membrane potential fluctuations that generate spiking (Meng et al. 2011). Here, the goal was to estimate multiple parameters of a stochastic Hodgkin-Huxley model directly using only the observed spike times. This model has non-linear dynamics and possesses multiple unknown parameters and unobserved dynamic variables, making the model fitting problem particularly challenging.

To solve this problem, a state-space model was defined with a multivariate state process representing the unobserved dynamical variables (membrane potential and ionic conductances of sodium Na^+ and potassium K^+) as well as the fixed, unknown model parameters. The state model reflected the nonlinear dynamics described by the Hodgkin-Huxley differential equations. The observation model was given by a conditional intensity that remained small until the membrane potential component of the state approached a threshold value, at which point it increased rapidly.

Since this model used a high dimensional state process with nonlinear dynamics, the integral on the right-hand side of Eq. (2.3) is not simple to compute numerically. In this case a sequential Monte Carlo (SMC) algorithm, or particle filter, was used to estimate the filter probability distribution of the dynamic components of the state as well as the unknown model parameters. Particle filters use random samples in order to estimate probability distributions and perform computations on them. Each particle represents a possible value of the state at a particular time, and has a weight related to the likelihood of the observations given that state value. The particles are repeatedly resampled based on these weights so that their distribution reflects the filter probability distribution. The SMC method used here incorporated both future and past spiking information to calculate the weight of each particle and identify sets of model parameters that were consistent with the spiking observations.

Figure 2.3 shows an example of the estimation procedure. Here six variables were estimated at each time point: the dynamic variables included the membrane

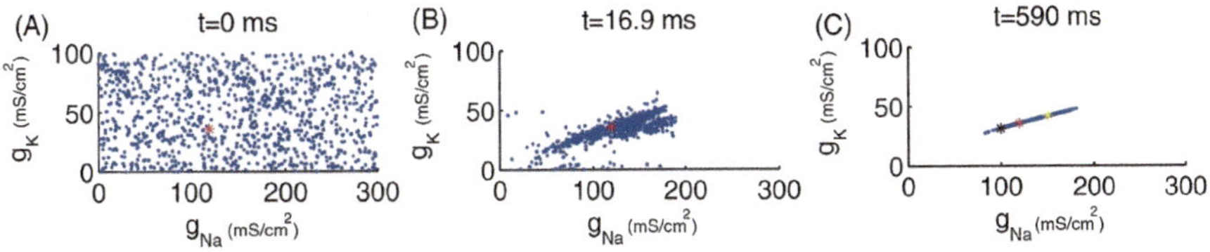

Fig. 2.3 Sequential parameter estimates for conductance parameters g_K and g_{Na}. The red asterisk denotes the true values for g_K and g_{Na}. The blue dots denote the parameter values for all of the particles. (**a**) The initial particle estimates are uniformly distributed in the parameter space. (**b**) Distribution of particles after the second observed spike. (**c**) Distribution of particles after 40 spikes. Adapted from Meng et al. (2011)

potential, and the ionic currents for Na^+ and K^+; the fixed parameters included the input current and the maximum conductance values for Na^+ and K^+, called g_K and g_{Na} respectively. In Fig. 2.3, the particle values for g_K and g_{Na} are shown at three time points: at the start, after two observed spikes, and after 40 observed spikes. Initially the particle estimates are uniformly distributed in the parameter space. After the second observed spike, the parameter values of the particles begin to concentrate in a region that contains the true values of for g_K and g_{Na}. After 40 spikes, the parameter estimates have converged to a narrow linear subspace of parameter values that are consistent with the spike data.

2.4 Identifying Replay Event from Multimodal Information

In Sect. 2.3, we have highlighted some examples of state-space models being used to describe complex neural phenomena and solve challenging estimation problems. In both instances, the challenges were overcome by carefully selecting the structure of the state and observation models so that the desired statistical relationships could be captured. Here we illustrate the development of a new state-space algorithm, which addresses some similar challenges as well as some new ones, associated with a neural phenomenon of recent interest, the detection and characterization of hippocampal replay events. As described in the Introduction section, replays are the sequential firing patterns of hippocampal place cells that represent previous experience and occur frequently during periods of awakeness (Buzsáki 1986; Wilson and McNaughton 1994; Diekelmann et al. 2011; Carr et al. 2011). They are multifaceted phenomena that involve features of multiple signals, including the rat's behavior (replay is thought to occur primarily during so-called sharp-wave ripple events that are most prevalent during low speed movement and immobility), hippocampal LFP (the presence of power in the ripple band of 150–250 Hz is often used to detect candidate replay events), and ensemble spiking activity. During active exploration of an environment, we might expect to see the rat's movement velocity fluctuate, the LFPs in hippocampus maintain a 8–12 Hz theta rhythm, and neurons fire with place receptive field structure based on the rat's current position. However, during replay events, we would expect the rat to remain still, the hippocampal LFPs to show low frequency sharp waves and high frequency ripples between 150–250 Hz, and to see patterns of spiking that resemble patterns that occur during active exploration but which are not directly related to the rat's actual position (Karlsson and Frank 2009; Davidson et al. 2009).

While in some cases, it may be easy to identify clear replay events with these properties by eye, in other cases these events may be hard to detect. From this description alone, it can be challenging to express one's confidence that a particular event is or is not an example of a replay event. Similarly, it can be challenging to define mathematically the degree to which each component signal supports or undermines that categorization.

Recent research has highlighted the potential role of hippocampal replay in learning, memory consolidation, and decision-making (Buzsáki 1986; Wilson and McNaughton 1994; Diekelmann et al. 2011; Carr et al. 2011). Thus the ability to detect these replay events, to define the periods over which they occur, and to express confidence about these estimates is critical. Additionally, researchers are often interested in decoding the information content of ripple events, that is, reconstructing a movement trajectory along which the observed spiking patterns might occur.

2.4.1 *Defining the State-Space Model*

The process of developing a state-space model to tackle this problem begins by coming up with a structure for the state variable. In this case, we are interested in knowing whether the brain is currently in a replay state or not, and if so, what kind of movement trajectory might correspond to the observed replay spike pattern. This suggests that our state variable should include two components, one binary indicator variable, call it I_k, which defines whether a replay is occurring at time t_k, and a second continuous variable, call it x_k, which will be used to express how neurons fire during replay.

We define the binary replay indicator state I_k so that

$$I_k = \begin{cases} 1, & \text{if a replay is occurring at time } t_k \\ 0, & \text{if a replay is not occurring at time } t_k \end{cases} \tag{2.4}$$

This is a latent state process in that we cannot directly observe the value of I_k at any moment. Instead, we will define observation models for the LFP and spiking activity as a function of this state, and then try to estimate the probability distribution of being in a replay state at any time.

We also need to define a continuous state variable, x_k, to describe the factors that influence spiking both in and out of the replay state. When out of the replay state, the neurons have place fields which fire as a function of the rat's position, m_k, at time t_k. Therefore, whenever $I_k = 0$, we set $x_k = m_k$. During periods when $I_k = 1$, we treat x_k as an unobserved variable such that the spiking intensity as a function of x_k during replay is equivalent to the spiking intensity as a function of position m_k during movement. Therefore, the observation model will be the same function of x_k, whether a replay state is occurring or not. x_k is observed whenever $I_k = 0$, but unobserved, or latent, whenever $I_k = 1$. Therefore, we call x_k a semi-latent state process. One goal is to estimate the trajectory of x_k through time during replay periods.

Now that we have defined the state variables, the next step is to define a state model for the temporal evolution of the states. For the discrete replay indicator state, we assume that the probability of being in a replay state at time t_k only depends on the values of the state and the rat's movement velocity at the previous time step, $\Pr(I_k = 1|I_{k-1}, v_{k-1})$. We model this probability using a pair of logistic models as a function of the rat's velocity for both possible values of $I_{k-1} \in \{0, 1\}$.

For the semi-latent state x_k, whenever $I_k = 0$, the state just follows the observed movement trajectory of the rat. Whenever the replay indicator state switches into a replay, we chose to make the distribution of the now unknown value of x_k uniform over the full space. Finally, when a replay state persists from one time step to the next, we assume the state update follows a zero mean random walk with a covariance based on the movement statistics of m_k, sped up by a factor of 20 (Nádasdy et al. 1999; Lee and Wilson 2002; Davidson et al. 2009). Mathematically, we define the semi-latent state equation as:

$$p(x_k|x_{k-1}, I_k, I_{k-1}) = \begin{cases} \delta(m_k), & \text{if } I_k = 0 \\ U(0, 200) & \text{if } I_k = 1, \text{ and } I_{k-1} = 0 \\ N(x_{k-1}, \hat{\sigma}) & \text{if } I_k = 1, \text{ and } I_{k-1} = 1 \end{cases} \tag{2.5}$$

With the state variables and state evolution model defined, the final step is to build models for all of the observed signals as functions of the states. The observations processes are the short time Fourier transform of the LFP in hippocampal area CA1, y_k, the rat's velocity, v_k, and the hippocampal neural spiking activity, $\Delta N_k^{(1:C)}$, at time t_k, where C is the total number of recorded neurons. We assume that the hippocampal LFP is influenced by the binary replay indicator state, but not the semi-latent state and define a multivariate Gaussian model, $p(y_k|I_k) \sim \mathcal{N}(\mu(I_k), \Sigma)$, where $\mu(I_k)$ is the mean power at each frequency in and out of the replay state, and Σ is a model covariance. We assume that the rat's velocity follows a random walk with a covariance that depends on the binary replay indicator state, $p(v_k|v_{k-1}, I_k) \sim \mathcal{N}(v_{k-1}, \varsigma(I_k))$, where $\varsigma(I_k)$ is the variability of the movement velocity in and out of the replay state. Finally, the spiking activity of each neuron is assumed to be a doubly stochastic Poisson process with a firing rate that depends on the value of the semi-latent state x_k, $p(\Delta N_k^c|x_k) \sim \texttt{Poisson}(\lambda^c(x_k)\Delta t_k)$, where $\lambda^c(x_k)$ is the firing rate function for neuron c as a function of x_k. Recall that when $I_k = 0$, x_k is the rat's position, and the firing model describes the neuron's place field; when $I_k = 0$, x_k is an unobserved state, but the neural firing as a function of this unknown value of x_k remains the same (Nádasdy et al. 1999; Lee and Wilson 2002; Davidson et al. 2009).

2.4.2 *A Filter to Identify and Decode Replay Events*

In Sect. 2.2, we discussed a general solution to the filter problem with spike train observations. Here, we work out the specific solution for the replay state and multimodal observation models discussed above. One goal is to compute at each instant the probability that a replay state is occurring, given all of the observed signals up to the current time, $\Pr(I_k|y_{1:k}, \Delta N_{1:k}^{(1:C)}, v_{1:k})$, where the subscript $1:k$, indicates the set of observations up to and including time t_k. Another goal is to compute the distribution of the trajectory of the continuous replay state, x_k, over replay periods, $p(x_k|I_k = 1, y_{1:k}, \Delta N_{1:k}^{(1:C)}, v_{1:k})$. Both of these can be computed directly from the joint filter distribution, $p(I_k, x_k|y_{1:k}, \Delta N_{1:k}^{(1:C)}, v_{1:k})$. The replay state probability is

$$\Pr(I_k|y_k, \Delta N_k^{(1:C)}, v_k, H_k) = \int p(I_k, x_k|y_k, \Delta N_k^{(1:C)}, v_k, H_k)dx_k \tag{2.6}$$

where $H_k = \{y_{1:k-1}, \Delta N_{1:k-1}^{(1:C)}, v_{1:k-1}\}$ is the history of observation up to, but not including the current time step. The continuous state density is given by

$$p(x_k|I_k = 1, y_k, \Delta N_k^{(1:C)}, v_k, H_k) = \frac{p(x_k, I_k = 1|y_k, \Delta N_k^{(1:C)}, v_k, H_k)}{\Pr(I_k = 1|y_k, \Delta N_k^{(1:C)}, v_k, H_k)} \tag{2.7}$$

We can compute the desired joint filter distribution using Bayes' rule

$$\begin{aligned} &p(I_k, x_k|y_k, \Delta N_k^{(1:C)}, v_k, H_k) \\ &\quad\propto p(y_k|I_k, x_k, \Delta N_k^{(1:C)}, v_k, H_k)p(\Delta N_k^{(1:C)}|I_k, x_k, v_k, H_k)p(v_k|I_k, x_k, H_k)p(I_k, x_k|H_k) \end{aligned} \tag{2.8}$$

The first three terms on the right-hand side of Eq. (2.8) are the likelihoods from each of the observations models, the hippocampal LFP, population spiking data, and the rat's velocity at the current time t_k conditioned on the replay states and the observation history, respectively. These terms can be simplified using the assumptions about our observations models discussed in the previous section.

$$\begin{aligned} &p(I_k, x_k|y_k, \Delta N_k^{(1:C)}, v_k, H_k) \\ &\quad\propto p(y_k|I_k, y_{1:k-1})p(\Delta N_k^{(1:C)}|x_k, \Delta N_{1:k-1}^{(1:C)})p(v_k|I_k, v_{1:k-1}))p(I_k, x_k|H_k) \end{aligned} \tag{2.9}$$

The last term on the right-hand side is the one-step prediction density, which once again can be expanded using the Chapman-Kolmogorov equation,

$$\begin{aligned} p(I_k, x_k|H_k) = \sum_{I_{k-1}} \int_{x_{k-1}} &p(x_k|x_{k-1}, I_k, I_{k-1}) \Pr(I_k|I_{k-1}, v_{k-1}) \\ &\times p(I_{k-1}, x_{k-1}|y_{k-1}, \Delta N_{k-1}^{(1:C)}, v_{k-1}, H_{k-1})dx_{k-1} \end{aligned} \tag{2.10}$$

The first term on the right-hand side of Eq. (2.10), $p(x_k|x_{k-1}, I_k, I_{k-1})$, is the semi-latent state transition density given by Eq. (2.5). The second term, $\Pr(I_k|\, I_{k-1}, v_{k-1})$, is the replay state transition density given the animal's most recent velocity. The third term, $p(I_{k-1}, x_{k-1}|y_{k-1}, \Delta N_{k-1}^{(1:C)}, v_{k-1}, H_{k-1})$, is the joint filter distribution of the replay state and semi-latent state from previous time step. This equation tells us how to combine the filter distribution from the previous time step with the two components of the state model to compute the distribution of both state variables given all but the most recent observations.

Substituting Eq. (2.10) into Eq. (2.9) we get the joint filter density of the replay state I_k and semi-latent state x_k:

$$\begin{aligned} &p(I_k, x_k|y_k, \Delta N_k^{(1:C)}, v_k, H_k) \\ &\quad \propto p(y_k|I_k, y_{1:k-1})p(\Delta N_k^{(1:C)}|x_k, \Delta N_{1:k-1}^{(1:C)})p(v_k|I_k, v_{1:k-1})p(I_k, x_k|H_k) \\ &\quad \times \sum_{I_{k-1}} \int_{x_{k-1}} p(x_k|x_{k-1}, I_k, I_{k-1}) \Pr(I_k|I_{k-1}, v_{k-1}) \\ &\quad \times p(I_{k-1}, x_{k-1}|y_{k-1}, \Delta N_{k-1}^{(1:C)}, v_{k-1}, H_{k-1})dx_{k-1} \end{aligned} \tag{2.11}$$

Equation (2.11) provides the solution to the filter problem for this state-space model. The last term on the right-hand side of the equation is the filter density from the previous time step. That gets multiplied by the two components of the state model and integrated and summed over the previous state values to produce the one-step prediction distribution. We then multiply by the likelihood of each of our observations at the current time step, based on the observation models, to compute the filter distribution at the current time step. Thus, we have an iterative method to compute the filter distribution at each time. If we select an initial distribution for the states at the beginning of the experiment, by iterating through Eq. (2.11) we can compute the filter distribution at all times.

The fact that each observation likelihood contributes in a multiplicatively separable manner means that it is easy to determine the degree to which each data modality is contributing to the estimate at each time step. This also makes it simple to deal with any missing data at any time, as the corresponding likelihood term can be removed and the information from the other data sources will still be maintained.

2.4.3 Replay Identification and Decoding Example

We applied the filter algorithm developed above to data from a rat performing a memory guided navigation task on a W-shaped maze. The data consisted of a 15.5 min trial during which the rat was required to alternate going down the center arm and then turning left or right on subsequent trials. Six LFP channels were used, and short time Fourier transforms of the past 20 ms were computed at each time point. Sorted spiking activity from 17 neurons was recorded. The spike and LFP data were down-sampled to 500 Hz resolution and the filter was computed with a 2 ms time step. For simplicity, we linearized the maze by assigning a value to each location based on its distance in cm to the tip of the center arm in the W-shaped maze. We then defined each of the state and observations models as functions of this linearized position.

As described above, we defined the binary indicator state model by setting the probability of being in a replay state at time t_k to be a logistic function related to the rat's movement velocity and the previous indicator state. Specifically, the form of this model is,

$$\texttt{logit}\,\Pr(I_k = 1|I_{k-1} = i, v_{k-1}) = \beta_0^{(i)} + \sum_{j=1}^{M} \beta_j^{(i)} g_j(v_{k-1}), \quad \text{for } i \in \{0, 1\}. \tag{2.12}$$

Where the $g_j(s)$ are a set of spline basis functions that ensure a smooth relationship between velocity and the probability of being in a replay state (Hearn et al. 2010). The model parameters, $\beta_j^{(i)}$, define the shape of this relationship, and are easily estimated using maximum likelihood (Truccolo et al. 2005). An example of the fit model is shown in Fig. 2.4. The left panel shows $\Pr(I_k = 1|I_{k-1} = 0, v_{k-1})$, the probability of switching into a replay state from a non-replay state as a function of velocity, and the right panel shows $\Pr(I_k = 1|I_{k-1} = 1, v_{k-1})$, the probability of remaining in a replay state at the next time as a function of velocity. We see that the probability of switching into a replay state in one discrete time step of 2 ms is always small, but is highest when the rat is near 0 cm/s. There is another local peak

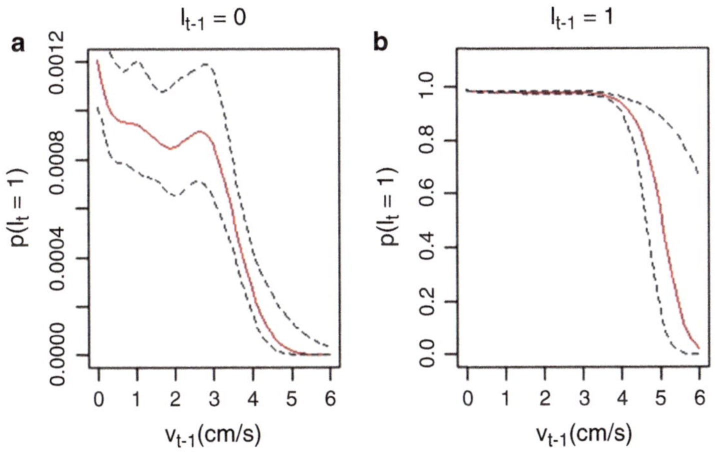

Fig. 2.4 Spline-based logistic model for the binary replay indicator state. (**a**) The probability of switching into a replay state in a single time step as a function of v_{k-1}, $\Pr(I_k = 1|I_{k-1} = 0, v_{k-1})$. (**b**) The probability of remaining in a replay state for a single time step as a function of v_{k-1}, $\Pr(I_k = 1|I_{k-1} = 1, v_{k-1})$. The red lines are the maximum likelihood estimates of the probability and the black dotted lines are the upper and lower 95% confidence levels

just below 3 cm/s, above which the probability drops off precipitously. Similarly, the probability of staying in a replay state from one time step to the next is close to, but below, 1 for low velocities and drops off quickly at velocities above 4 cm/s. The state model for the continuous, semi-latent state x_k was defined in Eq. (2.5), with a value of $\hat{\sigma} = 1.56$ used for the random walk variance term. Together, Eqs. (2.5) and (2.12) define the full state model.

Next the observation models for the hippocampal spiking, LFP, and rat's velocity were fit. The spiking for each neuron was modeled as a point process, as described in Sect. 2.2, with conditional intensity function

$$\lambda^{(i)}(t_k|H_k) = \exp\left(\alpha_0^{(i)} + \sum_{j=1}^{J} \alpha_j^{(i)} g_j(x_k)\right) \tag{2.13}$$

where $\lambda^{(i)}(t_k|H_k)$ is the intensity function for the i-th neuron and $g_j(x)$ are again a set of spline functions, this time taken as a function of position. This choice of model structure establishes a smooth relationship between the rat's position and place field firing during active exploration and a smooth relation between spiking and the unobserved continuous state during replay. The model parameters $\alpha_j^{(i)}$ are estimated by maximum likelihood. Note that since this model depends on a stochastic state process but does not depend on past spiking, it is also called a doubly stochastic Poisson model (Grandell 2006).

An example of the model fit for a single neuron is shown in Fig. 2.5. Panel A shows the model fit in blue overlaid on an occupancy normalized histogram of spiking vs position. The place field has a large peak centered about 80 cm from

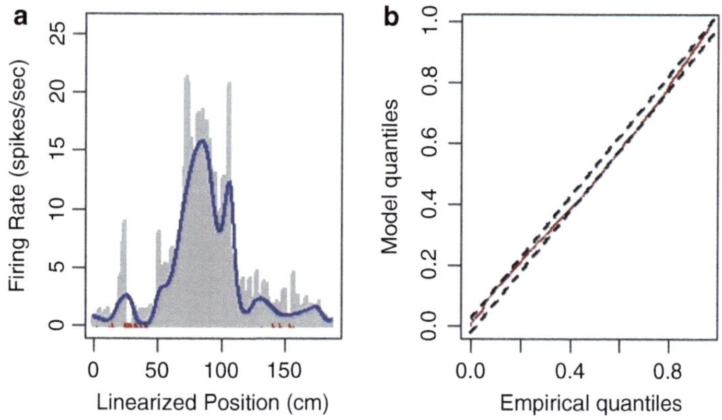

Fig. 2.5 Model fit and goodness-of-fit for one neuron. (**a**) Estimated conditional intensity of spiking as a function of the linearized position. The solid blue line is the maximum likelihood model fit. The gray bars are the occupancy-normalized histogram of the firing activity. (**b**) A KS plot for the rescaled inter-spike intervals (ISIs), with the dotted lines being the 95% confidence bounds for the KS statistic

the tip of the center arm. One advantage of writing down a point process model for neural spiking is that there are a number of natural goodness-of-fit methods to assess the quality of these models (Truccolo et al. 2005). In panel B we provide an example of one common goodness-of-fit assessment, a Kolmogorov-Smirnov (KS) plot. Details about the procedure to generate this plot are available (Brown et al. 2002; Truccolo et al. 2005), but briefly, the observed interspike intervals are rescaled according to the intensity model and compared to their expected distribution if the model were correct. A well-fit model should stay near the 45 degree line and should stay within the dotted confidence bounds. The KS plot in panel B shows a fairly well-fit model, though one where further refinement is still possible.

The remaining observation models are those for the short time Fourier transform of the LFP and for the rat's movement velocity. As described above, the observation model for the LFP is a multivariate Gaussian model, $p(y_k|I_k) \sim \mathcal{N}(\mu(I_k), \Sigma)$, where $\mu(I_k)$ is the mean power at each frequency in and out of the replay state, and Σ is the model covariance. Maximum likelihood estimates were obtained for $\mu(I_k = 0)$, $\mu(I_k = 1)$, and Σ. The observation model for the rat's movement velocity is also a Gaussian model, $p(v_k|v_{k-1}, I_k) \sim \mathcal{N}(v_{k-1}, \varsigma(I_k))$, where $\varsigma(I_k = 0)$ and $\varsigma(I_k = 1)$ are the variability of the movement velocity in and out of the replay state, and are estimated via maximum likelihood.

Finally, we define the initial condition for the states, $p(I_0, x_0)$. We assume that the binary state is initially known not to be in a replay state, $\Pr(I_0 = 0) = 1$. By definition, the density of $x_0|I_0 = 0$ is a delta function at the rat's position m_0. Together, these define the initial joint distribution of the states. We then iteratively compute the filter distribution at each time step based on Eq. (2.11), solving the integral numerically using Riemann summation.

An example segment of the decoding result is shown in Fig. 2.6. Figure 2.6a shows the posterior distribution for being in-replay state by the blue solid line. Figure 2.6b is a heat plot of the probability distribution of the semi-latent state jointly with the animal being in-replay state. The red solid line is the animal's actual linearized position. Figure 2.6c and d are a zoomed-in display of a smaller period of Fig. 2.6a and b highlighting a single decoded replay event. We see in panel A that the probability of being in a replay state tends to stay near 0 much of the time, and rapidly increases to values near 1 for short periods where the spiking, LFP, and movement observations are consistent with a replay event. During periods when the probability of a replay is small, the filter density of x_k is concentrated around the rat's actual position. As the probability of a replay starts to increase, the filter density of x_k is initially broadly distributed but very rapidly becomes more concentrated about a location, that is not necessarily at the location of the rat. The center of this filter distribution can then begin to move, as illustrated by the replay event in panel D, showing a replay with a population spike pattern that reflects movement from smaller to larger position values.

Figure 2.6 shows an example of a small section of time with a few clear replay events. Over the 15.5 min dataset we examined, we were able to detect 208 such events, approximately half of which showed a clear pattern corresponding to movement of the state variable x_k. Interestingly, over 80% of these reflected a state

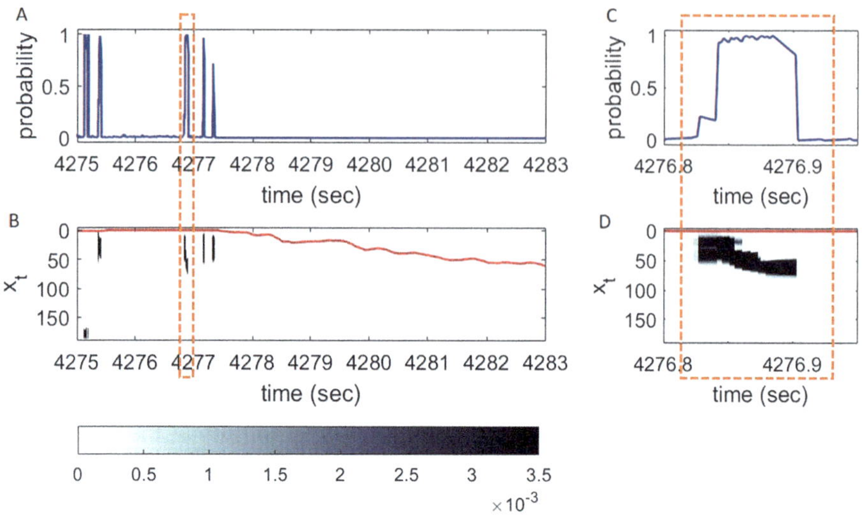

Fig. 2.6 Replay decoding example. (**a**) Shows the filter probability of a replay event at the current time. (**b**) Shows a heat plot of the joint distribution of the semi-latent state and of a replay event occurring. The red solid line is the animal's actual linearized position. (**c**) and (**d**) show a zoomed-in section from (**a**) and (**b**), respectively, illustrating a single decoded replay event that represents a trajectory beginning close to the animal's actual position and then proceeding away from the animal

trajectory that started close to the rat's actual location and moved away from the rat. We note that all of these detected replay events are detected using only past observed information, making them appropriate for closed loop experiments where replay events are detected during an experiment and rapidly disrupted.

2.5 Discussion

In this chapter, we considered the use of the state-space framework to help model complex neural phenomena. In particular, this class of models is well suited for analyzing high dimensional neural and behavioral signals that are noisy and have rich temporal structure. These types of problems have become increasingly prevalent in recent years, as our capacity to record neural activity and our interest in understanding complex relations between neural signals has expanded.

The fundamental structure of the state-space model, discussed in Sect. 2.2, provides some insight into the power of this framework. The state model is used to describe the dynamics of a stochastic signal or set of signals that influence neural activity or behavior. These signals can be very noisy or nearly deterministic, their dynamics simple or elaborate, depending on the structure of this model. The observation model is used to capture the statistical relation between these signals and any experimental measurements. These measurements may be scalar values or high-dimensional vectors, have distributions that are normal or not, and be related to the state variable through linear or highly nonlinear relationships. The challenge for many neural analysis problems lies in how to express its components using the state-space framework—what form the state variables should take, how they evolve, how noisy they are, and how they influence the observed data values. Once the model has been formulated, many tools for model fitting, model assessment, and estimation are immediately available.

The state-space framework also has important advantages in clarity; it forces the data analyst to define precisely what is meant by a particular state. In some cases these state definitions come directly from measurable quantities. The model describing the relationship between spiking and location defines a state-space process with an observable state: the present and future location of the animal. In other cases the state may not be directly observable, but nonetheless relates to potentially observable parameters; the model describing the relationship between spiking and membrane conductance falls into this category.

Alternatively, the state may represent something much more abstract, as in the analysis of replay events. Here the state that is estimated captures both whether the system is coding for the current local position (non-replay state) or a non-local position (replay state) as well as the current or non-local position that is being represented. The state-space approach forces us to define precisely what we mean by local and non-local activity and then makes it possible to ask, across all time, when the spiking activity is consistent with either state.

This approach has important advantages over standard analyses in the field. Putative replay events are typically detected using a combination of somewhat arbitrarily chosen criteria, which can include movement speed, LFP structure, and multiunit firing rates (Foster and Wilson 2006; Karlsson and Frank 2009; Gupta et al. 2010). Replay is then defined as events where a statistical test applied to the underlying spiking indicates that the spiking is similar in sequential structure (albeit on a compressed timescale) to that seen during behavior. While that approach was critical to the initial discovery of these events, it excludes both events where a single non-local representation might be activated as well as events where the sequential structure is not a good match for that seen during behavior. The state-space approach allows us to relax those constraints and to potentially discover new types of replay. Further, the two-step process of defining a set of candidate events and then applying a test to those events makes it difficult to know the extent to which the criteria chosen determine the results obtained. More broadly, further extensions of the state-space model will enable us to answer questions about the specific relationship of replay content to the animal's actual movement, and will help us identify potential differences in events that do or do not contain substantial power in the ripple band in addition to non-local spiking.

The examples in Sect. 2.3 also highlight the range of problems and applications that the state space paradigm can be used to address. The two examples we discussed used a similar point process filter to accomplish very different goals. In the first, the goal was to decode a movement signal, which was treated as unobserved, and predict a future turn direction, which was not directly observable. The challenge here was devising a state variable structure that could be estimated to inform both of these components. In the second example, the goal was not strictly to estimate the state variables corresponding to the membrane potential and ionic currents; this was just an ancillary step toward the objective of estimating parameters of the biophysical model. Here, the real challenge was not in setting up the state-space model, but in performing the computations to estimate the multiple state variables and parameters simultaneously. The first example used a very simple state model, and extracted most of the information about the state based on the observation model. The second example used a very simple observation model, and used a more complex state model, including multiple variables with interacting, nonlinear dynamics, to capture the structure in the data.

The power of the state-space approach is also illustrated by a wide array of other neuroscience applications of state-space modeling apart from decoding ensemble neural spike trains (Rieke et al. 1997; Brown et al. 1998; Barbieri et al. 2004; Wu et al. 2006). For instance, tracking the dynamics and plasticity of neural receptive field in general (Eden et al. 2004) and specifically in rat hippocampus and entorhinal cortex (Frank et al. 2002), looking at between-trial hippocampal neuronal dynamics in the primary motor cortex of monkeys (Czanner et al. 2008), and transitions in neural spiking dynamics in the subthalamic nucleus of Parkinson's patients (Deng et al. 2013). State-space models have also been successful at identifying specific states of neuronal ensembles, include stimulus-driven cortical states during behavior (Jones et al. 2007; Kemere et al. 2008) and intrinsic cortical UP/DOWN states during slow wave sleep (Chen et al. 2009). There have been many other extensions of state-space methods in neuroscience applications. For example, Calabrese and Paninski combined a mixture of Gaussians model, a Kalman filter and an EM algorithm to develop a computationally efficient method for online spike sorting (Calabrese and Paninski 2011); Pakman et al. developed a fast ℓ_1-penalized regression method for Kalman state-space models of the neuron voltage dynamics given noisy, subsampled voltage observations (Pakman et al. 2014); Archer et al. extended the standard Kalman filter-smoother with a structured Gaussian, variational posterior approximation to the posterior of much more general, nonlinear latent variable generative models (Archer et al. 2015); Linderman et al. combined multi-neuron point process models with flexible graph-theoretic priors, which characterize the relationship between latent features and neural connectivity patterns, to classify neurons and infer latent dimensions of circuit organization from correlated spike trains (Linderman et al. 2016); many other extensions of these methods have been explored, some of which are detailed in later chapters of this book.

In this chapter, we primarily focused on applications of the state-space paradigm to solve filtering problems, where only the data up to the current time are available to the model. Solving these problems is particularly useful for closed-loop experiments, where interventions can be triggered on the basis of the estimated state of the system. The utility of this approach extends to other types of problems, including cases where the entire dataset is available for state estimation (smoothing) or cases where the goal is to predict future states or use the state to control an external system. The identification of replay events, for example, is likely to be most accurate when a smoothing algorithm is applied. The relevant smoothing algorithm would use the same state model and observation model as the filtering algorithm, but it would calculate the smoothing probability density of the latent state by combining both past and future information, while the filtering algorithm only uses the observations up to current time.

There are several natural extensions of the state-space paradigm that are likely to be useful for future analyses. First, development of more efficient algorithms to compute non-Gaussian, multimodal posterior densities for large-scale neural data will be important. Individual neurons frequently have complex receptive field structures that are not well described by a single Gaussian distribution. Capturing these multimodal receptive fields and the potentially multimodal population-level representations requires more complex mathematical formulations and estimation algorithms. Second, the observed distribution could be extended to characterize more complex combinations of data types, including data from imaging experiments where calcium or voltage transients are measured. Third, more complex and potentially multi-layer formulations of the state model are likely to be important, as these could allow for more complicated dependence relationships between the observed signals and the hidden states.

More broadly, given that our eventual goal is a complete mathematical description of the state and information content of the system, the state-space approach offers a natural framework to begin to construct this sort of description. It can be applied to extract information about the internal state of neurons based on their spiking as well as about the representational state of neural ensembles. Further, applications to behavior allow for estimation of even more abstract variables like the learning state or attentional state of the animal. Combining models across all of these levels should eventually allow us to link events across single neuron, neural ensembles and behavior into a unified framework.

Acknowledgements We thank Hannah Joo for comments and writing suggestions.

References

Archer, E., Park, I. M., Buesing, L., Cunningham, J., & Paninski, L. (2015). Black box variational inference for state space models. arXiv preprint arXiv:1511.07367.

Archer, E. W., Koster, U., Pillow, J. W., & Macke, J. H. (2014). Low-dimensional models of neural population activity in sensory cortical circuits. In *Advances in neural information processing systems* (pp. 343–351). Red Hook: Curran.

Barbieri, R., Frank, L. M., Nguyen, D. P., Quirk, M. C., Solo, V., Wilson, M. A., et al. (2004). Dynamic analyses of information encoding in neural ensembles. *Neural Computation, 16*(2), 277–307.

Brown, E. N., Barbieri, R., Eden, U. T., & Frank, L. M. (2003). Likelihood methods for neural data analysis. In J. Feng (Ed.), *Computational neuroscience: A comprehensive approach* (pp. 253–286). London: Chapman and Hall/CRC Press.

Brown, E. N., Barbieri, R., Ventura, V., Kass, R. E., & Frank, L. M. (2002). The time-rescaling theorem and its application to neural spike train data analysis. *Neural Computation, 14*(2), 325–346.
Brown, E. N., Frank, L. M., Tang, D., Quirk, M. C., & Wilson, M. A. (1998). A statistical paradigm for neural spike train decoding applied to position prediction from ensemble firing patterns of rat hippocampal place cells. *Journal of Neuroscience, 18*(18), 7411–7425.
Brown, E. N., Ngyuen, D. P., Frank, L. M., Wilson, M. A., & Solo, V. (2001). An analysis of neural receptive field plasticity by point process adaptive filtering. *Proceedings of National Academy of Sciences USA, 98*, 12261–12266.
Bullmore, E., & Sporns, O. (2009). Complex brain networks: Graph theoretical analysis of structural and functional systems. *Nature Reviews Neuroscience, 10*(3), 186–198.
Buzsáki, G. (1986). Hippocampal sharp waves: Their origin and significance. *Brain Research, 398*(2), 242–252.
Buzsáki, G. (2015). Hippocampal sharp wave-ripple: A cognitive biomarker for episodic memory and planning. *Hippocampus, 25*(10), 1073–1188.
Calabrese, A., & Paninski, L. (2011). Kalman filter mixture model for spike sorting of non-stationary data. *Journal of Neuroscience Methods, 196*(1), 159–169.
Carr, M. F., Jadhav, S. P., & Frank, L. M. (2011). Hippocampal replay in the awake state: A potential substrate for memory consolidation and retrieval. *Nature Neuroscience, 14*(2), 147–153.
Chapin, J. (1986). Laminar differences in sizes, shapes, and response profiles of cutaneous receptive fields in the rat SI cortex. *Experimental Brain Research, 62*(3), 549–559.
Chapin, J. K. (2004). Using multi-neuron population recordings for neural prosthetics. *Nature Neuroscience, 7*(5), 452–455.
Chen, Z., Barbieri, R., & Brown, E. N. (2010). State space modeling of neural spike train and behavioral data. In K. Oweiss (Ed.), *Statistical signal processing for neuroscience and neurotechnology* (pp. 175–218). Amsterdam: Elsevier.
Chen, Z., Vijayan, S., Barbieri, R., Wilson, M. A., & Brown, E. N. (2009). Discrete-and continuous-time probabilistic models and algorithms for inferring neuronal UP and DOWN states. *Neural Computation, 21*(7), 1797–1862.
Coleman, T. P., Yanike, M., Suzuki, W., & Brown, E. N. (2006). A mixed filter algorithm for state estimation from simultaneously recorded continuous-valued point process and binary observations. In *Proceedings of 40th Asilomar Conference on Signals, Systems and Computers* (pp. 1949–1953). Piscataway: IEEE.
Czanner, G., Eden, U. T., Wirth, S., Yanike, M., Suzuki, W. A., & Brown, E. N. (2008). Analysis of between-trial and within-trial neural spiking dynamics. *Journal of Neurophysiology, 99*(5), 2672–2693.
Daley, D. J., & Vere-Jones, D. (2003). *An introduction to the theory of point processes* (Vol. 1). New York: Springer.
Davidson, T. J., Kloosterman, F., & Wilson, M. A. (2009). Hippocampal replay of extended experience. *Neuron, 63*(4), 497–507.
Deng, X., Eskandar, E. N., & Eden, U. T. (2013). A point process approach to identifying and tracking transitions in neural spiking dynamics in the subthalamic nucleus of Parkinson's patients. *Chaos, 23*(4), 046102.
Diekelmann, S., Büchel, C., Born, J., & Rasch, B. (2011). Labile or stable: Opposing consequences for memory when reactivated during waking and sleep. *Nature Neuroscience, 14*(3), 381–386.
Doucet, A., De Freitas, N., & Gordon, N. (Eds.) (2001). *Sequential Monte Carlo methods in practice*. New York: Springer.
Eden, U. T., Frank, L. M., Barbieri, R., Solo, V., & Brown, E. N. (2004). Dynamic analysis of neural encoding by point process adaptive filtering. *Neural Computation, 16*(5), 971–998.
Ergun, A., Barbieri, R., Eden, U. T., Wilson, M. A., & Brown, E. N. (2007). Construction of point process adaptive filter algorithms for neural systems using sequential monte carlo methods. *IEEE Transactions on Biomedical Engineering, 54*(3), 419–428.

Felleman, D. J., & Kaas, J. H. (1984). Receptive-field properties of neurons in middle temporal visual area (mt) of owl monkeys. *Journal of Neurophysiology, 52*(3), 488–513.

Foster, D. J., & Wilson, M. A. (2006). Reverse replay of behavioural sequences in hippocampal place cells during the awake state. *Nature, 440*(7084), 680–683.

Frank, L. M., Eden, U. T., Solo, V., Wilson, M. A., & Brown, E. N. (2002). Contrasting patterns of receptive field plasticity in the hippocampus and the entorhinal cortex: An adaptive filtering approach. *Journal of Neuroscience, 22*(9), 3817–3830.

Fuhs, M. C., & Touretzky, D. S. (2006). A spin glass model of path integration in rat medial entorhinal cortex. *Journal of Neuroscience, 26*(16), 4266–4276.

Girman, S. V., Sauvé, Y., & Lund, R. D. (1999). Receptive field properties of single neurons in rat primary visual cortex. *Journal of Neurophysiology, 82*(1), 301–311.

Grandell, J. (2006). *Doubly stochastic Poisson processes* (Vol. 529). Berlin: Springer.

Gupta, A. S., van der Meer, M. A., Touretzky, D. S., & Redish, A. D. (2010). Hippocampal replay is not a simple function of experience. *Neuron, 65*(5), 695–705.

Hafting, T., Fyhn, M., Molden, S., Moser, M.-B., & Moser, E. I. (2005). Microstructure of a spatial map in the entorhinal cortex. *Nature, 436*(7052), 801–806.

Hearn, D. D., Baker, M. P., & Carithers, W. (2010). *Computer graphics with open GL*. Upper Saddle River: Prentice Hall.

Huang, Y., Brandon, M. P., Griffin, A. L., Hasselmo, M. E., & Eden, U. T. (2009). Decoding movement trajectories through a T-maze using point process filters applied to place field data from rat hippocampal region CA1. *Neural Computation, 21*(12), 3305–3334.

Hubel, D. H., & Wiesel, T. N. (1962). Receptive fields, binocular interaction and functional architecture in the cat's visual cortex. *Journal of Physiology, 160*(1), 106–154.

Jones, J. P., & Palmer, L. A. (1987). An evaluation of the two-dimensional Gabor filter model of simple receptive fields in cat striate cortex. *Journal of Neurophysiology, 58*(6), 1233–1258.

Jones, L. M., Fontanini, A., Sadacca, B. F., Miller, P., & Katz, D. B. (2007). Natural stimuli evoke dynamic sequences of states in sensory cortical ensembles. *Proceedings of the National Academy of Sciences, USA, 104*(47), 18772–18777.

Karlsson, M. P., & Frank, L. M. (2009). Awake replay of remote experiences in the hippocampus. *Nature Neuroscience, 12*(7), 913–918.

Kemere, C., Santhanam, G., Byron, M. Y., Afshar, A., Ryu, S. I., Meng, T. H., et al. (2008). Detecting neural-state transitions using hidden Markov models for motor cortical prostheses. *Journal of Neurophysiology, 100*(4), 2441–2452.

Knierim, J. J., Kudrimoti, H. S., & McNaughton, B. L. (1995). Place cells, head direction cells, and the learning of landmark stability. *Journal of Neuroscience, 15*(3), 1648–1659.

Kuffler, S. W. (1953). Discharge patterns and functional organization of mammalian retina. *Journal of Neurophysiology, 16*(1), 37–68.

Lee, A. K., & Wilson, M. A. (2002). Memory of sequential experience in the hippocampus during slow wave sleep. *Neuron, 36*(6), 1183–1194.

Linderman, S., Adams, R. P., & Pillow, J. W. (2016). Bayesian latent structure discovery from multi-neuron recordings. In *Advances in neural information processing systems* (pp. 2002–2010).

McNaughton, B. L., Battaglia, F. P., Jensen, O., Moser, E. I., & Moser, M.-B. (2006). Path integration and the neural basis of the 'cognitive map'. *Nature Reviews Neuroscience, 7*(8), 663–678.

Meng, L., Kramer, M. A., & Eden, U. T. (2011). A sequential Monte Carlo approach to estimate biophysical neural models from spikes. *Journal of Neural Engineering, 8*(6), 065006.

Muller, R. U., Kubie, J. L., & Ranck, J. B. (1987). Spatial firing patterns of hippocampal complex-spike cells in a fixed environment. *Journal of Neuroscience, 7*(7), 1935–1950.

Nádasdy, Z., Hirase, H., Czurkó, A., Csicsvari, J., & Buzsáki, G. (1999). Replay and time compression of recurring spike sequences in the hippocampus. *Journal of Neuroscience, 19*(21), 9497–9507.

O'Keefe, J., & Burgess, N. (2005). Dual phase and rate coding in hippocampal place cells: Theoretical significance and relationship to entorhinal grid cells. *Hippocampus, 15*(7), 853–866.
O'Keefe, J., & Dostrovsky, J. (1971). The hippocampus as a spatial map. preliminary evidence from unit activity in the freely-moving rat. *Brain Research, 34*(1), 171–175.
O'Keefe, J., & Recce, M. L. (1993). Phase relationship between hippocampal place units and the eeg theta rhythm. *Hippocampus, 3*(3), 317–330.
Pakman, A., Huggins, J., Smith, C., & Paninski, L. (2014). Fast state-space methods for inferring dendritic synaptic connectivity. *Journal of Computational Neuroscience, 36*(3), 415–443.
Paninski, L., Shoham, S., Fellows, M. R., Hatsopoulos, N. G., & Donoghue, J. P. (2004). Superlinear population encoding of dynamic hand trajectory in primary motor cortex. *Journal of Neuroscience, 24*(39), 8551–8561.
Rao, R. P., & Ballard, D. H. (1997). Dynamic model of visual recognition predicts neural response properties in the visual cortex. *Neural Computation, 9*(4), 721–763.
Rieke, F., Warland, D., de Ruyter van Steveninck, R., & Bialek, W. (1997). *Spikes: Exploring the neural code*. Cambridge: MIT Press.
Rodieck, R. W. (1965). Quantitative analysis of cat retinal ganglion cell response to visual stimuli. *Vision Research, 5*(12), 583–601.
Shanechi, M. M., Hu, R. C., Powers, M., Wornell, G. W., Brown, E. N., & Williams, Z. M. (2012). Neural population partitioning and a concurrent brain-machine interface for sequential motor function. *Nature Neuroscience, 15*(12), 1715–1722.
Skaggs, W. E., & McNaughton, B. L. (1996). Theta phase precession in hippocampal neuronal populations and the compression of temporal sequences. *Hippocampus, 6*, 149–172.
Smith, A. C., & Brown, E. N. (2003). Estimating a state-space model from point process observations. *Neural Computation, 15*(5), 965–991.
Smith, A. C., Frank, L. M., Wirth, S., Yanike, M., Hu, D., Kubota, Y., et al. (2004). Dynamic analysis of learning in behavioral experiments. *Journal of Neuroscience, 24*(2), 447–461.
Solstad, T., Moser, E. I., & Einevoll, G. T. (2006). From grid cells to place cells: A mathematical model. *Hippocampus, 16*(12), 1026–1031.
Stephen, E. P., Lepage, K. Q., Eden, U. T., Brunner, P., Schalk, G., Brumberg, J. S., et al. (2014). Assessing dynamics, spatial scale, and uncertainty in task-related brain network analyses. *Frontiers in Computational Neuroscience, 8*, 31.
Truccolo, W., Eden, U. T., Fellows, M. R., Donoghue, J. P., & Brown, E. N. (2005). A point process framework for relating neural spiking activity to spiking history, neural ensemble, and extrinsic covariate effects. *Journal of Neurophysiology, 93*(2), 1074–1089.
Wilson, M. A., & McNaughton, B. L. (1994). Reactivation of hippocampal ensemble memories during sleep. *Science, 265*(5172), 676–679.
Wu, W., Gao, Y., Bienenstock, E., Donoghue, J. P., & Black, M. J. (2006). Bayesian population decoding of motor cortical activity using a Kalman filter. *Neural Computation, 18*(1), 80–118.
Zhang, K., Ginzburg, I., McNaughton, B. L., & Sejnowski, T. J. (1998). Interpreting neuronal population activity by reconstruction: unified framework with application to hippocampal place cells. *Journal of Neurophysiology, 79*(2), 1017–1044.

Chapter 3
Latent Variable Modeling of Neural Population Dynamics

Zhe Chen

3.1 Introduction

A fundamental task in neural data analysis is to discover the *regularity* or *structure* of measured (potentially high-dimensional) neural data. Measurements of neural activity, whether being discrete or continuous-valued, are noisy ("stochastic") and time-varying ("dynamic") at various spatiotemporal scales. In neuroscience experiments, neural activity is driven by both external behavior and internal brain dynamics. Any behavior has a temporal dynamics, thereby the derived neural activity is also dynamic. Even in the complete absence of behavior (e.g., during sleep or anesthesia), the neural activity is very likely to exhibit a rich neural dynamics. Notably, many experimental quantities are unobserved or unobservable. For instance, due to limited recording capacity, we are only able to measure partially observable neural populations in one or multiple brain regions. Therefore, neurons can receive a common input from an unobserved neuron or an unobservable modulatory input from other brain region.

Factor analysis and mixture models are two common statistical models that employ the concept of latent variables, one for continuous latent variables (Santhanam et al. 2009) and the other for discrete latent variables. To characterize temporal dynamic or dependency of time series data, it is natural to extend them using a state-space framework (Chen et al. 2010, 2013; Chen 2015b). The state space analysis provides a general framework for analyzing stochastic dynamical systems that are measured or observed through a stochastic process.

The latent process is often modeled as a Markov process (continuous-valued) or Markov chain (discrete-valued). In the continuous case, the most celebrated

Z. Chen (✉)
New York University School of Medicine, New York, NY, USA
e-mail: zhe.chen3@nyumc.org

Z. Chen, S.V. Sarma (eds.), *Dynamic Neuroscience*,
https://doi.org/10.1007/978-3-319-71976-4_3

continuous-valued latent process model is the linear dynamical system (LDS) or Kalman filter, which is an extension of the factor analysis. Another extension of factor analysis is the Gaussian process factor analysis (GPFA), which accommodates more smoothing power for characterizing neural population codes (Yu et al. 2009). In the discrete case, the most widely used latent process model is the hidden Markov model (HMM) and its many variants (depending on specific probabilistic assumptions).

Latent variable models have been successfully used in various analyses of neural data, such as *dimensionality reduction and visualization* (Cunningham and Yu 2014), *decoding* (Lawhern et al. 2010; Chen et al. 2014), deconvolution (Penny et al. 2005; Vogelstein et al. 2009, 2010), *denoising* (Wu et al. 2011, 2016), data exploration (Latimer et al. 2015), and interpretation of variability (Whiteway and Butts 2017).

In this chapter, bearing the goals mentioned above, we will propose an analysis paradigm for latent variable modeling of neural dynamics—with focus on neuronal population spike trains from freely behaving animals. Specifically, simultaneous recording of multiple spike trains from many neurons within one or several brain regions offers a window into how neural circuits work in concert to generate specific brain functions.

3.2 Latent Variable Models

3.2.1 Latent Variable

Latent variables are used to characterize *unobserved* random or unknown variables, which can represent quantitative or categorical (e.g., membership) information. For instance, in the linear regression model, the residual noise can be modeled as a latent variable. Latent variables can be either time-invariant or time-variant. For instance, the latent variables in the factor analysis and mixture models are time-invariant. When the latent variables evolve in time, they become time-variant.

We consider two types of latent variables: *discrete* and *continuous*. The discrete latent variable, denoted by S, can be either finite or infinite (but countable). The continuous latent variable, denoted as $\mathbf{z} \in \mathbb{R}^m$, can be either one or multi-dimensional. The latent variables are often assumed with specific probability distributions.

Latent variable models can be visualized by probabilistic graphical models (Jordan and Sejnowski 2001). Either node in the graph represents a random variable, the line (undirected) or arrow (directed) between nodes indicate the statistical dependency, which correspond to the undirected or directed graphical models, respectively. When the nodes are not connected, two random variables X and Y are considered to be conditionally independent or *factorial*; namely, $p(X, Y|\text{par}) = p(X|\text{par})p(Y|\text{par})$, where par denotes the parent nodes of X and Y.

Table 3.1 Various statistical assumptions lead to different variants of the HMM

Assumption	Variant
Semi-Markovian state transition	Hidden semi-Markov model (Yu 2010)
Dirichlet mixture model	Hierarchical Dirichlet process-HMM (Teh et al. 2006)
Hierarchical state transition	Hierarchical HMM (Fine et al. 1998)
Independent Markov chains	Factorial HMM (Ghahramani and Jordan 1997)
Nonlinear time warping	Markov processes on curve (Saul and Rahim 2000)
Decomposed large state space	Mixed memory Markov model (Saul and Jordan 1999)
Mixtures of factor analyzers	Hidden Markov factor analysis (Omigbodun et al. 2016)

3.2.2 *Latent State Dynamics*

When a latent variable is time-varying, we sometimes refer to it as the latent state. The latent state dynamics describes the evolution of latent process and how the future state depends on the present and the past states. In a general form, the latent state dynamics is characterized by a *probabilistic* mapping: $\Pr(S_t|S_{1:t-1})$ or $f(\mathbf{z}_{1:t-1}) \mapsto \mathbf{z}_t$.

The Markov process, named after the Russian mathematician Andrey Markov, is a stochastic process that satisfies the Markov property. Specifically, a process satisfies the Markov property if its future and past states are independent. In other words, the process is "memoryless." A Markov chain is a type of Markov process that has either discrete state space.

In many real-world examples, the latent process is not always Markovian in that the state durations may follow more specific probability distributions, such as Poisson, negative binomial, lognormal, and inverse Gaussian distribution. To model the history dependence of a Markov chain, one idea is to introduce a high-order Markov chain (Ching et al. 2015; Lee 2011). The other idea is to introduce an explicit-duration semi-Markov modeling for each state (Guédon 2003; Yu 2010; Chen 2015a). In addition, making different statistical assumptions will lead to different variants of the HMM (Table 3.1).

3.2.3 *Characterization of Neuronal Population Observations*

To characterize neuronal population responses, let $\mathbf{y}_t = [y_{1,t}, \ldots, y_{C,t}]^\top$ denote a C-dimensional neuronal population vector, with each element $y_{c,t} \geq 0$ denoting the c-th neuronal spike count at the t-th time bin; and let $\mathbf{y}_{1:T} = \{y_{c,t}\}_{C\times T}$ denote the time series of C-dimensional vector $\mathbf{y}_t$.

A common probability distribution for neuronal responses or spike count observations is the Poisson distribution. In addition, researchers have used the negative binomial distribution to model spike count observations with overdispersion (where the variance is greater than the mean statistic). In some cases, for the purpose

Table 3.2 Choices of probability distribution for the neuronal response

Distribution	Range of random variables	Degree of freedom
Gaussian	Real-valued	Mean μ, variance σ^2
Binomial	Binary	$0 < p < 1$
Poisson	Nonnegative integer	Rate $\lambda > 0$
Negative binomial	Nonnegative integer	$r > 0, 0 < p < 1$
Conway–Maxwell–Poisson (Stevenson 2016)	Nonnegative integer	$\lambda > 0, \nu \geq 0$
Generalized Gaussian (Gao et al. 2016)	Nonnegative integer	$\lambda, g(\cdot)$

of computational tractability, researchers often use a Gaussian approximation for Poisson spike counts through a variance stabilization transformation. Table 3.2 lists a few population probability distributions for modeling spike count observations.

3.3 Inference: Likelihood and Bayesian Approaches

The likelihood and Bayesian approaches are two fundamental methods to solve the inference problem. The likelihood approach computes a point estimate by maximizing the likelihood function and represents the uncertainty of estimate via confidence intervals (Pawitan 2001). The maximum likelihood estimate (m.l.e.) is asymptotically consistent, normal, and efficient; it is invariant to reparameterization (i.e., functional invariance). However, the m.l.e. is known to suffer from overfitting, and therefore model selection is required in statistical data analysis. In contrast, the Bayesian approach models the unknowns (parameters, latent variables, and missing data) and uncertainties with probabilities or probability densities (Gelman et al. 2004; Robert 2007).

The likelihood approach aims to optimize the likelihood function $\mathscr{L}(Y|Z,\theta)$ given the observed data Y and latent (missing) data Z and unknown parameter θ. To estimate the latent variable Z and parameter θ, the iterative *expectation-maximization* (EM) algorithm has been developed to maximize or increase the likelihood (Dempster et al. 1977; Smith and Brown 2003). In the E-step, conditional on the most recent estimate $\hat{\theta}$, update the latent variable $\hat{Z}$. In the M-step, conditional on the most recent estimate $\hat{Z}$, update the parameter $\hat{\theta}$. In each step, the optimization can be tackled by solving the equation $\frac{\partial \mathscr{L}}{\partial Z} = 0$ or $\frac{\partial \mathscr{L}}{\partial \theta} = 0$. When the analytic solution is unavailable, we can resort to Newton or gradient-based optimization methods.

The Bayesian approach computes the full posterior of the unknowns based on the rules of probability theory. The foundation of Bayesian inference is given by *Bayes' rule*, which consists of two rules: *product rule* and *sum rule*. Bayes' rule provides a way to compute the conditional, joint, and marginal probabilities. Specifically, let X and Y be two continuous random variables; the conditional probability $p(X|Y)$ is given by

$$p(X|Y) = \frac{p(X,Y)}{p(Y)} = \frac{p(Y|X)p(X)}{\int p(Y|X)p(X)dX} \tag{3.1}$$

In Bayesian language, $p(Y|X)$, $p(X)$, and $p(X|Y)$ are referred to as the likelihood, prior, and posterior, respectively. The Bayesian machinery consists of three types of basic operations: normalization, marginalization, and expectation, all of which involve integration. For a hierarchical Bayesian model, the prior $p(X)$ is further specified by a hyperparameter ρ: $p(X) = \int p(X|\rho)p(\rho)d\rho$. When $p(X|\rho)$ is sharply peaked, the integral can be replaced with the point estimate of its peak ρ^*. Therefore, Eq. (3.1) is rewritten as

$$p(X|Y) = \frac{p(Y|X)p(X)}{p(Y)} = \frac{p(Y|X)\int p(X|\rho)p(\rho)d\rho}{p(Y)} \approx \frac{p(Y|X)p(X|\rho^*)}{p(Y)} \tag{3.2}$$

This yields the empirical Bayes estimate.

In the presence of latent variables, for the joint estimation of latent state Z and unknown parameter θ, Bayesian inference aims to infer the joint posterior of the state and the parameter using Bayes' rule

$$\begin{aligned} p(Z,\theta|Y) &\approx p(Z|Y)p(\theta|Y) \\ &= \frac{p(Y|Z,\theta)p(Z)p(\theta)}{p(Y)} \\ &= \frac{p(Y|Z,\theta)p(Z)p(\theta)}{\int\int p(Y|Z,\theta)p(Z)p(\theta)dZd\theta} \end{aligned} \tag{3.3}$$

where the approximation in the first step has used the so-called "mean-field approximation."

Given the observed random variable X, latent variable Z, and unknown parameters θ, Bayesian inference attempts to maximize the marginal likelihood $p(Y)$ (also known as "evidence") as follows:

$$p(Y) = \int\int\int p(Y|X,Z,\theta)p(X)p(Z)p(\theta)dXdZd\theta \tag{3.4}$$

Direct optimization of those integrals in Bayesian estimation is often intractable. Therefore, many approximate Bayesian methods have been proposed. A detailed review can be found in Chen (2013). For instance, the variational Bayes (VB) approach employs the variational approximation (Jordan et al. 1999; Beal and Ghahramani 2006), and is also referred to as *ensemble learning*. Specifically, VB aims to maximize the marginal log-likelihood or its lower bound:

$$\begin{aligned} \log p(Y) &= \log \int d\theta \int dXp(\theta)p(Z,Y|\theta) \\ &= \log \int d\theta \int dZq(Z,\theta)\frac{p(\theta)p(Z,Y|\theta)}{q(Z,\theta)} \end{aligned}$$

$$\geq \int d\theta \int dZ q(Z,\theta) \log \frac{p(\theta)p(Z,Y|\theta)}{q(Z,\theta)}$$
$$= \Big\langle \log p(Z,Y,\theta) \Big\rangle_q + \mathscr{H}_q(Z,\theta) \equiv \mathscr{F}(q(Z,\theta)) \tag{3.5}$$

where $p(\theta)$ denotes the parameter prior distribution, $p(Z,Y|\theta)$ defines the complete data likelihood, and $q(Z,\theta)$ is called the variational posterior distribution which approximates the joint posterior of the unknown state and parameter $p(Z,\theta|Y)$. The term $\mathscr{H}_q$ represents the entropy of the variational posterior distribution q, and $\mathscr{F}(q(Z,\theta))$ is referred to as the free energy. Maximizing the free energy $\mathscr{F}(q(Z,\theta))$ is equivalent to minimizing the Kullback–Leibler (KL) divergence between the variational posterior and true posterior (denoted by $\mathrm{KL}(q\|p)$); since the KL divergence is nonnegative, we have $\mathscr{F}(q) = \log p(Y) - \mathrm{KL}(q\|p) \leq \log p(Y)$. The optimization problem in (3.5) can be resorted to the VB-EM algorithm (Beal and Ghahramani 2006) in a similar fashion as the standard EM algorithm (Dempster et al. 1977).

A common (but not necessary) VB assumption is a factorial form of the posterior $q(Z,\theta) = q(Z)q(\theta)$, although one can further impose certain structure within the parameter space. In the case of mean-field approximation, we have $q(Z,\theta) = q(Z)\prod_i q(\theta_i)$. With selected priors $p(Z)$ and $p(\theta)$, we may maximize the free energy by alternatively solving two equations: $\frac{\partial \mathscr{F}}{\partial Z} = 0$ and $\frac{\partial \mathscr{F}}{\partial \theta} = 0$. Specifically, VB-EM inference can be viewed as a natural extension of the EM algorithm, which consists of the following two steps:

- VB-E step: Given the available information of $q(\theta)$, maximize the free energy $\mathscr{F}$ with respect to the function $q(Z)$ and update the posterior $q(Z)$;
- VB-M step: Given the available information of $q(Z)$, maximize the free energy $\mathscr{F}$ with respect to the function $q(\theta)$ and update the posterior $q(\theta)$. The posterior update will have an analytic form provided that the prior $p(\theta)$ is conjugate to the complete-data likelihood function (the conjugate-exponential family).

Similar to the iterative EM algorithm, the VB-EM inference has local maxima in optimization. The EM algorithm can be viewed as a variant of the VB algorithm in that the VB-M step replaces the point estimate (i.e., $q(\theta) = \delta(\theta - \theta_{\mathrm{MAP}})$ from the traditional M-step with a full posterior estimate. Another counterpart of the VB-EM is the *maximization-expectation* (ME) algorithm (Kurihara and Welling 2009), in which the VB-E step uses the MAP point estimate $q(Z) = \delta(Z - Z_{\mathrm{MAP}})$), while the VB-M step updates the full posterior.

The Markov chain Monte Carlo (MCMC) approach is referred to a class of algorithms for drawing random samples from probability distributions by constructing a Markov chain that has the equilibrium distribution as the desired distribution. The designed Markov chain is reversible and has detailed balance. For example, given a transition probability P, the detailed balance holds between each pair of state i and j in the state space if and only if $\pi_i P_{ij} = \pi_j P_{ji}$ (where $\pi_i = \Pr(Z_{t-1} = i), P_{ij} = \Pr(Z_{t-1} = i, Z_t = j)$). The appealing use of MCMC methods for Bayesian inference is to numerically calculate high-dimensional integrals based on the samples drawn from the equilibrium distribution (Robert 2007).

The most common MCMC methods are the random walk algorithms, such as the *Metropolis-Hastings* (MH) algorithm and Gibbs sampling. The MH algorithm is the simplest yet the most generic MCMC method to generate samples using a random walk and then to accept them with a certain acceptance probability. For example, given a random-walk proposal distribution $g(Z \rightarrow Z')$ (which defines a conditional probability of moving state Z to Z′), the MH acceptance probability $\mathscr{A}(Z \rightarrow Z')$ is

$$\mathscr{A}(Z \rightarrow Z') = \min\left(1, \frac{p(Z')g(Z' \rightarrow Z)}{p(Z)g(Z \rightarrow Z')}\right)$$

which yields a simple MCMC implementation. Gibbs sampling is another popular MCMC method that requires no parameter tuning. Given a high-dimensional joint distribution $p(Z) = p(z_1, \ldots, z_n)$, Gibbs sampler draws samples from the individual conditional distribution $p(z_i|Z_{-i})$ in turn while holding others fixed (where Z_{-i} denote the $n-1$ variables in Z except for z_i).

Another important class of Bayesian methods is Bayesian nonparametrics (Gershman and Blei 2012; Müller et al. 2015). Since nonparametric Bayesian models accommodate a large number of degrees of freedom (infinite-dimensional parameter space) to exhibit a rich class of probabilistic structure, such approaches are very powerful in terms of data representation. The fundamental building blocks are two stochastic processes: Dirichlet process (DP) and Gaussian process (GP). In application of data clustering, partitioning and segmentation, such as spike sorting (Wood and Black 2008), Bayesian nonparametric models define a prior distribution over the set of all possible partitions, in which the number of clusters or partitions may grow as the increase of the data samples in both static and dynamic settings. The model selection issue is resolved implicitly in the process of infinite mixture modeling, such as the Dirichlet mixture model and infinite HMM. In the application of data smoothing or prediction, the GP defines priors for the mean function and covariance function, where the covariance kernel determines the smoothness and stationarity between the data points. Although Bayesian nonparametrics offer greater flexibility for modeling complex data structures, most inference algorithms for Bayesian nonparametric models rely on MCMC methods, which may be computationally prohibitive for large-scale neural data analysis.

3.4 Latent Variable Models in Neural Data Analysis

3.4.1 Uncovering Neural Representations of Rodent Hippocampal-Neocortical Population Codes

3.4.1.1 Background

Population codes derived from simultaneous recordings of ensembles of neurons have been studied in the representation of sensory or motor stimuli and in their relationship to behavior. Uncovering the internal representation of such codes remains a fundamental task in systems neuroscience. In practice, this is usually formulated as a neural coding or decoding problem.

The rodent hippocampus plays a key role in episodic memory, spatial navigation, and memory consolidation (O'Keefe and Dostrovsky 1971; O'Keefe and Nadel 1978). Pyramidal cells in the CA1 area of the rodent hippocampus have localized receptive fields (RFs) that are tuned to the animal's spatial location during navigation in one-dimensional (1D) or two-dimensional (2D) environments. These cells are referred to as place cells, and their RFs are referred to as place fields. Place field can be used for topographic map representation, in which a topographic map contains metric information (such as distance and orientation) between two locations in the map. In contrast to topographic map, the hippocampus has also been suggested in topological coding of space (Curto and Itskov 2008; Dabaghian et al. 2011, 2012; Chen et al. 2014). From a neural data analysis point of view, the question of our interest is that: *How do we transform the temporal patterns of spiking activity in the form of multiple time series into a spatial representation with pattern of place fields?*

3.4.1.2 Modeling Methodology

To discover latent structures of large-scale population codes, we have developed a hidden Markov model (HMM) for analyzing hippocampal population codes during spatial navigation (Chen et al. 2012, 2014). In a basic model (Fig. 3.1), we assume that the latent state process follows a first-order discrete-state Markov chain $\{S_t\} \in \{1, 2, \ldots, m\}$. We use a finite m-state HMM to characterize the temporal population spike activity from a population of C hippocampal neurons. We assume that the spike counts of individual place cells at discrete time index t, conditional on the latent state S_t, follow a Poisson probability distribution with associated tuning curve functions $\boldsymbol{\Lambda} = \{\boldsymbol{\lambda}_c\} = \{\lambda_{c,i}\}$:

$$p(\mathbf{y}_{1:T}, S_{1:T}|\boldsymbol{\pi}, \boldsymbol{P}, \boldsymbol{\Lambda}) = p(S_1|\boldsymbol{\pi}) \prod_{t=2}^{T} p(S_t|S_{t-1}, \boldsymbol{P}) \prod_{t=1}^{T} p(\mathbf{y}_t|S_t, \boldsymbol{\Lambda}) \tag{3.6}$$

where $\boldsymbol{P} = \{P_{ij}\}$ denotes an m-by-m state transition matrix, with P_{ij} representing the transition probability from state i to j; $\boldsymbol{\pi} = \{\pi_i\}$ denotes a probability vector for the initial state S_1; and $p(\mathbf{y}_t|S_t, \boldsymbol{\Lambda}) = \prod_{c=1}^{C} \texttt{Poisson}(y_{c,t}|\lambda_{c,S_t})$ defines products of Poisson distributions given the individual rate parameters.

To derive a Bayesian inference procedure, we further introduce the following prior distributions over the parameters $\{\boldsymbol{\pi}, \boldsymbol{P}, \boldsymbol{\Lambda}\}$ (Chen et al. 2012):

$$\boldsymbol{\pi} \sim \texttt{Dirichelet}(\alpha_0 \mathbf{1}),$$
$$\boldsymbol{P}_{i,:} \sim \texttt{Dirichelet}(\alpha_0 \mathbf{1}),$$

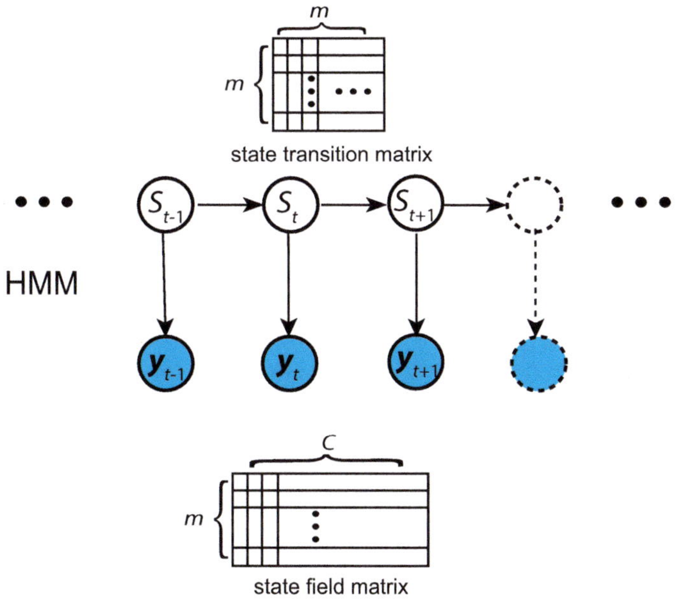

Fig. 3.1 Finite-state hidden Markov model (HMM). The latent state $\{S_t\}$ follows a Markov process. The state transition is characterized by an $m \times m$ state transition matrix. Conditional on the latent state, the population spike activity of C neurons is characterized by a Poisson distribution characterized by a mean firing rate represented by an $m \times C$ state field matrix

$$\alpha_0 \sim \texttt{Gamma}(a_{\alpha_0}, 1),$$
$$\lambda_{c,i} \sim \texttt{Gamma}(a_c^0, b_c^0).$$

where `Dirichelet` denotes the Dirichlet distribution, and $\texttt{Gamma}(a_c^0, b_c^0)$ denotes the gamma distribution with shape parameter a_c^0 and scale parameter b_c^0.

To accommodate automatic model selection for the unknown parameter m, we have further developed a Bayesian nonparametric version of the HMM, the so-called hierarchical Dirichlet process-HMM (HDP-HMM), which extends the finite-state HMM with a nonparametric HDP prior, and inherits a great flexibility for modeling complex data (Linderman et al. 2016). Specifically, we sample a distribution over latent states, G_0, from a DP prior, $G_0 \sim \texttt{DP}(\gamma, H)$, where γ is the concentration parameter and H is the base measure. We also place a prior distribution over the concentration parameter, $\gamma \sim \texttt{Gamma}(a_\gamma, 1)$. Given the concentration, we sample from the DP via the "stick-breaking process (STP)": the stick-breaking weights, $\boldsymbol{\beta}$, is drawn from a beta distribution:

$$\tilde{\beta}_i \sim \texttt{Beta}(1, \gamma), \quad \beta_i = \tilde{\beta}_i \prod_{j=1}^{i-1} (1 - \tilde{\beta}_j) \tag{3.7}$$

where $\beta_1 = \tilde{\beta}_1$, $\sum_{i=1}^{\infty} \beta_i = 1$, and $\mathtt{Beta}(a, b)$ defines a beta distribution with two shape parameters $a > 0$ and $b > 0$. For inference, we use a "weak limit" approximation in which the DP prior is approximated with a symmetric Dirichlet prior

$$
\begin{aligned}
\gamma &\sim \mathtt{Gamma}(a_\gamma, 1) \\
\alpha_0 &\sim \mathtt{Gamma}(a_{\alpha_0}, 1) \\
\boldsymbol{\beta}|\gamma &\sim \mathtt{Dirichelet}(\gamma/M, \ldots, \gamma/M), \\
\boldsymbol{\pi}|\alpha_0, \boldsymbol{\beta} &\sim \mathtt{Dirichelet}(\alpha_0\beta_1, \ldots, \alpha_0\beta_M), \\
\boldsymbol{P}_{i,:}|\alpha_0, \boldsymbol{\beta} &\sim \mathtt{Dirichelet}(\alpha_0\beta_1, \ldots, \alpha_0\beta_M).
\end{aligned}
$$

where M denoted a truncation level for approximating the distribution over the countably infinite number of states.

For the Poisson likelihood, we use a Gibbs sampler for parameter $\boldsymbol{\Lambda}$. Since we are using conjugate gamma priors, the posterior can be updated in a closed form

$$
\lambda_{c,i}|\boldsymbol{y}, S_{1:T} \sim \mathtt{Gamma}\left(\alpha_c^0 + \sum_{t=1}^{T} y_{c,t}\mathbb{I}[S_t = i],\ \beta_c^0 + \sum_{t=1}^{T} \mathbb{I}[S_t = i]\right).
$$

Furthermore, since the priors on $\boldsymbol{P}_{i,:}$ and $\boldsymbol{\pi}$ reduce to Dirichlet distributions, we can derive conjugate Gibbs updates for these parameters as follows:

$$
\begin{aligned}
\boldsymbol{\pi}|\alpha_0, \boldsymbol{\beta} &\sim \mathtt{Dirichelet}\left(\alpha_0\boldsymbol{\beta} + \mathbf{1}_{S_1}\right), \\
\boldsymbol{P}_{i,:}|\alpha_0, \boldsymbol{\beta} &\sim \mathtt{Dirichelet}\left(\alpha_0\boldsymbol{\beta} + \boldsymbol{n}_i\right), \\
n_{i,j} &= \sum_{t=1}^{T-1} \mathbb{I}[S_t = i, S_{t+1} = j],
\end{aligned}
$$

where $\mathbf{1}_j$ is a unit vector with a one in the j-th entry. Conditioned upon the firing rates, the initial state distribution, and the transition matrix, we can jointly update the latent states using a *forward filtering, backward sampling* algorithm to obtain a full sample from $p(S_{1:T}|\boldsymbol{P}, \boldsymbol{\pi}, \boldsymbol{\Lambda})$.

Regarding the firing rate hyperparameters $\{\alpha_c^0, \beta_c^0\}$ for the c-th neuron, we have proposed three methods for update (Linderman et al. 2016): (1) empirical Bayesian, which aims to maximize the marginal likelihood of the spike counts; (2) Hamiltonian Monte Carlo (HMC) sampling for joint posterior $\{\log\alpha_c^0, \log\beta_c^0\}$; and (3) sampling the scale hyperparameter β_c^0 (using a gamma prior) while fixing the shape hyperparameter, $\alpha_c^0 = 1$. In practice, the second and third methods are found to work very well.

To summarize, we have constructed a hierarchical probabilistic model for characterizing population spike trains, consisting of model parameters, hyperparameters, and hyper prior parameters. This model is sufficiently flexible and hierarchical Bayesian inference also enables us to impose different prior information onto the model. More importantly, we have tested (using synthetic datasets) the robustness of this model with respect to those hyper priors or hyperparameter optimization.

3.4.1.3 Application to Rat Hippocampal Population Codes

Hippocampal functions have been widely investigated in various rodent spatial and nonspatial experimental tasks. In spatial tasks, freely behaving rodents are instructed to navigate in specific spatial environments. The typical experimental paradigm is spatial navigation followed by post-behavior sleep.

In our first illustrated example, the rat was freely foraging in an open field arena (Fig. 3.2a). The micro-drive arrays containing multiple tetrodes were implanted above the right dorsal hippocampus of male Long-Evans rats. The tetrodes were slowly lowered into the brain reaching the cell layer of CA1 2–4 weeks following the date of surgery. Recorded spikes were manually clustered and sorted to obtain single units using a custom software. In this example, we apply the unsupervised population decoding analysis to about 9-min recording of 49 well-isolated rat hippocampal CA1 units. The details of experimental setup are referred to Linderman et al. (2016). From the inferred state trajectory, we infer a two-dimensional (2D) state space map (Fig. 3.2b), from which we further quantify the median decoding error in time (Fig. 3.2c). In this case, the median decoding error is around 0.10 m).

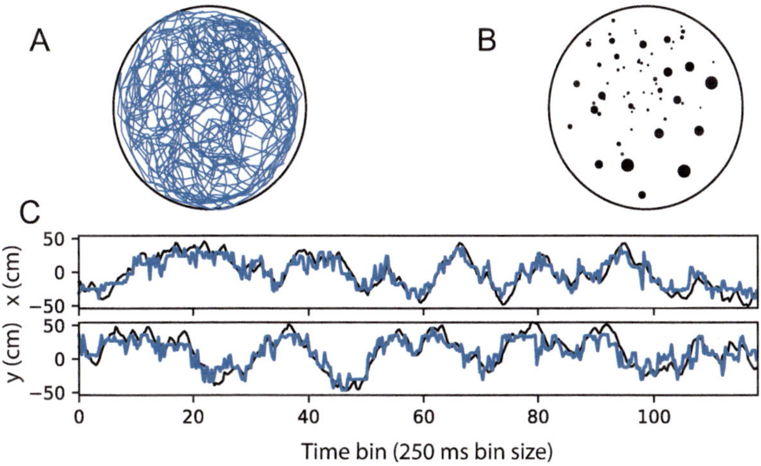

Fig. 3.2 Uncovering topological representation of rat hippocampal population codes. (**a**) Rat's run trajectory in an open field arena. (**b**) Inferred state space map, where the mean value of the spatial position for each latent state is shown by a black dot. The size of the dot is proportional to the occupancy of the state. (**c**) Snapshots of decoded trajectories (blue) in x and y coordinates (black: animal's true trajectory). Figure adapted from Linderman et al. (2016)

In our second illustrated example, a naive rat was freely foraging a circular maze. The recording session consisted of a long (∼4 h) pre-RUN sleep epoch home-cage recording performed in a familiar room, followed by a RUN epoch (∼45 min) in a novel circular maze (https://crcns.org/data-sets/hc/hc-11/) (Grosmark and Buzsaki 2016). After the RUN epoch, the animal was transferred back to its home cage in the familiar room where a long (∼4 h) post-RUN sleep was recorded. Upon off-line spike sorting, we observed clear place receptive fields from 77 hippocampal CA1 neurons.

In this example, we first apply unsupervised decoding analysis to the RUN data and infer the state transition matrix $\boldsymbol{P}$ and firing rate matrix $\boldsymbol{\Lambda}$. The number of the latent states is automatically identified and the choice of hyperparameter priors is determined from the predictive likelihood of held on data. Next, we can apply the inferred structure to detect sleep-associated hippocampal memory reactivation (or "replay") during quiet wakefulness (QW) on the circular track, as well as during post-RUN slow wave sleep (SWS) in a sleep box (Chen et al. 2016a). From post-SWS epochs, we further identify candidate events for hippocampal spatial memory reactivation. The events are selected based on hippocampal local field potential (LFP) ripple power and hippocampal multi-unit activity (threshold > mean + 3SD). We also impose a minimum cell activation criterion (>10% of cell population). Among those candidate events, we temporally bin the hippocampal neuronal spikes (20 ms) and run a population decoding analysis to detect replay events. Statistically significant reactivation events are determined by established criteria, followed by random shuffling operations (Chen et al. 2016a). A few representative significant (Monte Carlo $p < 0.01$ or Z-score greater than 2.33) examples are shown in Fig. 3.3.

Notably, detecting the statistical significance of a trajectory sequence often relies on the line fitting procedure (Davidson et al. 2009). To overcome the limitation of linear weighted correlation metric (Wu and Foster 2014), we adapt the "distance correlation" metric (Székely and Rizzo 2009), and derive a new metric called "weighted distance correlation." The motivation is to address the deficiency of Pearson's correlation in the presence of discontinuity of the trajectory (e.g., due to linearization of a circular track or T-maze) or nonlinear relationship between two variables (Liu et al. 2018). In comparison with the standard receptive-field based population decoding analysis, our unsupervised population decoding allows non-even spacing with the environment according to the sampling occupancy or spiking data. In addition, the analysis paradigm "*memory first, meaning later*" (or "*structure first, content later*") provides an unbiased assessment of neural population codes (Chen et al. 2016a; Chen and Wilson 2017).

Thus far, we have only tested our method on RUN→RUN and RUN→SLEEP. However, our analysis paradigm is purely unsupervised and is independent of the templates. For instance, in principle we can test the method on SLEEP→RUN or SLEEP→SLEEP (different sleep epochs or sleep stages). Specifically, to test our method on SLEEP→RUN, we use the detected significant memory replay events as the training data and run the unsupervised population decoding analysis (Chen et al. 2016a). We use the inferred firing rate matrix $\boldsymbol{\Lambda}$ from SLEEP epochs and to infer the

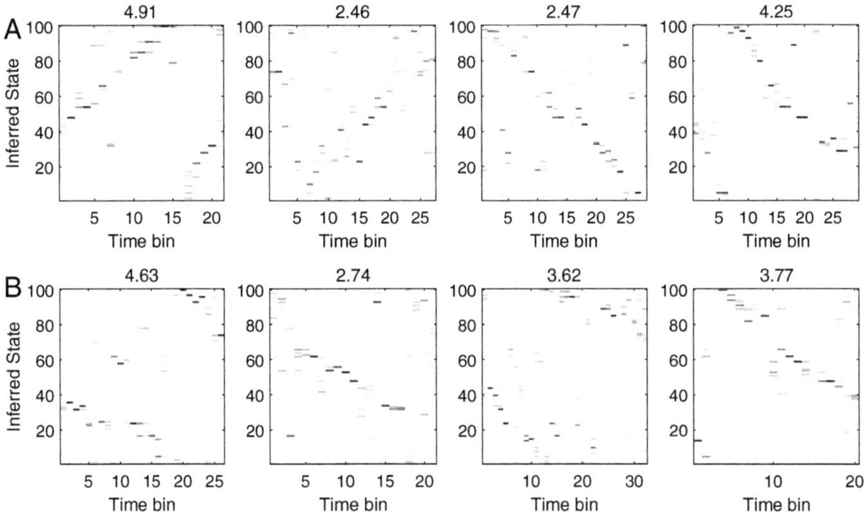

Fig. 3.3 Detecting hippocampal memory replays in offline states. (**a**) Quiet wakefulness (QW). (**b**) Post-SWS. Temporal bin size: 20 ms. The Z-score of weighted distance correlation is shown on the top of each panel

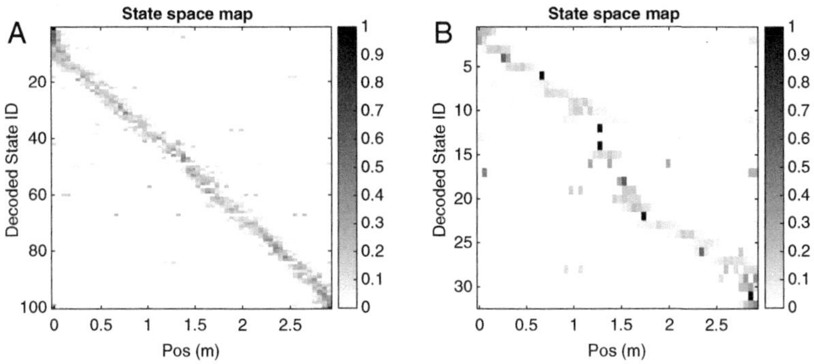

Fig. 3.4 Inferred state space correspondence maps. (**a**) Using population spike data from RUN epochs alone (effective number of state: 100, median decoding error: 0.05 m). (**b**) Using population spike data from post-SWS epochs alone (effective number of state: 34, median decoding error: 0.16 m)

state sequence during RUN. We then compare the inferred latent state sequence with the animal's position ("ground truth") to assess the state space map (Fig. 3.4) and decoding error (median error: 0.16 m). To our surprise, despite the fact that sleep-associated population spike data are sparse and fragmental, we are still capable of extracting meaningful latent structure of the "place code."

3.4.1.4 Application to Rat Hippocampal-Neocortical Population Codes

In rodent hippocampal neural representations, the latent states correspond to spatial sequences. In general, the latent states can be abstract or undefined a priori. Therefore, a similar analysis can be applied to uncover neocortical population representations, except that the derived latent states may correspond to a different behavioral correlate. The visual cortex provides a crucial sensory input to the hippocampus, and is a key component for the creation of spatial memories (Chen et al. 2013). Specifically, visual and spatial inputs provide two dominant cues in *visuospatial* information processing during spatial navigation (Haggerty and Ji 2015a). In this case, the latent states derived from the visual population codes may correspond to visuospatial sequences.

In the third illustrated example, the rat navigated in a figure-"8" environment (Fig. 3.5a), and neuronal ensemble spike activity was recorded from the rat hippocampus CA1 and primary visual cortex (V1) simultaneously (Ji and Wilson 2007; Haggerty and Ji 2015b). One research question is to investigate the content dependency between two neuronal populations. Since large percentage of V1 neurons contain high spatial information rate (Haggerty and Ji 2015a), it is expected that the spike activity from V1 population can code spatial information.

We apply the unsupervised decoding analysis to the simultaneous CA1-V1 ensemble recording from one rat during spatial navigation in one session. Based on two independent analyses, we infer the latent sequences $S_{1:T}^{\mathrm{CA1}}$ and $S_{1:T}^{\mathrm{V1}}$ separately from respective (sorted) ensemble spikes (16 CA1 units and 21 V1 units). First, we assess the results by the median decoding error. Surprisingly, the decoding accuracy derived from both data sets without using behavioral measures are very good (median decoding error: 3.68 cm for the hippocampus, 2.01 cm for the V1; illustrations in Fig. 3.5b). The inferred number of states are 60 from V1 and 39 from CA1, respectively.

Next, we assess the statistical dependency of neural representations between CA1 and V1. The mutual information (MI) is a measure that quantify statistical dependency between two discrete random variables. Comparing two inferred latent sequences: $S_{1:T}^{\mathrm{CA1}}$ vs. $S_{1:T}^{\mathrm{V1}}$ (Fig. 3.5c), we compute the Shannon entropy, conditional entropy (unit: bits): $H_{\mathrm{CA1}} = 4.5171, H_{\mathrm{V1}} = 5.1851, H_{\mathrm{CA1|V1}} = 1.5514$ and normalized mutual information (NMI): $\frac{H_{\mathrm{CA1}}-H_{\mathrm{CA1|V1}}}{\sqrt{H_{\mathrm{CA1}}H_{\mathrm{V1}}}} = 0.6128$ (bootstrapped SD: 0.0059). A large value of NMI ($0 \leq \mathrm{NMI} \leq 1$) indicates high statistical dependency between two latent state sequences, which contributed by their place coding. The additional information might be contributed by visual coding.

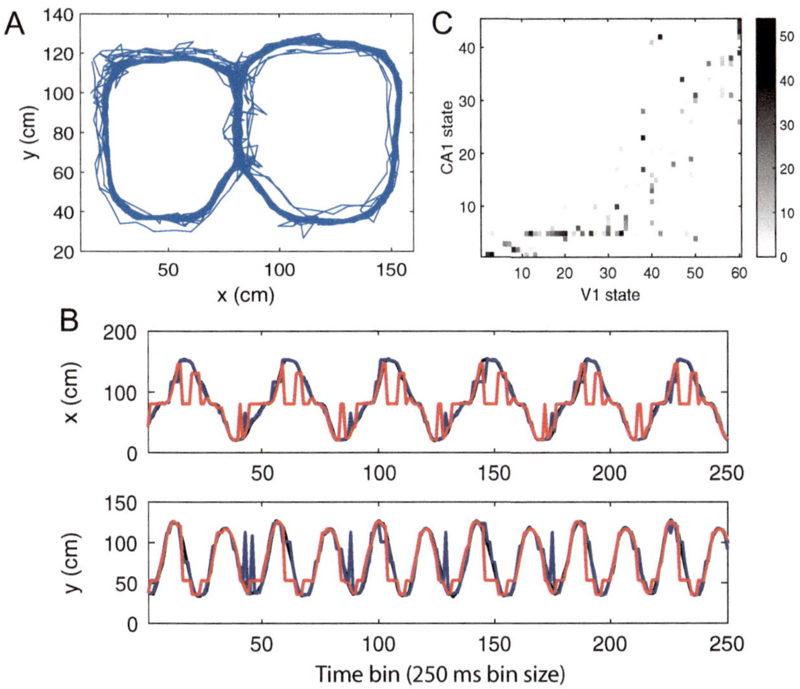

Fig. 3.5 Neural representations of rat CA1-V1 population codes. (**a**) Rat's run trajectory in a figure-"8" maze. (**b**) Snapshots of decoded trajectories from CA1 (red) and V1 (blue) population codes in x and y coordinates. The black curve denotes animal's actual position. (**c**) Confusion matrix map between two inferred latent state sequences $S_{1:T}^{\mathrm{CA1}}$ and $S_{1:T}^{\mathrm{V1}}$

3.4.2 Detecting Onset of Acute Pain Signals

3.4.2.1 Background

In the neuroscience literature, neuroimaging and neurophysiological studies have identified circuit changes in the primary somatosensory cortex (S1) and anterior cingulate cortex (ACC) during pain states (Bushnell et al. 2013; Wagner et al. 2013; Fuchs et al. 2014; Kuo and Yen 2005; Zhang et al. 2011; Vierck et al. 2013). The use of multiple microwire arrays has enabled us to record many neurons from the S1 and ACC in freely behaving rats during an experimental protocol of acute thermal pain (Chen et al. 2017a).

A central computational goal is to reliably detect on the onset of acute thermal pain signals encoded by ensemble S1 and ACC responsive neurons. This can be formulated as a change-point detection problem. Depending on the nature of application, the detection can be off-line or on-line (sequential). To illustrate the idea of latent variable modeling, here we discuss the off-line setting.

3.4.2.2 Modeling Methodology

Let the latent univariate variable $z_t \in \mathbb{R}$ represent the unobserved common input (e.g., pain stimulus) that drives the neuronal population firing. The data-driven model can be described by a latent-state Poisson linear dynamical system (PLDS) (Chen et al. 2017a)

$$z_t = az_{t-1} + \epsilon_t \tag{3.8}$$

$$\mathbf{y}_t \sim \texttt{Poisson}\big(\exp(\mathbf{c}z_t + \mathbf{d})\Delta\big) \tag{3.9}$$

where the state equation (3.8) is a first-order autoregressive (AR) model driven by a zero-mean Gaussian noise process $\epsilon_t \in \mathcal{N}(0, \sigma^2)$. The parameters $\mathbf{d}$ and $\mathbf{c}$ are unconstrained. Previously, we developed a variational EM algorithm to estimate the posterior of $\hat{z}_{t|T}$ as well as the unknown parameters (Chen et al. 2017a).

Alternatively, we transform the spike count vector using the change-of-variable, such as $\tilde{\mathbf{y}}_t = \sqrt{\mathbf{y}_t}$ or $\tilde{\mathbf{y}}_t = \log(\mathbf{y}_t + 1)$ such that $\tilde{\mathbf{y}}_t \geq 0$. Assuming that the transformed variable $\tilde{\mathbf{y}}_t$ is Gaussian, we have the linear dynamical system (LDS):

$$z_t = az_{t-1} + \epsilon_t \tag{3.10}$$

$$\tilde{\mathbf{y}}_t = \mathbf{c}z_t + \mathbf{d} + \mathbf{w}_t \tag{3.11}$$

where $\mathbf{w}_t \in \mathcal{N}(\mathbf{0}, \boldsymbol{\sigma}_\mathbf{w})$ is a Gaussian noise process with zero mean and covariance $\boldsymbol{\sigma}_\mathbf{w}$. Similarly, we use the EM algorithm or spectral learning algorithm to estimate the unknown parameters $\{a, \mathbf{c}, \mathbf{d}, \sigma_\epsilon, \boldsymbol{\sigma}_\mathbf{w}\}$ (Buesing et al. 2012a). In practice, we may fix $\mathbf{d}$ in Eqs. (3.9) and (3.11) with the baseline firing rate.

From the estimated latent state, we further compute the Z-score related to the baseline: Z-score $= \frac{z\text{-mean of } z_{\text{baseline}}}{\text{SD of } z_{\text{baseline}}}$. Under the assumption that the Z-score is standard normally distributed, we convert it to the one-tailed P-value:

$$P(\text{Z-score} > \hat{z}_{t|T}) = 1 - P(\text{Z-score} \leq \hat{z}_{t|T}) = 1 - \int_{-\infty}^{\hat{z}_{t|T}} \frac{1}{\sqrt{2\pi}} e^{-u^2/2} du \tag{3.12}$$

The criterion of Z-score change (or equivalent P-value) is determined by a critical threshold for reaching statistical significance. For instance, using the significance criterion with one-sided P-value 0.05, it is concluded that a change point occurs when Z-score $-$ CI > 1.65 or Z-score $+$ CI < -1.65, where CI denotes the scaled confidence interval derived from the standard deviation of latent variable.

3.4.2.3 Inference

Given the PLDS, we can estimate the latent variables and unknown model parameters by an iterative EM algorithm based on maximum likelihood estimation. In the E-step, we compute the Gaussian smoothed posterior for the latent state $z_t \sim$

$\mathcal{N}(\hat{z}_{t|T}, Q_{t|T})$; in the M-step, we update the parameters using the most recent state estimate. The iteration continues until the log-likelihood value reaches to the local maximum. Because of the non-Gaussian likelihood, the E-step in the EM algorithm is intractable. Therefore, Laplace or variational approximation methods can be considered (Smith and Brown 2003; Buesing et al. 2012b; Macke et al. 2012). In our experimental investigation, we found that variational approach yielded slightly better results in predictive likelihood, but the Laplace approach converged much faster and obtained similar results in the state estimate (Chen et al. 2017a).

Note that we have assumed a univariate latent variable here. In general, the latent variable can be multivariate and we can apply model selection procedures to determine the optimal model dimensionality (Chen et al. 2017a). However, in this specific application, our central goal is the quickest detection of change in neural activity; therefore, a univariate latent variable is sufficient.

In off-line applications, the sufficient statistics $\mathcal{N}(\hat{z}_{t|T}, Q_{t|T})$ of the smoothed state are used for assessing change points. In on-line applications, we can derive recursive filtering algorithm for spike count observations (Smith and Brown 2003; Eden et al. 2004)

$$\hat{z}_{t|t-1} = a\hat{z}_{t-1|t-1} \tag{3.13}$$

$$Q_{t|t-1} = a^2 Q_{t-1|t-1} + \sigma_\epsilon^2 \tag{3.14}$$

$$\hat{\mathbf{y}}_{t|t-1} = \exp(\mathbf{c}\hat{z}_{t|t-1} + \mathbf{d})\Delta \tag{3.15}$$

$$Q_{t|t}^{-1} = Q_{t|t-1}^{-1} + \mathbf{c}^\top \mathrm{diag}(\hat{\mathbf{y}}_{t|t-1})\mathbf{c} \tag{3.16}$$

$$\hat{z}_{t|t} = \hat{z}_{t|t-1} + Q_{t|t}\mathbf{c}^\top(\mathbf{y}_t - \hat{\mathbf{y}}_{t|t-1}) \tag{3.17}$$

where $Q_{t|t} = \mathrm{Var}[\hat{z}_{t|t}]$ denotes the filtered state variance.

3.4.2.4 Application to Rat ACC-S1 Population Codes

In the experiment, noxious stimulation via a blue laser was applied to plantar surface of the hind paw contralateral to the brain recording site in freely moving male Sprague-Dawley rats (Chen et al. 2017a). The onset of noxious pain was identified from videos (60 frame/s), indicated by the paw withdrawal. Spikes were sorted off-line to obtain well-isolated single units from either stereotrodes or tetrodes. All sorted single units (putative pyramidal neurons and interneurons) are included in population analysis.

In the first illustrated example, we apply our proposed models (PLDS and LDS) to one recording session of rat ACC during 150 mW laser stimulations. All analyses are on the single-trial basis. The algorithm converges very fast, typically within 50–100 iterations for 15-s data. For each trial, we define the baseline as 5 s prior to the laser onset as the baseline period. Notably, the ACC population consists of both positive and negative responders, which show increased and decreased firing

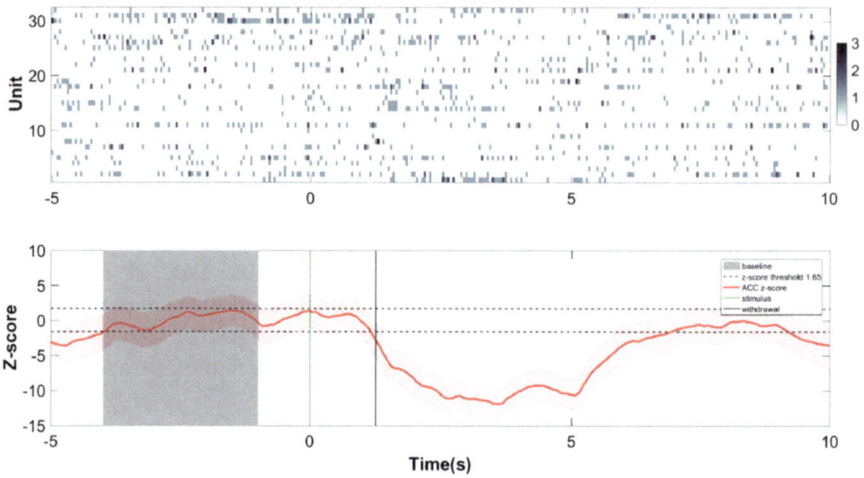

Fig. 3.6 Detecting acute thermal pain signals. (**a**) Rat recording of neuronal ensemble spike counts from 30 ACC units under 150 mW laser stimulation (on-line sorting). Time 0 denotes the laser onset. Bin size 50 ms. Color bar indicates spike count, with dark color representing large spike count. (**b**) Estimated mean *Z*-score (blue curve) from the latent state $\hat{z}_t$. Vertical red line indicates the animal's paw withdrawal—an indicator of acute pain behavior. Horizontal dashed lines mark the significant thresholds of ± 1.65. Shaded area around the red curve marks the 95% confidence intervals. Baseline period: $[-4, -1]$ s

in response to pain stimuli, respectively. Some of units do not show significant modulation with respect to pain stimuli. As illustrated in this example (Fig. 3.6), our approach successfully detected the "neuronal threshold for acute pain" from ACC ensemble neurons, and the change point is identified around the onset of paw withdrawal.

In the second illustrated example, we apply our proposed models (PLDS and LDS) to one session with simultaneous recordings of rat S1 and ACC units during 250 mW laser stimulations. Two independent decoding analyses are carried out for S1 and ACC recordings at each single trial, and the final decision is made based upon integrating the detection results from two separate analyses (Fig. 3.7). In this example, note that first, the change point is identified before the onset of paw withdrawal using either S1 and ACC population spikes; and it is faster to cross the significance threshold in the case of S1 population. Second, the significant period detected from the S1 population is much longer than the significant period detected from the ACC population. This may be due to the reason that S1 and ACC neurons have different sensitivity or specificity to pain stimuli. To accommodate this difference, we can design a suboptimal decision rule and adjust the detection threshold according to their sensitivity or specificity (Hu et al. 2017).

In summary, the state space analysis provides a principled paradigm to detect the change in neuronal ensemble spike activity. In our observations, the model-based approach is more robust than the model-free approach, and the PLDS model

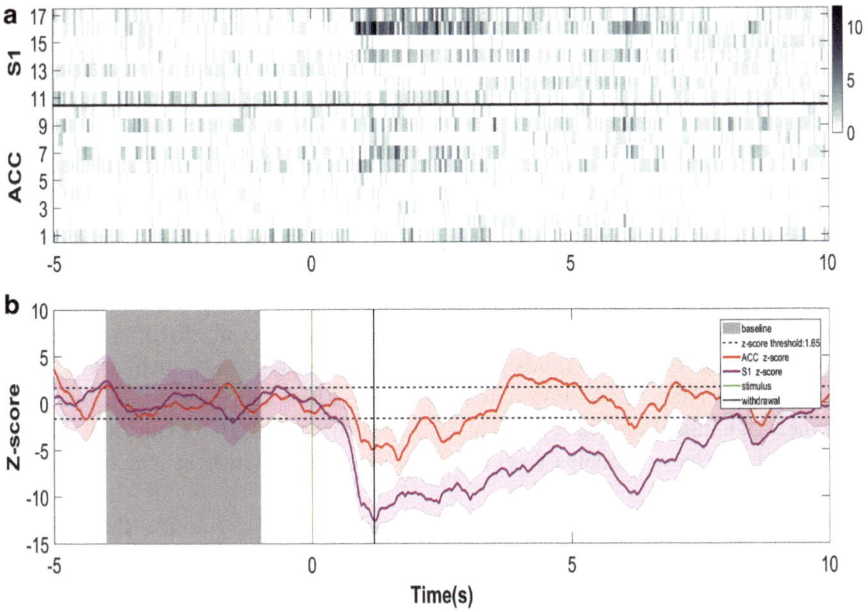

Fig. 3.7 Detecting acute thermal pain signals. (**a**) Simultaneous recordings of neuronal ensemble spike counts from the 7 S1 and 10 ACC units under 250 mW laser stimulation. Time 0 denotes the laser onset. Bin size 50 ms. Color bar indicates spike count, with dark color representing large spike count. (**b**) Estimated mean *Z*-score (blue curve) from the latent state $\hat{z}_t$. Vertical red line indicates the animal's paw withdrawal—an indicator of acute pain behavior. Horizontal dashed lines mark the significant thresholds of ± 1.65. Shaded area around the red or blue curve mark the 95% confidence intervals. Baseline period: $[-4, -1]$ s

performs better than the LDS (Chen et al. 2017b). For real-time applications, we can further optimize the filtering algorithms to improve the detection speed and accuracy (Hu et al. 2018).

3.4.3 Unfolding Motor Population Dynamics

3.4.3.1 Background

Single-neuron responses in motor cortex are complex, and there is marked disagreement regarding which movement parameters are represented (Churchland et al. 2012). Therefore, it is important to discover latent structure of motor population dynamics using statistical methods (Cunningham and Yu 2014; Aghagolzadeh and Truccolo 2016; Feeney et al. 2017). One popular approach is to use *supervised learning* for establishing encoding models for individual motor neurons, and then use the assumed encoding model in population decoding. However, this approach has several drawbacks: First, we need to make strong statistical assumptions for the neuronal encoding model. In addition, the measured movement behavior (e.g., kinematics) is high-dimensional, the model fitting may easily suffer from overfitting. Second, there is often strong heterogeneity among neuronal populations. Without prior information, it may be unwise to assume that every neuron share the same encoding model. In contrast, the alternative approach is to use *unsupervised learning* for unbiased assessment of neuronal population codes.

The latent variable approach can be viewed as a subclass of "neural trajectory" methods, with the goal to uncover the latent neural trajectory (Buonomano and Laje 2010). One type of methods is based on trial averaging, such as principal component analysis (PCA) or other subspace methods (Churchland et al. 2012; Ames et al. 2014). The other type of methods focus on single-trial dynamics, such as Gaussian process factor analysis (GPFA) (Yu et al. 2009; Zhao and Park 2017), linear or nonlinear dynamical systems (Wu et al. 2009; Lawhern et al. 2010; Gao et al. 2016), mixture of trajectory models (Yu et al. 2007), and neural networks (Michaels et al. 2016).

Given the measured neuronal motor population spike activity, the goal of unsupervised learning is to unfold the latent state trajectory that drives the motor population dynamics, where the relationship between the inferred state sequence and measured behavior can be established a posteriori. Here we employ two unsupervised learning approaches: the first approach is based on the PLDS, where the state is assumed to be continuous with unknown dimensionality; the second approach is based on the HDP-HMM, where the state is assumed to be discrete and the state transition is assumed to be Markovian. In both cases, we can slightly differentiate assumptions on the latent state without explicitly defining what the latent state is a priori. For illustration purpose, we will not make comprehensive comparisons with other models here.

3.4.3.2 Simulation and Results

As an illustration, we construct the synthetic population spike data based on known underlying relationship between the motor movement and neural population dynamics. Specifically, the following four types of trajectory paths are assumed for movement kinematics $\boldsymbol{K} \stackrel{\Delta}{=} [x_t, y_t, \dot{x}_t, \dot{y}_t] \in \mathbb{R}^4$ (Wu and Srivastava 2011):

$$
\begin{aligned}
x_t^{(1)} &= -\cos(0.5\pi t) \\
y_t^{(1)} &= \sin(0.5\pi t) \\
x_t^{(2)} &= -\cos(0.5\pi t) \\
y_t^{(2)} &= -\sin(0.5\pi t) \\
x_t^{(3)} &= 0.5(\cos(\pi t) + 1)\Theta_t
\end{aligned}
$$

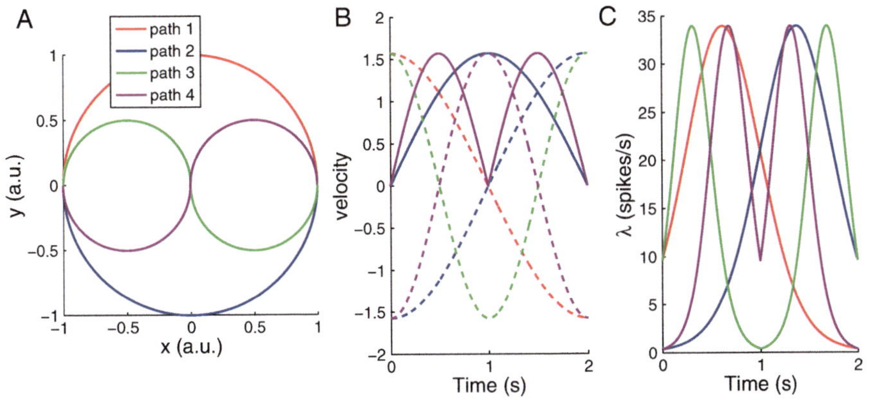

Fig. 3.8 Simulated trajectory and representative tuning curves. (**a**) Simulated movement trajectory (x_t, y_t), with four colors representing different paths. (**b**) Velocity of $(\dot{x}_t, \dot{y}_t)$ for different paths. Solid and dashed lines denote the x and y directions, respectively. (**c**) Profile of neuronal firing rate

$$y_t^{(3)} = 0.5\sin(\pi t)$$
$$x_t^{(4)} = 0.5(\cos(\pi t) + 1)\Theta_t$$
$$y_t^{(4)} = -0.5\sin(\pi t)$$

where Θ_t is a step function. The trajectory paths are shown in Fig. 3.8. In addition, we assume that the firing rate of the c-th neuron, $\lambda_{c,t}$, is *nonlinearly* modulated by a four-dimensional *instantaneous* kinematic vector $[x_t, y_t, \dot{x}_t, \dot{y}_t]$. Specifically, for the i-th trajectory path ($i = 1, 2, 3, 4$), the firing rate is represented as follows:

$$\lambda_{c,t}^{(i)} = b_c \exp(\mathbf{a}_c^\top [x_t^{(i)}, y_t^{(i)}, \dot{x}_t^{(i)}, \dot{y}_t^{(i)}]) \tag{3.18}$$

We use the following setup for simulation: 40 neurons, 50 trials, with each trial of 2-s duration. We use 40 trials for training and the remaining 10 trials for testing. Population spikes are binned into 100 ms to obtain spike count observations.

In the first method, we assume that the latent state is continuous valued. We set the dimensionality of latent state $\mathbf{z}$ to 4 and employ the PLDS and EM algorithm based on likelihood inference. Once the latent trajectory is inferred (Fig. 3.9a), we apply the canonical correlation analysis (CCA) to assess the statistical dependency between the linear subspaces of $\mathbf{z}$ and $\boldsymbol{K}$. For two random vector variables $\mathbf{z} = \{z_i\}$ and $\boldsymbol{K} = \{K_j\}$, CCA is aimed to find linear combinations of the $\{z_i\}$ and $\{K_j\}$ which have maximum correlation with each other. When using $\dim(\mathbf{z}) = 4$, the scatter plots of training data projected on two maximally correlated subspaces are shown in Fig. 3.9b, where the correlation coefficients are 0.90 and 0.81, respectively. Specifically, the maximum correlation between the latent state and the behavioral subspace for four trajectory paths are 0.91, 0.94, 0.91 and 0.85, respectively. This

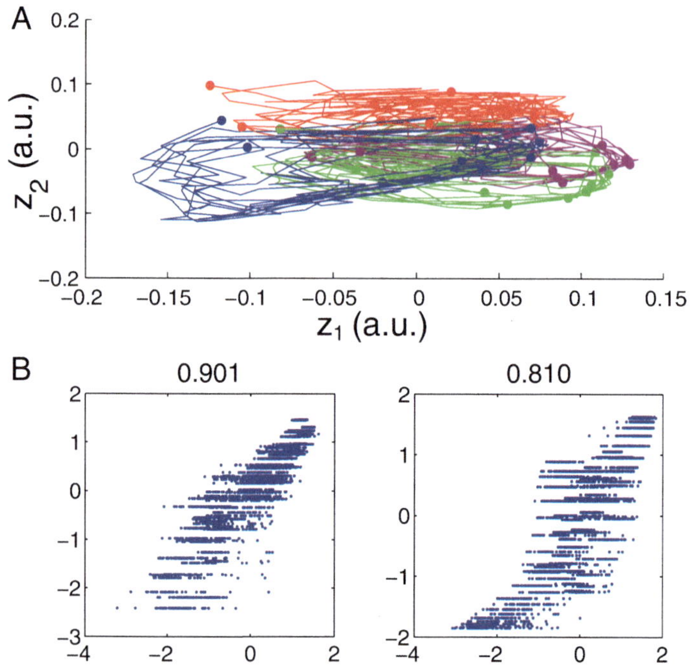

Fig. 3.9 Uncovering latent structures of motor population codes. (**a**) Inferred continuous latent state trajectory for four trajectory paths projected on the 2D state space. Circle denotes the trajectory onset. (**b**) Scatter plots of training data projected onto two dominant CCA subspaces. Correlation coefficients are shown in the top of panels

suggests that the inferred latent variables derived from the population spike activity capture the majority of variance of behavioral kinematics. By varying the latent state dimensionality, it is also found that the predictive log-likelihood of testing data is the highest when we use the true state dimensionality.

In the second method, we assume that the latent state space is discrete. We employ the HDP-HMM with MCMC inference to characterize the simulated population spike data. We assess the modeling performance by predicted data log-likelihood and the predicted state sequences. Examples are shown in Fig. 3.10. It appears that the state sequences captures the internal dynamics of periodic movement. Since the behavioral variables (i.e., kinematics) are relatively high-dimensional, we use a temporal clustering method known as *aligned cluster analysis* (ACA) and hierarchical ACA (HACA) to label the behavioral measurement (Zhou et al. 2013). The ACA provides a natural framework to find a low-dimensional embedding for time series by combining the concepts of *kernel k-means* with *dynamic time alignment kernel*. Next, from the inferred state trajectory, we compute the NMI between the inferred latent state sequence and the clustered behavioral sequence. Specifically, the NMI between the latent state and the clustered behavioral sequence is 0.71 (0.66, 0.48, 0.60, and 0.75 for four trajectory paths, respectively).

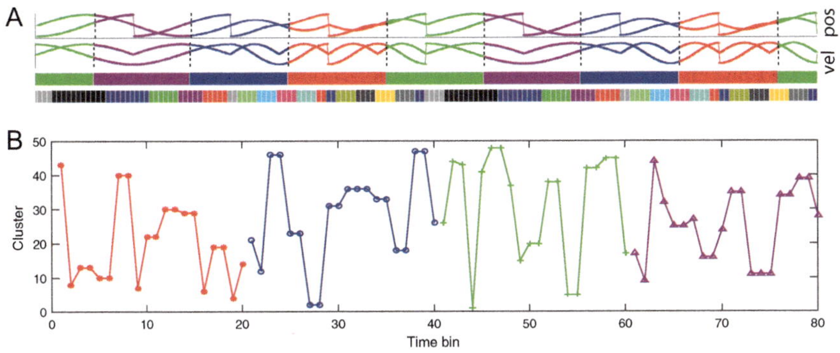

Fig. 3.10 Uncovering latent structures of motor population codes. (**a**) Illustration of hierarchically clustered behavioral sequences (four trajectory paths, each with two repetitions) at two levels. Different color represents different clusters. At the first level, there are 4 clusters, roughly corresponding to four trajectory paths. At the second level, there are a total of 48 clusters. (**b**) Examples of inferred discrete latent state sequences. Different colors or symbols represent different trajectory paths

By carefully examining the correspondence map (result not shown), we find that the majority of behavioral states are captured by 1–2 dominant HMM latent states, whereas the most occupied HMM states represent multiple clustered behavioral states, suggesting that it employs a conductive coding scheme to approximate the continuous firing rate mapping function. Furthermore, from the inferred state-firing rate matrix, we can develop new goodness-of-fit measures to characterize the "dissimilarity" between latent state sequences induced by trial variability. Specifically, it is found that the within-type trial variability is significantly lower than the between-type trial variability. Detailed results will be presented elsewhere due to space limitation.

3.5 Discussion

In this chapter, we have presented a class of latent variable models to characterize spatiotemporal neural dynamics. Although we have illustrated our methodology using neuronal population spike trains, this framework is rather general and can be extended to other neurophysiological recordings, such as the LFP, EEG, and calcium imaging data. To do so, one need to establish accurate probability distributions or likelihood models to characterize those neural signals.

In the context of neuroscience applications, the latent variable can be viewed as a proxy that represents a continuous variable of a specific task, whether it is referred to animal's spatial locations or movement kinematics in time. More generally, the latent variable can also be used to represent an abstract memory space that drives complex behavior. The complexity of the latent variable model is determined by many factors, such as the model architecture (e.g., directed graph, hierarchical and recurrent structure), the probability distribution and statistical dependency between random variables, temporal embedding of observations, and local or global nonlinear mapping.

3.5.1 Model Extension

3.5.1.1 Discrete State Case

Recently, we have extended the HDP-HMM method into two directions. First, we relax the common Markovian assumption and introduced a hidden semi-Markov model (HSMM) that allows for greater modeling flexibility of behavioral or neuronal dynamics. The HDP-HSMM also accommodates the HDP-HMM as a special case. Specifically, we assume that the sojourn duration for state i, denoted by $p(d_t|S_t = i)$, follows a parametric distribution (Chen et al. 2016b):

$$\begin{aligned} d_t|S_t = i &\sim \texttt{NegBin}(r, p) \\ &= \binom{d+r-2}{d-1} (1-p)^r p^{d-1} \quad (d = 1, 2, \ldots) \end{aligned}$$

where $\texttt{NegBin}(r, p)$ denotes a negative binomial distribution (discrete analog of the gamma distribution), which reduces to the geometric distribution when $r = 1$ as a special case (i.e., Markovian); namely, the HMM has a geometric sojourn time distribution such that the probability of staying in state i for d steps is $P_{ii}^d(1 - P_{ii})$.

Second, we introduce the concept of temporal embedding of population vector observations to enhance the representation power of HMM or HSMM (Chen 2017). Specifically, we construct a time-delay firing rate vector and augment the population vector from size C to $C\tau$, where τ denotes the temporal embedding length (which can contain forward or backward direction, or both). The inference procedure with such temporal embedding remains unchanged. At the end of inference, the newly derived firing rate matrix $\tilde{\boldsymbol{\Lambda}} = \{\tilde{\boldsymbol{\lambda}}_c\}$ is of size $C \times \tau m$, which can be reorganized and interpreted as an $m \times \tau$ spatiotemporal receptive field for the c-th neuron. When the embedding length is 1, it reduces to the standard setup.

3.5.1.2 Continuous State Case

The LDS and PLDS models employ a continuous latent state space. There are a few possible ways to extend the model representation. First, the log link function in the Poisson generalized linear model can be replaced by a complex nonlinear embedding, such as a nonlinear feedforward neural network, and the Poisson distribution can be replaced by a generalized count distribution (Gao et al. 2016).

Second, the network can be introduced a recurrent structure to incorporate internal memory, which yields a wide class of recurrent latent variable model (Pachitariu et al. 2013; Chung et al. 2016). Third, we can introduce a hierarchical structure to yield a deep neural network (DNN) (LeCun et al. 2015; Goodfellow et al. 2016). In addition, the unsupervised latent variable method can be combined with supervised learning to decode behavioral measures such as movement kinematics (Aghagolzadeh and Truccolo 2016).

3.5.2 *Challenges and Future Direction*

3.5.2.1 From Neural Space to Behavior, and to Neural Codes

Our unsupervised learning methods provide a paradigm to identify a meaningful latent "neural space" based on neuronal ensemble spike activity, which is further linked to behavioral measures at a specific timescale, such as animal's position or movement kinematics. However, one great challenge is the lack of prior information for choosing the representation of behavioral measures (e.g., the choice of coordinate system, potential nonlinear transformation or interaction among the behavioral measures), especially in the presence of high-dimensional behavioral measurements. How to establish the "behavioral space" that is mostly relevant to neural data remains an open question. Once the bridge is established between the neural space and behavior, we will ultimately reveal important neural coding principles.

3.5.2.2 Generative Models vs. Neural Networks

The stochastic HMM-type models have fundamental limitations due to its limited representation power. First, the states are mutually exclusive. Even with infinite number of states, it must select one of its hidden states at each time step; therefore with m hidden states it can only remember $\log_2 m$ bits about what it has generated. Second, they have relatively constrained latent state transition structures (characterized by $\boldsymbol{P}$).

Unlike the probabilistic generative models (e.g., HMM, LDS and PLDS), the recurrent or deep neural networks are mostly deterministic (with a very few exception such as the Boltzmann machine); namely, the hidden state of the neural network is computed as a deterministic version of probability mode. However, because of recurrent and hierarchical structure, the deterministic neural networks can characterize a rich distributed internal state representation and accommodate flexible nonlinear transition operations. For instance, they can be used to model oscillation, fixed-point or chaotic attractors (Rivkind and Barak 2017). In a special case of one-hidden-layer restricted Boltzmann machine (RBM) network (with Poisson observations), the model is very similar to the HMM: the binary hidden nodes correspond to the latent state, and the hidden-to-input connection weight matrices correspond to (unconstrained) mean firing rate per state. Presumably, adding more hidden layers or recurrent connections will further augment the model's representation power. One future research direction is to integrate the strengths of probabilistic generative model and neural network to build a rich repertoire of latent variable models. The neural network can be pre-trained, such as in the DNN-HMM (Dahl et al. 2012).

Acknowledgements I would like to thank Emery N. Brown for the intellectual inspiration during my postdoctoral career and all coauthors who have contributed to previously published work. I also thank S.L. Hu and S.Z. Liu for assistance in preparing some figures. Reproduction of some copyrighted materials is granted from publishers. The work was partially supported by an NSF-CRCNS award IIS-1307645 from the US National Science Foundation and an NIH-CRCNS award R01-NS100065 from the NINDS.

References

Aghagolzadeh, M., & Truccolo, W. (2016). Inference and decoding of motor cortex low-dimensional dynamics via latent state-space models. *IEEE Transactions on Neural Systems and Rehabilitation Engineering, 24*(2), 272–282.

Ames, K. C., Ryu, S. I., & Shenoy, K. V. (2014). Neural dynamics of reaching following incorrect or absent motor preparation. *Neuron, 81*(2), 438–451.

Beal, M., & Ghahramani, Z. (2006). Variational Bayesian learning of directed graphical models. *Bayesian Analysis, 1*(4), 793–832.

Buesing, L., Macke, J. H., & Sahani, M. (2012a). Learning stable, regularized latent models of neural population dynamics. *Network: Computation in Neural Systems, 23*, 24–47.

Buesing, L., Macke, J. H., & Sahani, M. (2012b). Spectral learning of linear dynamics from generalised-linear observations with application to neural population data. In *Advances in neural information processing systems* (Vol. 25, pp. 1682–1690). New York: Curran Associates.

Buonomano, D. V., & Laje, R. (2010). Population clocks: Motor timing with neural dynamics. *Trends in Cognitive Science, 14*, 520–527.

Bushnell, M. C., Ceko, M., & Low, L. A. (2013). Cognitive and emotional control of pain and its disruption in chronic pain. *Nature Review Neuroscience, 14*, 502–511.

Chen, Z. (2013). An overview of Bayesian methods for neural spike train analysis. *Computational Intelligence and Neuroscience, 2013*, 251905.

Chen, Z. (2015a). Estimating latent attentional states based on simultaneous binary and continuous behavioral measures. *Computational Intelligence in Neuroscience, 2015*, 493769.

Chen, Z. (Ed.). (2015b). *Advanced state space methods in neural and clinical data*. Cambridge: Cambridge University Press.

Chen, Z. (2017). Unfolding representations of trajectory coding in neuronal population spike activity. In *Proceedings of Conference on Information Sciences and Systems (CISS'17)*.

Chen, Z., Barbieri, R., & Brown, E. N. (2010). State-space modeling of neural spike train and behavioral data. In K. Oweiss (Ed.), *Statistical signal processing for neuroscience and neurotechnology* (pp. 175–218). Amsterdam: Elsevier.

Chen, Z., Gomperts, S. N., Yamamoto, J., & Wilson, M. A. (2014). Neural representation of spatial topology in the rodent hippocampus. *Neural Computation, 26*(1), 1–39.

Chen, Z., Grosmark, A. D., Penagos, H., & Wilson, M. A. (2016a). Uncovering representations of sleep-associated hippocampal ensemble spike activity. *Scientific Reports, 6*, 32193.

Chen, Z., Hu, S., Zhang, Q., & Wang, J. (2017b). Quickest detection of abrupt changes in neuronal ensemble spiking activity using model-based and model-free approaches. In *Proceedings of 8th International IEEE/EMBS Conference on Neural Engineering (NER)*.

Chen, G., King, J. A., Burgess, N., & O'Keefe, J. (2013). How vision and movement combine in the hippocampal place code. *Proceedings of National Academy of Sciences USA, 110*, 378–383.

Chen, Z., Kloosterman, F., Brown, E. N., & Wilson, M. A. (2012). Uncovering spatial topology represented by rat hippocampal population neuronal codes. *Journal of Computational Neuroscience, 33*(2), 227–255.

Chen, Z., Linderman, S., & Wilson, M. A. (2016b). Bayesian nonparametric methods for discovering latent structures of rat hippocampal ensemble spikes. In *Proceedings of IEEE Workshop on Machine Learning for Signal Processing* (pp. 1–6).

Chen, Z., & Wilson, M. A. (2017). Deciphering neural codes of memory during sleep. *Trends in Neurosciences, 40*(5), 260–275.

Chen, Z., Zhang, Q., Tong, A. P. S., Manders, T. R., & Wang, J. (2017a). Deciphering neuronal population codes for acute thermal pain. *Journal of Neural Engineering, 14*(3), 036023.

Ching, W.-K., Huang, X., Ng, M. K., & Siu, T.-K. (2015). *Markov chains: Models, algorithms and applications* (2nd ed.). Berlin: Springer.

Chung, J., Kastner, K., Dinh, L., Goel, K., Courville, A., & Bengio, Y. (2016). *A Recurrent Latent Variable Model for Sequential Data*. Technical report. https://arxiv.org/pdf/1506.02216.pdf

Churchland, M. M., Cunningham, J. P., Kaufman, M. T., Foster, J. D., Nuyujukian, P., Ryu, S. I., et al. (2012). Neural population dynamics during reaching. *Nature, 487*, 51–56.

Cunningham, J. P., & Yu, B. M. (2014). Dimensionality reduction for large-scale neural recordings. *Nature Neuroscience, 17*(11), 1500–1509.

Curto, C., & Itskov, V. (2008). Cell groups reveal structure of stimulus space. *PLoS Computational Biology, 4*(10), e1000205.

Dabaghian, Y., Cohn, A. G., & Frank, L. M. (2011). Topological coding in the hippocampus. In *Computational modeling and simulation of intellect: Current state and future prospectives* (pp. 293–320). Hershey: IGI Global.

Dabaghian, Y., Memoli, F., Frank, L. M., & Carlsson, G. (2012). A topological paradigm for hippocampal spatial map formation using persistent homology. *PLoS Computational Biology, 8*(8), e1002581.

Dahl, G., Yu, D., Deng, L., & Acero, A. (2012). Context-dependent, pre-trained deep neural networks for large vocabulary speech recognition. *IEEE Transactions on Audio, Speech & Language Processing, 20*(1), 30–42.

Davidson, T. J., Kloostserman, F., & Wilson, M. A. (2009). Hippocampal replay of extended experience. *Neuron, 63*, 497–507.

Dempster, A., Laird, N., & Rubin, D. (1977). Maximum likelihood from incomplete data via the EM algorithm. *Journal of the Royal Statistical Society, Series B, 39*(1), 1–38.

Eden, U. T., Frank, L. M., Barbieri, R., Solo, V., & Brown, E. N. (2004). Dynamic analysis of neural encoding by point process adaptive filtering. *Neural Computation, 16*(5), 971–998.

Feeney, D. F., Meyer, F. G., Noone, N., & Enoka, R. M. (2017). A latent low-dimensional common input drives a pool of motor neurons: A probabilistic latent state-space model. *Journal of Neurophysiology, 117*, 1690–1701.

Fine, S., Singer, Y., & Tishby, N. (1998). The hierarchical hidden Markov model: Analysis and applications. *Machine Learning, 32*, 41–62.

Fuchs, P. N., Peng, Y. B., Boyette-Davis, J. A., & Uhelski, M. L. (2014). The anterior cingulate cortex and pain processing. *Frontiers in Integrative Neuroscience, 8*, 35.

Gao, Y., Archer, E., Paninski, L., & Cunningham, J. P. (2016). Linear dynamical neural population models through nonlinear embeddings. In *Advances in Neural Information Processing Systems*. New York: Curran Associates.

Gelman, A., Carlin, J. B., Stern, H. S., & Rubin, D. B. (2004). *Bayesian data analysis* (2nd ed.). Boca Raton, FL: Chapman & Hall/CRC Press.

Gershman, S., & Blei, D. M. (2012). A tutorial on Bayesian nonparametric models. *Journal of Mathematical Psychology, 56*, 1–12.

Ghahramani, Z., & Jordan, M. I. (1997). Factorial hidden Markov models. *Machine Learning, 29*(2), 245–273.

Goodfellow, I., Bengio, Y., & Courville, A. (2016). *Deep learning*. Cambridge, MA: MIT Press.

Grosmark, A. D., & Buzsaki, G. (2016). Diversity in neural firing dynamics supports both rigid and learned hippocampal sequences. *Science, 351*, 1440–1443.

Guédon, Y. (2003). Estimating hidden semi-Markov chains from discrete sequences. *Journal of Computational and Graphical Statistics, 12*, 604–639.

Haggerty, D. C., & Ji, D. (2015a). Activities of visual cortical and hippocampal neurons co-fluctuate in freely-moving rats during spatial behavior. *eLife, 4*, e08902.

Haggerty, D. C., & Ji, D. (2015b). Coordinated sequence replays between the visual cortex and hippocampus. In M. Matsuno (Ed.), *Analysis and modeling of coordinated multi-neuronal activity* (pp. 183–206). New York: Springer.

Hu, S., Zhang, Q., Wang, J., & Chen, Z. (2017). A real-time rodent neural interface for deciphering acute pain signals from neuronal ensemble spike activity. In *Proceedings of the 51st Asilomar Conference on Signals, Systems and Computers*.

Hu, S., Zhang, Q., Wang, J., & Chen, Z. (2018). Real-time particle filtering and smoothing algorithms for detecting abrupt changes in neural ensemble spike activity. *Journal of Neurophysiology*, in press.

Ji, D., & Wilson, M. A. (2007). Coordinated memory replay in the visual cortex and hippocampus during sleep. *Nature Neuroscience, 10*, 100–107.

Jordan, M. I., Ghahramani, Z., Jaakkola, T. S., & Saul, L. K. (1999). An introduction to variational methods for graphical models. *Machine Learning, 37*, 183–233.

Jordan, M. I., & Sejnowski, T. J. (Eds.). (2001). *Graphical models: Foundations of neural computation*. Cambridge, MA: MIT Press.

Kuo, C. C., & Yen, C. T. (2005). Comparison of anterior cingulate and primary somatosensory neuronal responses to noxious laser-heat stimuli in conscious, behaving rats. *Journal of Neurophysiology, 94*, 1825–1836.

Kurihara, K., & Welling, M. (2009). Bayesian *k*-means as 'maximization-expectation' algorithm. *Neural Computation, 21*, 1145–1172.

Latimer, K. L., Yates, J. L., Meister, M. L. R., Huk, A. C., & Pillow, J. W. (2015). Single-trial spike trains in parietal cortex reveal discrete steps during decision-making. *Science, 349*, 184–187.

Lawhern, V., Wu, W., Hatsopoulos, N. G., & Paninski, L. (2010). Population decoding of motor cortical activity using a generalized linear model with hidden states. *Journal of Neuroscience Methods, 189*, 267–280.

LeCun, Y., Bengio, Y., & Hinton, G. E. (2015). Deep learning. *Nature, 521*, 436–444.

Lee, L.-M. (2011). High-order hidden Markov model and application to continuous mandarin digit recognition. *Journal of Information Science and Engineering, 27*(13), 1919–1930.

Linderman, S., Johnson, M. J., Wilson, M. A., & Chen, Z. (2016). A Bayesian nonparametric approach for uncovering rat hippocampal population codes during spatial navigation. *Journal of Neuroscience Methods, 263*, 36–47.

Liu, S., Grosmark, A. D., & Chen, Z. (2018). Methods for assessment of memory reactivation. *Neural Computation*, to appear.

Macke, J. H., Buesing, L., Cunningham, J. P., Yu, B. M., Shenoy, K. V., & Sahani, M. (2012). Empirical models of spiking in neural populations. In *Advances in neural information processing systems* (Vol. 24). New York: Curran Associates.

Michaels, J. A., Dann, B., & Scherberger, H. (2016). Neural population dynamics during reaching are better explained by a dynamical system than representational tuning. *PLoS Computational Biology, 12*(11), e1005175.

Müller, P., Quintana, F. A., Jara, A., & Hanson, T. (2015). *Bayesian nonparametric data analysis*. Cham: Springer.

O'Keefe, J., & Dostrovsky, J. (1971). The hippocampus as a spatial map: preliminary evidence from unit activity in the freely-moving rat. *Brain Research, 34*(1), 171–175.

O'Keefe, J., & Nadel, L. (1978). *The hippocampus as a cognitive map*. London: Oxford University Press.

Omigbodun, A., Doyle, W. K., Devinsky, O., & Gilja, V. (2016). Hidden-Markov factor analysis as a spatiotemporal model for electrocorticography. In *Proceedings of IEEE Engineering in Medicine and Biology Conference* (pp. 1632–1635).

Pachitariu, M., Petreska, B., & Sahani, M. (2013). Recurrent linear models of simultaneously-recorded neural populations. In L. Bottou, C. J. C. Burges, M. Welling, Z. Ghahramani & K. Q. Weinberger (Eds.), *Advances in neural information processing systems* (Vol. 26). New York: Curran Associates.

Pawitan, Y. (2001). *In all likelihood: Statistical modeling and inference using likelihood*. Oxford: Clarendon Press.

Penny, W., Ghahramani, Z., & Friston, K. (2005). Bilinear dynamical systems. *Philosophical Transactions of Royal Society of London B: Biological Sciences, 360*, 983–993.

Rivkind, A., & Barak, O. (2017). Local dynamics in trained recurrent neural networks. *Physics Review Letter, 118*, 258101.

Robert, C. P. (2007). *The Bayesian choice: From decision-theoretic foundations to computational implementation* (2nd ed.). Berlin: Springer.

Santhanam, G., Yu, B. M., Gija, V., Ryu, S. I., Afshar, A., Sahani, M., et al. (2009). Factor-analysis methods for higher-performance neural prostheses. *Journal of Neurophysiology, 102*(2), 1315–1330.

Saul, L. K., & Jordan, M. I. (1999). Mixed memory Markov models: Decomposing complex stochastic processes as mixtures of simpler ones. *Machine Learning, 37*, 75–86.

Saul, L. K., & Rahim, M. G. (2000). Markov processes on curves. *Machine Learning, 41*, 345–363.

Smith, A. C., & Brown, E. N. (2003). Estimating a state-space model from point process observations. *Neural Computation, 15*(5), 965–991.

Stevenson, I. H. (2016). Flexible models for spike count data with both over- and under-dispersion. *Journal of Computational Neuroscience, 41*, 29–43.

Székely, G. J., & Rizzo, M. L. (2009). Brownian distance covariance. *Annals of Applied Statistics, 3/4*, 1233–1303.

Teh, Y. W., Jordan, M. I., Beal, M. J., & Blei, D. M. (2006). Hierarchical Dirichlet processes. *Journal of American Statistical Association, 101*, 1566–1581.

Vierck, C. J., Whitsel, B. L., Favorov, O. V., Brown, A. W., & Tommerdahl, M. (2013). Role of primary somatosensory cortex in the coding of pains. *Pain, 154*, 334–344.

Vogelstein, J., Packer, A., Machado, T. A., Sippy, T., Babadi, B., Yuste, R., et al. (2010). Fast nonnegative deconvolution for spike train inference from population calcium imaging. *Journal of Neurophysiology, 104*, 3691–3704.

Vogelstein, J., Watson, B., Packer, A., Yuste, R., Jedynak, B., & Paninski, L. (2009). Spike inference from calcium imaging using sequential Monte Carlo methods. *Biophysical Journal, 97*(2), 636–655.

Wagner, T. D., Atlas, L. Y., Lindquist, M. A., Roy, M., Woo, C.-W., & Kross, E. (2013). An fMRI-based neurologic signature of physical pain. *New England Journal of Medicine, 368*, 1388–1397.

Whiteway, M. R., & Butts, D. A. (2017). Revealing unobserved factors underlying cortical activity with a rectified latent variable model applied to neural population recordings. *Journal of Neurophysiology, 117*, 919–936.

Wood, F., & Black, M. J. (2008). A nonparametric Bayesian alternative to spike sorting. *Journal of Neuroscience Methods, 173*(1), 1–12.

Wu, X., & Foster, D. (2014). Hippocampal replay captures the unique topological structure of a novel environment. *Journal of Neuroscience, 34*, 6459–6469.

Wu, W., Chen, Z., Gao, S., & Brown, E. N. (2011). A hierarchical Bayesian approach for learning sparse spatio-temporal decompositions of multichannel EEG. *Neuroimage, 56*(4), 1929–1945.

Wu, W., Kulkarni, J. E., Hatsopoulos, N. G., & Paninski, L. (2009). Neural decoding of hand motion using a linear state-space model with hidden states. *IEEE Transactions on Neural Systems and Rehabilitation Engineering, 17*, 370–378.

Wu, W., Nagarajan, S., & Chen, Z. (2016). Bayesian machine learning: EEG/MEG signal processing measurements. *IEEE Signal Processing Magazine, 33*(1), 14–36.

Wu, W., & Srivastava, A. (2011). An information-geometric framework for statistical inferences in the neural spike train space. *Journal of Computational Neuroscience, 31*, 725–748.

Yu, B. M., Cunningham, J. P., Santhanam, G., Ryu, S. I., Shenoy, K. V., & Sahani, M. (2009). Gaussian-process factor analysis for low-dimensional single-trial analysis of neural population activity. *Journal of Neurophysiology, 102*(1), 614–635.

Yu, B. M., Kemere, C., Santhanam, G., Ryu, S. I., Meng, T. H., Sahani, M., et al. (2007). Mixture of trajectory models for neural decoding of goal-directed movements. *Journal of Neurophysiology, 97*, 3763–3780.

Yu, S.-Z. (2010). Hidden semi-Markov models. *Artificial Intelligence, 174*(2), 215–243.

Zhang, Y., Wang, N., Wang, J.-Y., Chang, J.-Y., Woodward, D. J., & Luo, F. (2011). Ensemble encoding of nociceptive stimulus intensity in the rat medial and lateral pain systems. *Molecular Pain, 7*, 64.

Zhao, Y., & Park, I. M. (2017). Variational latent Gaussian process for recovering single-trial dynamics from population spike trains. *Neural Computation, 29*, 1293–1316.

Zhou, F., De la Torre, F., & Hodgins, J. K. (2013). Hierarchical aligned cluster analysis for temporal clustering of human motion. *IEEE Transactions Pattern Analysis and Machine Intelligence, 35*(3), 582–596.

Chapter 4
What Can Trial-to-Trial Variability Tell Us? A Distribution-Based Approach to Spike Train Decoding in the Rat Hippocampus and Entorhinal Cortex

Michael J. Prerau and Uri T. Eden

4.1 Introduction

4.1.1 The Neural Code

Neuroscientists have long sought to understand the *neural code* (Rieke et al. 1997), the cognitive Rosetta Stone defining the language of the brain—mapping neural spiking activity to the representation of world around and within us. In practice, the neural code is commonly studied through electrophysiological experiments, in which neural activity is recorded. Cells within the brain experience quick, stereotyped changes in membrane potential called *action potentials*, which are marked by a sharp peak in voltage due to a depolarization then repolarization of the cell membrane (Dayan and Abbott 2001). In recording the time at which this large depolarization reaches its peak, the continuous-valued voltage trace is converted into a single event called a *spike*, a series of which is termed a *spike train*. From these experiments, spiking data from individual neurons or ensembles of cells are collected simultaneously with measurements of behavioral, biological, or other factors (Moeliker 2001) germane to the experimental procedure. To explore the way in which different intrinsic and extrinsic factors are encoded and processed within the brain, experimental parameters can be varied in a controlled and predictable manner, so that the relationship between the parameters and neural activity can be

M.J. Prerau (✉)
Massachusetts General Hospital, Harvard Medical School, Boston, MA, USA
e-mail: prerau@nmr.mgh.harvard.edu

U.T. Eden
Boston University, Boston, MA, USA
e-mail: tzvi@bu.edu

Z. Chen, S.V. Sarma (eds.), *Dynamic Neuroscience*,
https://doi.org/10.1007/978-3-319-71976-4_4

characterized. In characterizing this relationship, we can form hypotheses about the neural code for a given system.

Measures of spiking from a single neuron under identical, repeated stimuli or conditions produce different observed spike trains at every iteration, or *experimental trial*. Computational analysis of spiking activity has therefore viewed spiking as a stochastic process (Rieke et al. 1997). Therefore, multiple experimental trials are typically performed to characterize the spiking activity in the aggregate or to characterize the distributional properties of the data.

4.1.2 The Neural Code in the Rat Hippocampus and Entorhinal Cortex

One area of neuroscience in which the neural code has been widely investigated is the representation within the medial temporal lobe of the brain of the space that an animal navigates. In particular, many studies of the neural code have focused on the activity of neurons within the hippocampus and entorhinal cortex (EC) of rats during spatial navigation tasks.

4.1.2.1 Place Cells and Grid Cells

The hippocampus is a region within the medial temporal lobe, and is comprised of the dentate gyrus and the cornu ammonis (CA), which itself is broken into subregions CA1 through CA3. Functionally, the hippocampus was initially implicated as a primary center for the processing of *declarative memory* (Cohen and Squire 1980; Eichenbaum et al. 1994), that is, the recollection of specific facts or events. The most famous example of the linkage between memory and the hippocampus was the case study of the Henry Gustav Molaison (known clinically as H.M.), who, due to epilepsy, underwent a bilateral temporal lobe resection, resulting in the removal the majority of his hippocampus in both cerebral hemispheres. Studies of H.M. showed severe anterograde amnesia and some retrograde amnesia—leaving him with severe deficits in the formation of new memories and remembrance of only his early childhood (Scoville and Milner 1957).

Subsequent studies of the hippocampus have suggested that, in addition to memory, the hippocampus represents and processes information relating to spatial location and navigation. In O'Keefe and Dostrovsky (1971), cells were identified within the rat hippocampus that exhibited firing activity related to the spatial location of the animal. These cells, called *place cells* (Eichenbaum et al. 1999; Muller 1996; O'Keefe 1979; O'Keefe and Dostrovsky 1971; O'Keefe and Nadel 1978; Wilson and McNaughton 1993), fire whenever the animal passes through the specific region of space to which that particular cell is tuned. The spatial receptive field of a place cell is called a *place field*. In their book "*The Hippocampus as a Cognitive Map*" (1978), O'Keefe and Nadel proposed that these cells provided a "cognitive map" for spatial knowledge (Tolman 1948). Subsequent research on the nuances of place cell activity, however, has provided evidence which suggests that aspects of hippocampal function are inconsistent with or insufficient for a true hippocampal cognitive map (Eichenbaum et al. 1999). Thus, the study of the function and dynamics of place cell activity remains an active area of research.

The EC is another region within the medial temporal lobe, and it provides the primary input to the hippocampus, which in turn has projections back to the EC. Initial studies of the EC showed some spatially tuned firing activity akin to that of place fields (Mizumori et al. 1992; Quirk et al. 1992). In Fyhn et al. (2004), however, May-Britt and Edvard Moser and colleagues discovered cells within the dorsomedial EC that possessed multiple place fields, which completely covered the experimental environment in a hexagonal tessellation pattern. The cells, called *grid cells*, suggest an organized, Euclidean representation of space within the medial EC (MEC), not present in the hippocampus.

4.1.2.2 Context-Dependent Activity: Understanding What's Going On

Beyond static representations of spatial location, cells in both the hippocampus and EC have been found to change their firing properties as a function information related to certain sets of behavioral or experimental parameters, called *contexts*. Such neurons can be said to have context-dependent neural activity, or to exhibit *differential firing*.

In Wood et al. (2000), rats were trained to perform a continuous spatial alternation task on a modified T-maze, in which they were required to alternate between left and right-turns on the T-maze. During this task, neural spiking data was recorded from neural ensembles in the CA1 region of the hippocampus. This study identified neurons fired that almost exclusively during trials where the rat would eventually turn to one of the directions and not at all during trials where the rat would turn to the other. These cells, termed *splitter cells*, represent the neural code for future turn direction through the presence or absence of spiking activity during a given trial. Further work has expanded on the representation of behavioral context in the hippocampus (Ainge et al. 2007; Ferbinteanu and Shapiro 2003; Frank et al. 2000; Griffin et al. 2007; Lee et al. 2006; Lipton et al. 2007; Smith and Mizumori 2006). Studies have also been able to identify cells within EC (Frank et al. 2000; Lipton et al. 2007) that exhibit context-dependent neural activity as a function of future turn-direction during continuous alternation tasks.

The findings from Wood et al. (2000) suggested that the neural code for future turn direction was essentially an on/off switch, with firing preceding only one turn-direction. Consequently, studies of context-dependent neural activity have identified splitter cells using statistical analyses, such as the ANOVA, that look for large differences in mean firing rate. From the standpoint of an accurate characterization of context, the mean may not be the most appropriate descriptor of neural activity over multiple trials for these particular regions of the brain.

4.1.2.3 Characterizing Sparse Encoding of Context

In order to demonstrate that the information necessary for a robust encoding of context is present within a population made of cells that sparsely encode context, we developed a novel spike train decoding procedure, which attempts to predict context given observed spiking data from a neural receptive field model that includes trial-to-trial variability (Fig. 4.2a). Predominantly, spike train decoding approaches have been used to estimate continuous-valued external stimuli, such as predicting the hand trajectory from motor cortical spike data (Wessberg et al. 2000; Serruya et al. 2002), or reconstructing the animal's spatial location for hippocampal population spike data (Brown et al. 1998; Zhang et al. 1998; Barbieri et al. 2004; Johnson and Redish 2007; Huang et al. 2009). Some previous methods for decoding discrete experimental contexts have used Poisson mixture models of inter-spike intervals (Wiener and Richmond 2003), likelihood ratios for mean firing rate differences (Lipton et al. 2007), subspace clustering (Lin et al. 2005), and state-space analyses (Johnson and Redish 2007; Huang et al. 2009).

In this chapter, we will develop likelihood-based decoding methodologies to predict turn direction from spiking data from individual cells and ensembles from real and simulated units. We will present two variants of the decoding procedure based on different models of context-dependent neural activity. The first model assumes that context is encoded only through differences in mean firing rate (Fig. 4.2b). This model is designed to capture the scenario in which there is context-dependent rate, and all variability comes from the stochastic nature of the spiking. The second model, on the other hand, is designed to capture the idea that context can be encoded through changes the distribution of the firing rate trajectories (Fig. 4.2c). By examining the performance of these methods on simulated and real populations of ICD neurons, we aim to gain a further understanding of the interaction between single cell and population representations of behavioral context within the hippocampus and EC.

4.1.3 An Inconsistent Language: Trial-to-Trial Variability in the Hippocampus and Entorhinal Cortex

Although the neural code is often conceptualized as a single, albeit stochastic, representation of the world in terms of a spike train, in practice it is not quite so straightforward. For example, spiking within the hippocampus has been demonstrated to be highly variable across experimental trials (Fenton and Muller 1998), with neural activity changing drastically from trial to trial during repeated tasks. In fact, even with classic splitter cells, there has yet to be reported a single cell that fires 100% of the time during one context and never fires during the other.

4.1.3.1 Characterizing Trial-to-Trial Variability

Given the complexity of spike-based representations, a proper characterization of neural coding properties is vital in accurately understanding a neural system. One prominent example of the importance of characterizing trial-to-trial variability was a study of the macaque lateral intraparietal (LIP) area, which has a population of neurons previously observed to smoothly increase its rate prior to a monkey making a decision in a behavioral task (Latimer et al. 2015). While these neurons appeared overall to smoothly be ramping up their firing rate, a more careful analysis of the data revealed that individual neurons were in fact making large discrete jumps in firing rate at random times across different trials.

To perform this analysis, the authors compared the fit of two competing models: a step model of firing rate increase and a model in which the rates of the neurons gradually ramp up. They were able to show that the step model provided a much better fit to the data, as well as provided significantly more information than the ramp model could. As such, properly characterizing the trial-to-trial variability of these neurons provided a whole new way of understanding the way information processing occurs prior to decision making.

Numerous other approaches have been for characterizing the trial-to-trial variability in the hippocampus and other regions, including parametric modeling (Brown et al. 2001, 2004; Eden et al. 2004; Truccolo et al. 2005), history-dependent models using descriptive statistics (Churchland et al. 2010), models of changes in network state, or "reference frame" (Touretzky and Redish 1996; Redish and Touretzky 1997; Touretzky and Muller 2006) with doubly stochastic Poisson state transitions (Lansky et al. 2001; Jackson and Redish 2007), and multiplicative models of specific components of the variability (Ventura et al. 2005).

4.1.3.2 Intermittent Context-Dependent Firing Activity in Parahippocampal Regions

In a previous study (Prerau et al. 2014), we identified place and grid cells from the rat CA1 and dorsocaudal medial EC (dcMEC) that exhibited statistically significant context-dependent differences in firing rate variance (Fig. 4.1b) or in the 95th percentile of the firing rate (Fig. 4.1c) across trials during a T-maze continuous spatial alternation task. What value could context-dependent changes in variability have to the brain in the encoding of information, especially since individuals typically don't have the luxury of observing multiple trials of a stimulus before making a decision?

In addition to thinking about these cells as having context-dependent changes in variance, these cells could also be said to have sparse encoding of context. This is because they possess firing activity that is predominantly invariant, save for a small set of trials with extreme firing rates, which occurs during only one context. Overall, these findings suggest a hypothesis that groups of cells that sparsely encode context could provide a robust population representation of context in the aggregate. In this

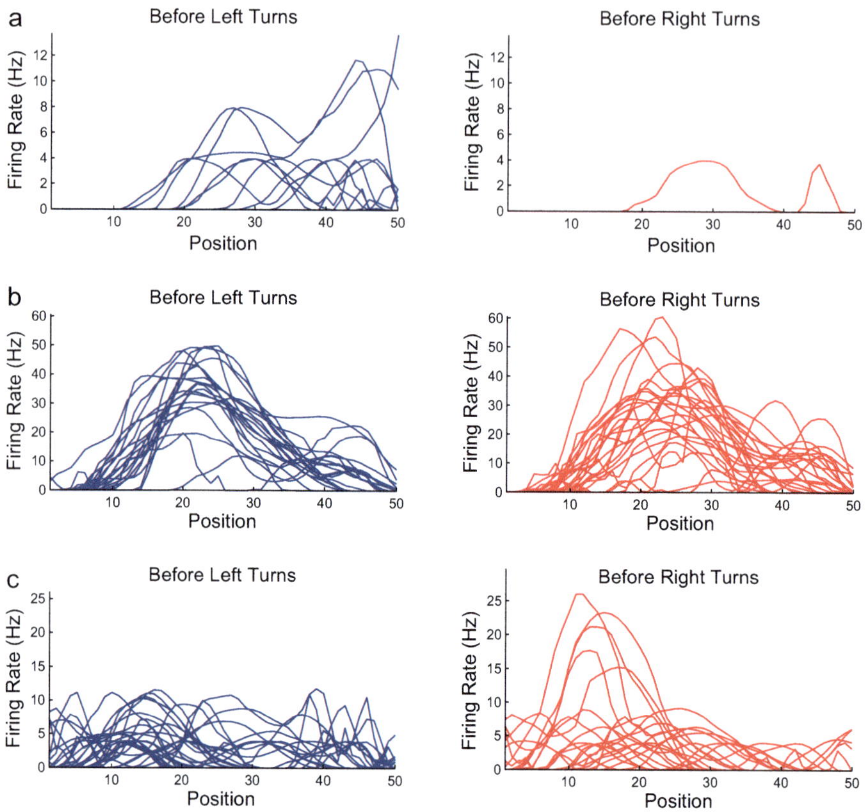

Fig. 4.1 Examples of different types of context-dependent neural activity in the rat hippocampus and entorhinal cortex. (**a**) An example of a splitter cell with context-dependent mean firing rate, firing almost exclusively on preceding left turn trials. Cells such as these can be identified by large context-dependent differences in the mean firing rate. (**b**) One example of cell that exhibits intermittent context-dependence. It shows context-dependent changes variance, with significantly greater trial-to-trial variability preceding right turns. (**c**) Another example of cell that exhibits intermittent context-dependence. It has nearly identical behavior preceding left and right turns, except for five elevated trials occurring preceding right turns only. These types of cells show statistically significant context-dependent differences in the 95th percentile of firing rate distribution. In all examples, the firing rates for three different cells are plotted as a function of position. The firing rate trajectories are separated by trials preceding left turns (left panels, blue curves) and preceding right turns (right panels, red curves)

scenario, each cell would cast its "vote" on context for only a few trials, and abstain otherwise. Given enough cells covering of all trials, the population could encode context throughout the entire experiment. We call the activity of these neurons *intermittent context-dependent* (ICD) firing, as they encode context, but only on a subset of trials.

4.2 Modeling Trial-to-Trial Variability

4.2.1 Decoding Behavioral Context from Neural Spiking Data

Neural spike train decoding is a mathematical procedure in which an external signal is estimated from spiking data. In this case, we wish to decode behavioral context, which is essentially the problem of selecting to which of two discrete states (left-turn context, right-turn context) a given spike train belongs. Previous methods for decoding context have used likelihood ratios (Lipton et al. 2007), parametric models (Wiener and Richmond 2003; Lin et al. 2005), or adaptations of continuous methods (Huang et al. 2009) to predict discrete state. Herein, we develop nonparametric models of the firing activity for each context, and use likelihood methods to predict the behavioral context given a single unit or population spiking activity.

To decode behavioral context from neural spiking data, we must compute Pr(context|spikes), the probability of a behavioral context given an observed spike train. It follows from Bayes' rule

$$\Pr(\text{context}|\text{spikes}) = \frac{\Pr(\text{spikes}|\text{context})\,\Pr(\text{context})}{\Pr(\text{spikes})} \tag{4.1}$$

Assuming each context is equally likely to occur at any given point during the experiment, the probability of a given context is

$$\Pr(\text{context}) = \frac{1}{\#\text{contexts}} \tag{4.2}$$

A better estimate of Pr(context) can be computed by using a model of the probability of a correct response at each trial (Smith et al. 2004, 2005, 2007; Prerau et al. 2008, 2009). While this estimate would be highly informative, especially if the animal performs the task accurately. Our goal here is to explore the contextual information contained exclusively within the spiking activity, and thus we assume equal probabilities for all contexts.

In Eq. (4.1), Pr(spikes) is a normalization factor, and Pr(context) is equal across all contexts. Thus we can group those terms as a constant, and say

$$\Pr(\text{context}|\text{spikes}) \propto \Pr(\text{spikes}|\text{context}) \tag{4.3}$$

indicating that the probability of the context given the spikes is proportional the probability or likelihood of the spikes as a function of the context. Thus, given a fixed probability of Pr(context), the decode is determined exclusively through the computation of Pr(spikes|context). We estimate this likelihood using the theory of point process, which has been used successfully to analyze the activity of neural data from single cells (Brillinger 1988; Barbieri et al. 2001; Kass and Ventura 2001; Brown et al. 2003) and from populations (Brillinger 1992). We present two methods for computing Pr(spikes|context) based on models of differential firing derived from two different types of differential firing activity. The first model is based on splitter cells (Wood et al. 2000), which fire almost exclusively during a single context (Fig. 4.1a), and assumes that context can be characterized through a single trial-invariant firing rate (Fig. 4.2b). The second model is based on cells that exhibit intermittent context-dependent activity (Fig. 4.1b, c), and assumes that context may be characterized by capturing the trial-to-trial variability through the distribution of possible firing rate trajectories (Fig. 4.2c).

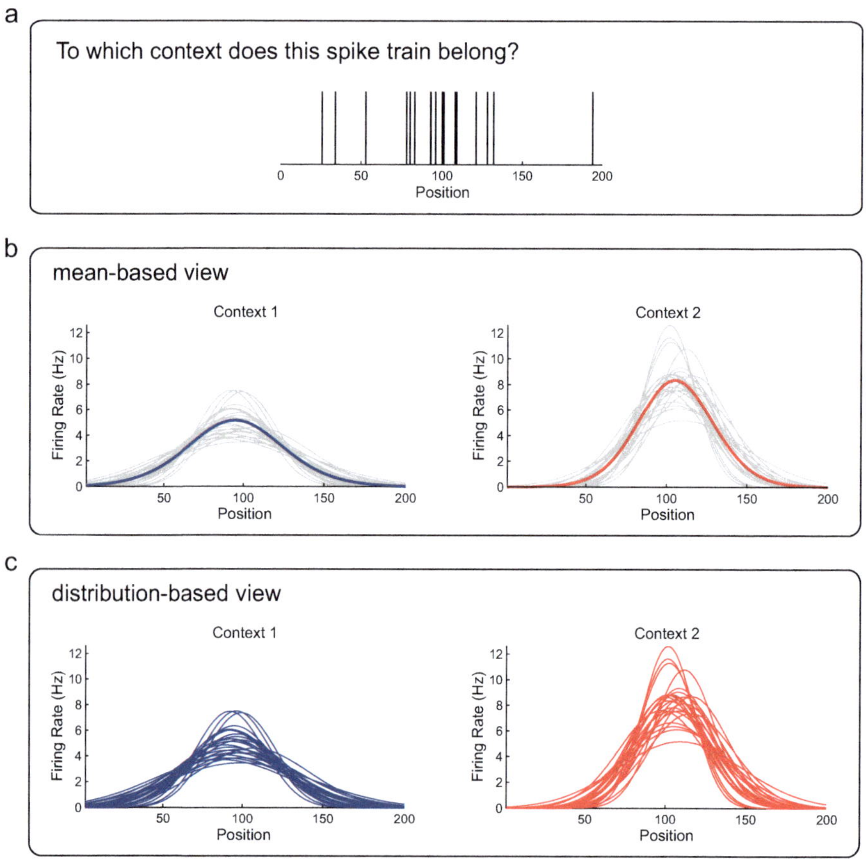

Fig. 4.2 A schematic of two distinct views on decoding procedures, both of which attempt to classify a spike train into one of contexts. (**a**) A context decoding question for an observed neuronal spike train. (**b**) The first view assumes a mean-based coding of context, and thus aims to decode the spike train given one fixed mean firing rate for each context. All variation is assumed to be noise from the stochasticity of the spike train realization. (**c**) The second view assumes that there is information within the trial-to-trial variation and uses the firing rate trajectory empirical distribution to decode context from a given spike train

We performed our analyses in discrete time, defined as $t_i = i\Delta t$, where Δt is the sampling interval, and i is a positive integer ranging from 1 to T. ΔN_i denotes the spike count in the time interval $[t_{i-1}, t_i)$. For both models, we analyzed neural activity as a function of x_i, the animal's linear position on the T-maze at time t_i, and divided the T-maze center into 50 equally sized spatial bins.

4.2.2 *Single-Cell Decoding*

4.2.2.1 Mean Firing Rate Model

The predominant view of differential firing is that there is a distinct, trial-invariant probability of spiking for each behavioral context, which may vary as a function of space or time (Fig. 4.2b). This viewpoint suggests that any variation observed is due to the stochasticity in the realizations of the spiking from the probability function, and that consequently, a good estimator of the underlying firing probability is the mean firing rate across trials. Given this assumption, we can estimate Pr(spikes|context) by modeling this trial-invariant spiking probability function for each context. Then, using the likelihood of the spikes given that function, we can determine the context from which the spike train was most likely generated.

There are several parametric modeling techniques, such as generalized-linear models (McCullagh 1984; Truccolo et al. 2005) and state-space models (Brown et al. 1998, 2001; Eden et al. 2004; Czanner et al. 2008; Kulkarni and Paninski 2008), which may be used to build an estimate for the spiking probability function for each context. As an intuitive and computationally accessible first approach, we estimate $\hat{\lambda}_C$, the spiking probability for a given context C, by computing the mean firing rate across the trials from that context. Using the spiking from each trial, we use smoothed estimates of the firing rate in the time domain. Since we have simultaneously recorded positional data for the rats, we compute the mean firing rate for that context in each of 50 linear spatial bins on the T-maze. Thus, for any position x on the T-maze, we define $\hat{\lambda}_C(x)$ as the average firing rate for context C for the spatial bin that contains x.

To ensure that overfitting is avoided, we perform a leave-one-out cross-validation (Efron and Gong 1983) to compute $\hat{\lambda}_C(x)$. This is done in the calculation of the mean across trials by omitting the firing rate data from the trial of the spike train to be decoded. In this way, we avoid using any data related to the spike train in the decoding procedure.

Pr(spikes|context) can also be written as $\mathscr{L}(\text{spikes}|C)$, the inhomogeneous Poisson likelihood (Snyder and Miller 1991) of the spikes given a position-dependent firing rate $\hat{\lambda}_C(x)$, defined as

$$\mathscr{L}(\text{spikes}|C) = \exp\left(\sum_{i=1}^{T}\left\{\Delta N_i \log\left[\hat{\lambda}_C(x_i)\Delta t\right] - \hat{\lambda}_C(x_i)\Delta t - \log(\Delta N_i!)\right\}\right) \quad (4.4)$$

This likelihood integrates over time within the trial, combining the spike train data with the firing probability, as determined by the mean firing rate, at every position that the animal traverses.

4.2.2.2 Empirical Firing Rate Distribution Model

Another method for modeling Pr(spikes|context) stems from our previous work (Prerau et al. 2014), which suggests that, for certain cells, the structure of the trial-to-trial firing rate variability can be highly informative about behavioral context (Fig. 4.1b, c). An analysis of ensembles from CA1 and dcMEC found many cells with statistically significant differences in the distribution of the firing rate structure between left-turn and right-turn trials. Cells in both regions that failed tests of significance for the mean firing rate exhibited significant context-dependent differences in other statistics—most notably the rate variance or the 95th percentile. For such cells, decoding from the mean rate would fail to capture the aspects of the firing activity in which the contextual information may be encoded. Thus, it is useful to devise an estimator for Pr(spikes|context) that can incorporate a characterization of the trial-to-trial variability (Fig. 4.2c).

To model the variability, we use essentially the same idea as in the mean decoding model, but instead of having a single spiking probability function for a context, we create a mixture model of inhomogeneous Poisson processes representing the range of trial-to-trial variability for that context. We achieve this by marginalizing Pr(spikes|context) over the observed firing rate trajectories in that given context

$$\Pr(\text{spikes}|\text{context}) = \sum_{r}^{\#\text{ context trials}} \Pr(\text{spikes}|\text{context}, \text{rate}_r) \Pr(\text{rate}_r) \tag{4.5}$$

which creates a mixture model of spiking probabilities based on the observed firing rates. This equation, in effect, can be said to compute the probability of the spike train given an empirical distribution of firing rate trajectories. The quality of this estimate is based on how well the previously observed rates span the space of possible rates. The animals ran roughly between 25 and 35 non-error trials per context, and our previous work (Prerau et al. 2014) suggests can produce estimates of empirical densities sufficiently accurately to compute context-dependent differences in trial-to-trial variability.

Assuming no a priori knowledge of the distribution of firing rate as a function of spatial trajectories across the T-maze, we set the probability of each observed firing rate as

$$\Pr(\text{rate}) = \frac{1}{\#\text{trials in context}}. \tag{4.6}$$

To estimate Pr(spikes|context, rate_r), we create a mixture model of inhomogeneous Poisson processes. We combine Eqs. (4.5) and (4.6) with the inhomogeneous

Poisson process likelihood (Snyder and Miller 1991), and compute the log-likelihood of a spike train from trial k given a context C, denoted as $\mathscr{L}(\text{spikes}_k|C)$, which is defined by

$$\mathscr{L}(\text{spikes}_k|C) = \frac{1}{R}\sum_{r\neq k}\exp\left(\sum_{i=1}^{T}\left\{\Delta N_i \log\left[\hat{\lambda}_r(x_i)\Delta t\right] - \hat{\lambda}_r(x_i)\Delta t - \log(\Delta N_i!)\right\}\right) \tag{4.7}$$

where R is the number of trials in the context, and T is the number of time bins in the trial. We define $\hat{\lambda}_r(x_i)$ as the average firing rate of trial r within the spatial bin into which the animal's linear position falls at time t_i. We compute $\hat{\lambda}_r$ by first estimating the firing rate in the time domain using a Hanning kernel smoother and the spikes on trial r. We then use the position of the animal during trial r to translate the rate into the position domain, and calculate the mean firing rate for trial r in each of the 50 spatial bins across the T-maze.

Similarly, we use a leave-one-out cross-validation to compute $\mathscr{L}(\text{spikes}_k|C)$, omitting the firing rate data from trial k, from which the spikes are observed.

4.2.2.3 Decoding Context Probabilities

For both mean firing rate and empirical firing rate distribution models, we can use the formulation of $\mathscr{L}(\text{spikes}|C)$ to compute the probability of each context given the observed spike train. For two different behavioral contexts C_1 and C_2, assuming that each context is equally likely, we use the results from Eq. (4.4) or (4.7) and calculate $\mathscr{L}(\text{spikes}|C_1)$and $\mathscr{L}(\text{spikes}|C_2)$ as

$$\Pr(\text{spikes}|C_1) = \frac{\mathscr{L}(\text{spikes}|C_1)}{\mathscr{L}(\text{spikes}|C_1) + \mathscr{L}(\text{spikes}|C_2)} \tag{4.8}$$

$$\Pr(\text{spikes}|C_2) = \frac{\mathscr{L}(\text{spikes}|C_2)}{\mathscr{L}(\text{spikes}|C_1) + \mathscr{L}(\text{spikes}|C_2)} \tag{4.9}$$

We make the prediction based on the higher likelihood ratio

$$\text{Prediction} = \begin{cases} C_1, & \text{if } \Pr(\text{spikes}|C_1) > \Pr(\text{spikes}|C_2) \\ C_2, & \text{if } \Pr(\text{spikes}|C_2) > \Pr(\text{spikes}|C_1) \end{cases} \tag{4.10}$$

which essentially uses the maximum rule:

$$\text{context} = \arg\max\{\Pr(\text{spikes}|C_1), \Pr(\text{spikes}|C_2)\} \tag{4.11}$$

It is possible to calculate a "running decode" of a context, which is $\Pr(C|\text{spikes})$ computed at each time point in the trial. This is useful for observing how incoming spiking data provides information over the entire course of the trial. A running decode is achieved by progressively computing the summation in Eqs. (4.4) and (4.7), varying the maximum trial time from 1 through T for each context, and then computing the probabilities Pr(spikes|C) as a function of time or as a function of the rat's position at any point in time.

In summary, the process for decoding context from the spikes from a single neuron can be structured as follows:

- Define behavioral contexts C_1 and C_2, two sets of experimentally related trials and corresponding spike trains.
- Estimate firing rates for the experimental trials in each context.
- Compute Pr(spikes|content) using Eq. (4.4) for the mean firing rate model or Eq. (4.7) for the empirical firing rate model.
- The maximum trial time in Eqs. (4.4) and (4.7) may be progressively increased from $i = 1$ to T to create a running decode.
- Compute Pr(spikes|C_1) and Pr(spikes|C_2) using the maximum rule (Eq. (4.11)).

4.2.3 Population Decoding

We can take advantage of the ability to record simultaneously from multiple neurons and perform a population decode, combining information from the spiking data from all cells to determine behavioral context. The probability for the population, assuming an ensemble of U independent cells, is the product of the likelihoods

$$\mathscr{L}_{\text{pop}}(\text{spikes}|C) = \prod_{u=1}^{U} \mathscr{L}(\text{spikes}^u|C) \tag{4.12}$$

where spikes^u denotes the spikes from a given cell u.

From the mean firing rate model (Eq. (4.4)), we rewrite the population likelihood as

$$\mathscr{L}_{\text{pop}}(\text{spikes}|C) = \prod_{u=1}^{U} \left(\exp\left[\sum_{i=1}^{T} \left\{ \Delta N_i^u \log\left[\hat{\lambda}_C^u(x_i)\Delta t\right] - \hat{\lambda}_C^u(x_i)\Delta t - \log(\Delta N_i^u!) \right\} \right] \right) \tag{4.13}$$

where ΔN_i^u denotes the spike count for cell u at time t_i, and $\hat{\lambda}_C^u(x_i)$ is the leave-one-out cross-validated mean firing rate for context C at the position of the animal at time t_i.

From the empirical firing rate model (Eq. (4.7)), we rewrite the population likelihood as

$$\mathscr{L}_{\text{pop}}(\text{spikes}|C) = \prod_{u=1}^{U} \left(\frac{1}{R} \sum_{r \neq k} \exp \left[\sum_{i=1}^{T} \left\{ \Delta N_i^u \log \left[\hat{\lambda}_r^u(x_i) \Delta t \right] \right. \right. \right.$$
$$\left. \left. \left. - \hat{\lambda}_r^u(x_i) \Delta t - \log(\Delta N_i^u!) \right\} \right] \right) \tag{4.14}$$

where $\hat{\lambda}_r^u(x_i)$ is the mean firing rate for trial r at the animal's position at time t_i.

Analogous to the single cell decoding, the population probability for each context is

$$\Pr(\text{spikes}|C_1) = \frac{\mathscr{L}_{\text{pop}}(\text{spikes}|C_1)}{\mathscr{L}_{\text{pop}}(\text{spikes}|C_1) + \mathscr{L}_{\text{pop}}(\text{spikes}|C_2)} \tag{4.15}$$

$$\Pr(\text{spikes}|C_2) = \frac{\mathscr{L}_{\text{pop}}(\text{spikes}|C_2)}{\mathscr{L}_{\text{pop}}(\text{spikes}|C_1) + \mathscr{L}_{\text{pop}}(\text{spikes}|C_2)} \tag{4.16}$$

To make prediction, we use the same maximum rule based on the likelihood rate ratio.

4.3 Experimental Data

To test these algorithms, we used data from two previously published datasets. In each experiment, Long-Evans rats were trained to perform a continuous spatial alternation task on a modified T-maze, in which they were required to alternate between left-turns and right-turns on the T-maze. For both datasets, we examined the neural activity while the rat traversed the center portion of T-maze stem.

The first dataset was from the experiment described in Lipton et al. (2007). Eight male Long-Evans rats were trained on the continuous spatial alternation task, and spiking data was acquired from 111 cells from 10 sets of simultaneously recorded neurons from five rats with six tetrodes aimed at dorsal CA1, and 210 cells from 10 sets of simultaneously recorded neurons from three rats with 13 tetrodes aimed at dcMEC. Each of the 20 total datasets was recorded during a separate experimental session. The number of non-error experimental trials per session ranged from 31 to 69, with an average of 46 trials per session. The second dataset was from the experiment described in detail elsewhere (Huang et al. 2009; Lee et al. 2006). We used data from ensembles of simultaneously recorded neurons CA1 of the hippocampus from one rat during two sessions of a T-maze continuous alternation task.

For all cells, we calculated the firing rate for the spiking activity over the span of the T-maze for each trial with a 500 ms Hanning smoothing kernel (Parzen 1962; Dayan and Abbott 2001). Rather than select bandwidth in the traditional ad hoc manner, the size of the Hanning window was chosen using an established cross-validation bandwidth estimation framework (Prerau and Eden 2011). In this procedure, the temporal variability was determined using a cross-validation scheme (Turlach 1993), which computes the smoother bandwidth that best predicts the missing data. A representative set of neurons from both CA1 and dcMEC were selected from the first dataset, and the cross-validated kernel smoother was used to calculate the bandwidth for each neuron for each trial. In examining the distribution of the selected bandwidths, the largest mode was found close to 500 ms. We use a single fixed bandwidth parameter for the neural activity in these regions.

4.3.1 *Single Cell Decoding*

To demonstrate the single cell decoding process, we will present illustrative examples of both the mean and empirical models as they predict context from single trials from different cell types. The aim is to highlight the specific utility of each model, which, by design, is tailored to decode best from the cells from which the underlying mathematical assumptions are drawn. In each case, given the spiking data along with the decode results, we are able to reveal the contribution each spike or non-spike data point to the representation of context.

4.3.1.1 Decoding from a Cell with Context-Dependent Changes in Mean Firing Rate

Figure 4.3 shows an example of a "splitter" cell recorded from the rat dcMEC during a single trial of the continuous alternation task. This cell clearly exhibits context-dependent differences in mean firing rate. The firing rates for 46 non-error trials comprised of 22 from the left-turn context (Fig. 4.3a, thin blue curves) and 24 from right-turn context (Fig. 4.3b, thin red curves) are displayed as a function of T-maze stem position. The rat's direction of motion towards the T-maze choice-point is represented from left to right. This particular cell fires predominantly during the right-turn context (Fig. 4.3b), with only a 3 out of 22 non-error left-turn trials containing any spikes (Fig. 4.3a). Consequently, the mean firing rate for the right-turn trials (Fig. 4.3a, b, thick red curve) shows a clear receptive field, while the mean firing rate for the left-turn trials (Fig. 4.3a, b, thick blue curve) remains very close or equal to zero.

Figure 4.3c, d illustrates the running decode for two representative spike trains from this cell, from the left-turn and right-turn contexts, respectively. For each spike train, we compute running estimates of Pr(left-turn|spikes) (blue curves) and Pr(right-turn|spikes) (red curves) as a function of time, mapped onto stem position, for both the mean (dotted) and empirical (solid) models. The trial from the left-turn context (Fig. 4.3c) has no spikes, and, for both models, Pr(left-turn|spikes) increases and Pr(right-turn|spikes) decreases as time progresses. For the mean firing rate model, this is because the left-turn context mean is much lower than the right-turn context mean or zero at all points on the stem. Thus, non-spiking implies a

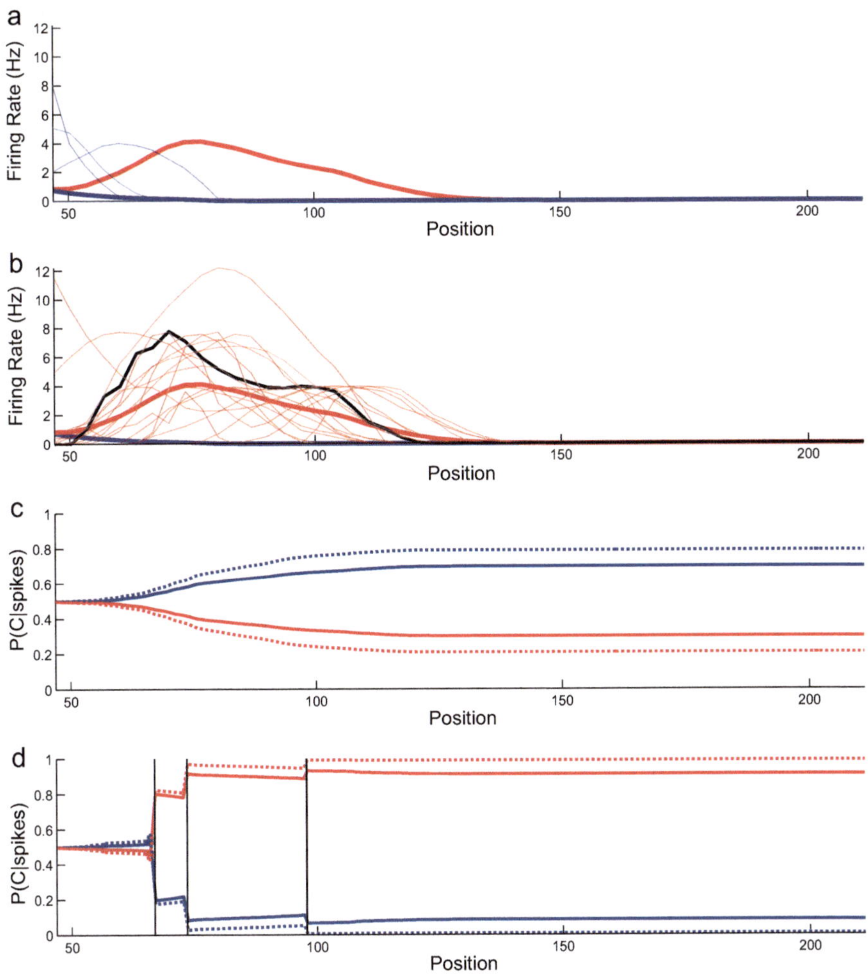

Fig. 4.3 Mean and empirical model-based decoding of spike trains from a "splitter" cell from the rat dcMEC, possessing context-dependent changes in mean firing rate. (**a**, **b**) For both left-turn (**a**, blue) and right-turn (**b**, red) contexts, we use the trial-to-trial firing rates (thin curves) to compute the mean firing rates (thick curves, superimposed on both contexts). (**c**, **d**) The results of the decoding algorithms for spiking data from a left-turn trial (black curve at 0 Hz) and a right-turn trial (black curve), respectively. For both trials, Pr(left-turn|spikes) (blue curves) and Pr(right-turn|spikes) (red curves) are computed using the mean (dotted) and empirical (solid) model decoding algorithms. In this case, both decoding models correctly identify the context for both trials

firing rate far closer to the left-turn context mean than to the right, and will drive Pr(left-turn|spikes) up and Pr(right-turn|spikes) down every time no spikes are observed. The same follows for the empirical model, with a non-spiking trial being more likely in the distribution of left-turn firing rate trajectories.

The second example spike train (Fig. 4.3d, vertical lines) comes from a right-turn trial, and the estimated firing rate is shown (Fig. 4.3b, black curve). For both models, Pr(left-turn|spikes) and Pr(right-turn|spikes) both start off at chance, and as time/position progresses and no spike has been observed, Pr(left-turn|spikes) increases as Pr(right-turn|spikes) decreases, just as in the left-turn example with no spiking. At this point, the decodes are predicting the incorrect context for the spike train. As soon as the first spike appears, Pr(right-turn|spikes) jumps up to around 0.8 and Pr(left-turn|spikes) jumps down to around 0.2 for both models. This spike provides strong instant evidence to greatly increase the likelihood of a right-turn trial, thus switching the probabilities such that both decodes predict the correct outcome. Directly after the spike, the decoding algorithms are presented with non-spike evidence, which starts to drive Pr(right-turn|spikes) down and Pr(left-turn|spikes) up, because the more time observed without a spike, the more the firing resembles like the left- turn context. The subsequent two spikes show the same pattern of a sharp increase in Pr(right-turn|spikes) at the spike times followed by a very slow decrease after the spikes. With each spike, however, the decodes become more and more certain that it is the right turn context, until at the very end, both pick the correct context.

In both of these examples, the decodes from the mean and empirical models each returned the correct answer. The mean firing rate model decode, was, however, more certain than the empirical firing rate model at virtually all points. That is to say, that the probability of the correct response was consistently higher for the mean firing rate model than for the empirical firing rate model. For the non-spike trial (Fig. 4.3c), the final values for Pr(left-turn|spikes) were 0.84 and 0.72 for the mean and empirical models, respectively. For the spiking trial (Fig. 4.3d), the final values for Pr(right-turn|spikes) were 0.99 and 0.94 for the mean and empirical models, respectively. These are precisely the results we expect by design. In the case of cells that exhibit gross changes in average firing activity, the trial-to-trial variability provides little or no information. Therefore, for cells with gross changes in activity between context, a difference between the context means is a more robust and proper estimator of context probability than the differences between the overall firing rate distributions, the tails of which may unduly inject uncertainty into the decoding estimate.

4.3.1.2 Decoding from a Cell with Context-Dependent Changes in Firing Rate Variance

Figure 4.4 shows an example of CA1 cell that displays context-dependent changes in firing rate variance. The firing rates for the left-turn context (Fig. 4.4a, thin blue curves) and right-turn context (Fig. 4.4b, thin red curves) are displayed as a function of T-maze stem position. The rat's direction of motion towards the T-maze choice-point is represented from left to right. For both left-turn and right-turn contexts, the peak of the field is around position 75 followed by a gradual decline until position 150. It is between positions 75 and 150 that the context-modulated variance is

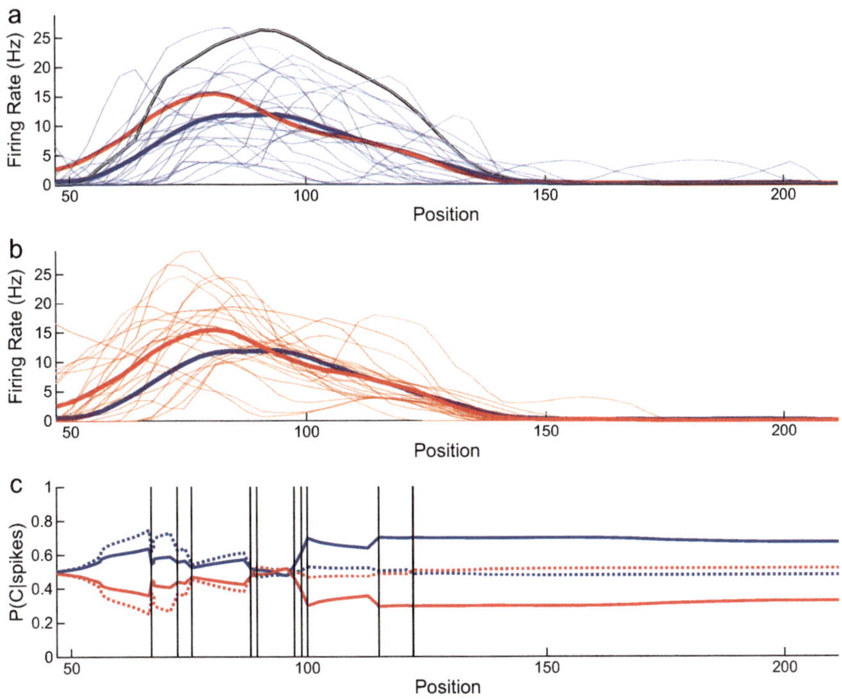

Fig. 4.4 Mean and empirical firing rate model-based decoding of a spike train from a cell with context-dependent firing rate variance in the rat CA1. (**a**, **b**) For both left-turn (**a**, blue) and right-turn (**b**, red) contexts, we use the trial-to-trial firing rates (thin curves) to compute the mean firing rates (thick curves, superimposed on both contexts). (**c**) Results of the decoding algorithms for spiking data from a left-turn trial (black curve) and a right-turn trial. For this trial, Pr(left-turn|spikes) (blue curves) and Pr(right-turn|spikes) (red curves) are computed using mean (dotted) and empirical (solid) model decoding algorithms. In this case, the empirical firing rate model correctly identifies the context, whereas the mean firing rate model fails to differentiate between two contexts

evident. The variance of firing rate is much larger for the left-turn context (Fig. 4.4a) than for the right-turn context (Fig. 4.4b), in which the majority of the observed firing rates closely track the mean.

Figure 4.4c illustrates the running decode for a representative spike train from this cell, which comes from the left-turn context. The corresponding rate estimate (Fig. 4.4a, black curve) is shown. We compute Pr(left-turn|spikes) (blue curves) and Pr(right-turn|spikes) (red curves) as a function of time, mapped onto stem position, for both the mean (dotted) and empirical (solid) models. For both models, the Pr(left-turn|spikes) and Pr(right-turn|spikes) waver in and out for the first 5 spikes, with the probabilities getting less and less certain over time. At the 6th spike, at position 97 the model estimates begin to greatly diverge. The estimate from the mean firing rate model (dotted) continues to get more and more uncertain, until the last spike drives Pr(right-turn|spikes), incorrectly, to become greater than

Pr(left-turn|spikes). Conversely, the empirical model estimates (solid) become drastically more certain in the correct direction after the sixth spike. The reason for the difference in the model estimates is that both context means are practically identical between positions 75 and 150. Consequently, the certainty in the mean decode is extremely low. The final values of Pr(left-turn|spikes) and Pr(right-turn|spikes) are 0.48 and 0.52, respectively. These results suggest that for this cell, the mean provides very little information in distinguishing between behavioral contexts, and that for this trial, the neuron provides almost no contextual information to the population.

On the other hand, the empirical model is readily able to capture the change in trial-to-trial variability that differentiates the neural activity between contexts, as there is a marked difference in the between variances of the left-turn and right-turn trial sets. Thus, the empirical firing rate model chooses the correct context more confidently by taking advantage of information that the mean firing rate model discards. The final values of Pr(left-turn|spikes) and Pr(right-turn|spikes) are 0.67 and 0.33. It follows that for trials with high or low firing rates in this region of high variability, this neuron will predict left-turns correctly with high certainty, intermittently providing contextual information to the population. This confirms the findings from the bootstrap analysis (Prerau et al. 2014) providing single trial evidence of the presence of intermittent context-dependent activity in cells in both CA1 and dcMEC.

4.3.2 Population Decoding Analysis

To explore how information from single cells of different types can be combined to produce a representation of context in a neural population, we used our proposed methods to decode contexts from the recorded population spike trains. We present results from neural population recordings from two different experimental sessions.

4.3.2.1 Full Population Decoding from Experimental Data

We examined the ability of these decoding algorithms to estimate behavioral context from neural spike recordings from actual experimental data. To these ends, we used both the mean and empirical decoding algorithms on the data from an ensemble of 47 units from CA1 from Huang et al. (2009). Both the mean and empirical models decoded context from the population with very high accuracy. The mean firing rate model's prediction accuracy was 100% (65/65 trials), and the empirical firing rate model's prediction accuracy was 98.5% (64/65 trials). For both models the average single cell decoding accuracy was about 53% and the maximum single-cell decoding accuracy was 93.9% (61/65) for the mean firing rate model and 92.3% (60/65) for the empirical firing rate model. These results indicate a neural population with a few strongly predictive cells and many more weakly predictive cells, which was indeed the case. Since in both cases the population result exceeded the maximum single cell results, this strongly indicates that the mean and empirical firing rate models can adequately combine information from many cells to accurately decode from the experimental data. From a physiological perspective, these results confirm that robust contextual information exists within hippocampal neural populations, and that there exists a small population of strongly differential firing cells, which encode most of the contextual information.

4.3.2.2 Decoding from Populations of ICD Cells from CA1 and dcMEC

We next used the decoding models to explore the how multiple cells with ICD neural activity could work together within a population to represent behavioral context. To these ends, we selected a subpopulation of 10 ICD cells from both CA1 and dcMEC and performed a population decoding analysis. These neurons were selected from the entire population by removing all of the strong splitter cells (prediction accuracy $\geq$85%), as well as the cells for which there was little or no spiking activity. From this subset, we chose cells that appeared to exhibit intermittent context-dependent activity, to examine how they would combine in a population analysis.

Analysis of CA1 Population

We selected a 10 unit CA1 subpopulation (Fig. 4.5a) from another session of the same rat performing the experiment described in Huang et al. (2009). For this session, there were 77 total non-error trials in the session, with 38 from the left-turn context (Fig. 4.5a, left subpanels, red curves) and 39 from the right-turn context (Fig. 4.5a, right subpanels, blue curves). We computed the single cell and population decodes using the mean and empirical firing rate models, and computed the corresponding prediction rates. For the mean firing rate model, the mean single-cell prediction accuracy was 59%, and the maximum single-cell prediction accuracy was 71% (55/77). For the empirical model, the mean single-cell prediction accuracy rate was 56.5%, and the maximum single-cell prediction accuracy was 71% (55/77). Both results confirm populations of individually uninformative neurons with respect to turn direction. The population decodes produced prediction accuracy rates of 83% (64/77) and 79% (61/77). For both models, the population prediction rate exceeded the maximum single-cell decode, suggesting that a strong population representation is present in CA1 within ensembles of individually poorly predicting, non-splitter cells.

To further examine the way in which the individual cells contribute information to population, we created mean (Fig. 4.5b) and empirical (Fig. 4.5c) firing rate model trial-by-trial decode certainty matrices, as in the simulation, for the individual cell decodes (left matrices), as well as for the population (right single columns). Trial number is represented in the rows, and the cell numbers correspond to the columns. In both matrices, there are no neurons that show high certainty at every trial, as one would expect to see with a population of splitter cells. Rather, for each neuron,

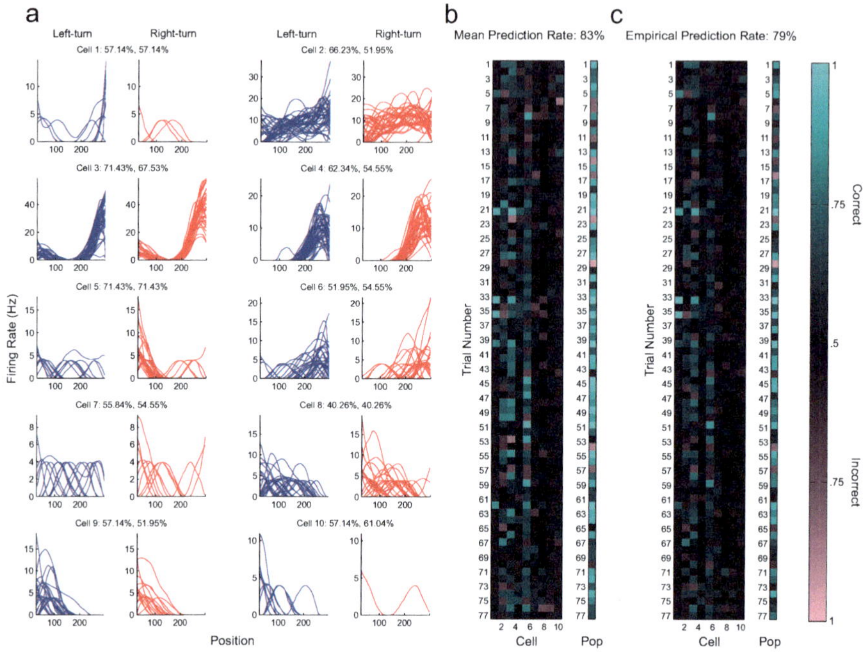

Fig. 4.5 Population decoding analysis from a group of simultaneously recorded CA1 cells from the rat hippocampus. (**a**) The raw firing rate traces are shown from a population of 10 cells for the left-turn (left panel, blue curves) and right turn (right panel, red curves), along with the single cell prediction rates for the mean and empirical firing rate models, respectively. (**b**) Using the spike data, we create a matrix of single cell trial prediction results for the mean (left) and empirical (right) model decoding procedures, where each row represents a trial, and each column represents a cell. Correct predictions of context are indicated by cyan and incorrect predictions by red, with brightness representing the certainty. Next to each matrix, is a column showing the population prediction of context. For this group of cells, both the mean and empirical population decoding algorithms produced approximately the same performance

there are several trials for which the neuron is certain and correct (bright blue). This result is precisely what we expect to see for a population of cells with ICD activity, and mirrors the output of the simulated population analysis on the two types of ICD cells (Fig. 4.5b, middle and right panels). This provides compelling evidence that non-splitter neurons within CA1 show ICD through subsets of trials during which they are certain and correct, and that populations of these individually uninformative neurons can provide an improved representation of behavioral context.

Analysis of dcMEC Population

To analyze ICD population dynamics in dcMEC, we selected a subpopulation of 10 non-splitter cells from an ensemble from the experiment described in Lipton et al.

(2007). For this session, there were 40 total non-error trials in the session, with 20 from the left-turn context (Fig. 4.6a, left subpanels, red curves) and 28 from the right-turn context (Fig. 4.6a, right subpanels, blue curves). We computed the single cell and population decodes using the mean and empirical models, and computed the corresponding prediction rates. For the mean firing rate model, the mean single-cell prediction accuracy was 58%, and the maximum single-cell prediction accuracy was 72.5% (29/40). For the empirical firing rate model, the mean single-cell prediction accuracy was 58%, and the maximum single-cell prediction accuracy was 80% (32/40). In this case, the population decodes differed greatly between models. The mean firing rate model performed with a moderate 75% (30/40) prediction accuracy, and gained only 2.5% accuracy improvement over the best single-cell decoding result. These results suggest a noninformative or redundant population encoding of behavioral context. In contrast, the empirical firing rate model's accuracy had 90% (36/40) prediction accuracy. For this ensemble, not only did the population

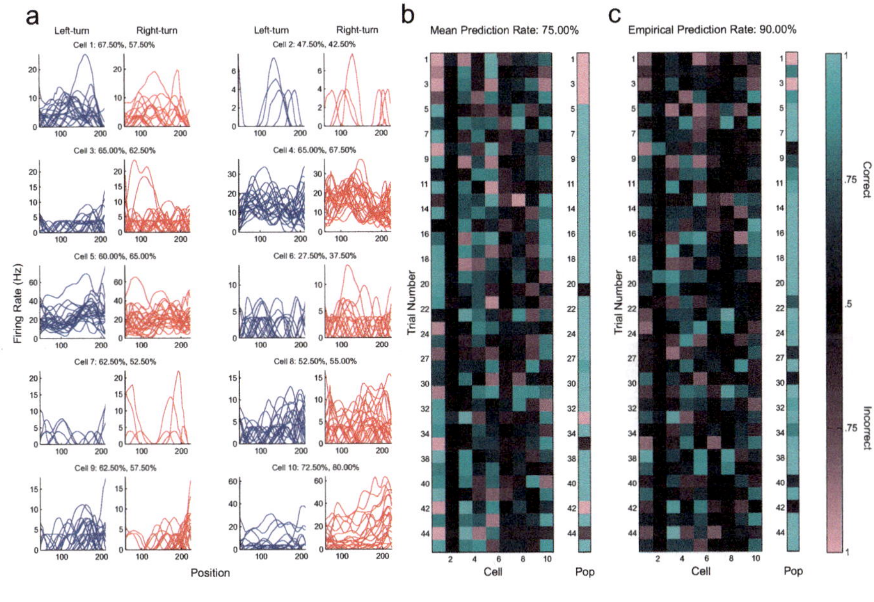

Fig. 4.6 Population decoding analysis from a group of simultaneously recorded ICD cells from the rat dcMEC. (**a**) The raw firing rate traces are shown from a population of 10 cells for the left-turn (left panel, blue curves) and right turn (right panel, red curves), along with the single cell prediction rates for the mean and empirical firing rate models, respectively. (**b**) Using the spiking data, we created a matrix of single cell trial prediction results for the mean (left) and empirical (right) model decoding procedures, where each row represents a trial, and each column represents a cell. Correct predictions of context are indicated by cyan and incorrect predictions by red, with brightness representing certainty. Next to each matrix, is a column showing the population prediction of context. For this group of cells, the empirical population model greatly outperformed the mean firing rate model, suggesting that a population of ICD cells is more robust in representing the context

prediction rate show a marked increase over that of the mean firing rate model, but the difference between population and the maximum single cell decoding accuracy was 10% as opposed to 2.5%. These results strongly suggest that, for this dcMEC population, there is information contained within the firing rate distribution beyond the mean statistic that contributes to a robust encoding of behavioral context on a population level.

To examine the way in which the individual dcMEC ICD cells contribute information to population, we created mean (Fig. 4.6b) and empirical (Fig. 4.6c) model trial-by-trial decode certainty matrices for the single-cell decodes (left matrices), and for the population (right single columns). Trial number is represented in the rows, and the cell numbers correspond to the columns. As in the CA1 population, there are no neurons that show high certainty and correctness at every trial, as one would expect to see with a population of splitter cells. In general, there are several trials for which each cell strongly predicts context. This provides compelling evidence that populations of individually uninformative ICD neurons within dcMEC can also provide a robust encoding of behavioral context.

4.4 Discussion

In this chapter, we have shown that trial-to-trial variability in neural responses does not prevent a population from maintaining a robust representation of the signals being encoded. Specifically, we were able to use empirical models of trial-to-trial variability to develop decoding methodologies that predict turn direction from spiking data from individual cells, as well as from populations of cells. The goal of these methods was to examine the way ICD cells, which provide information only for a few trials, work individually, as well as to understand how ensembles of ICD cells interact to produce a population representation of behavioral context. On a single trial level, decoding provides a story for how and when contextual information is conveyed by a particular spike train or sets of simultaneously recorded spike trains. Over multiple trials, decoding not only provides trial-by-trial predictions of context, but also the certainty of those predictions. Using these computational tools, we decoded from populations of simulated and real neurons, and examined the prediction certainty of each individual cell, and the mechanics of the contributions of multiple cells to the population representation of behavioral context.

Our findings confirm that certain ICD cells in CA1 and dcMEC can act as experts on a subset of trials, and that groups of these cells can provide a consistent population representation of turn direction during continuous spatial alternation. Thus, instead of the full burden of contextual information resting upon just a few robustly encoding individual cells, the information can also be found distributed throughout the population. Therefore, a much larger proportion of cells in these regions encode behavioral context than was previously assumed, indicating a potentially more prominent role of CA1 and dcMEC in processing and transmitting spatial trajectory information. We are confident that the decoded ICD behavior is not simply a result of noise for two reasons. First, because we use leave-one-out cross-validation in the decoding procedure, which requires repetition and consistency of the neural activity beyond a single anomalous trial in order to predict correctly. Second, in both population examples, there are many more correct and certain trials than there are incorrect and certain trials, indicating a consistent encoding of context.

The two models used in this analysis were designed to capture the different methods with which splitter and ICD cells theoretically transmit information, rather than to create an optimal decoding algorithm. For splitter cells, both models decoded well, but the prediction certainty of the mean model was generally higher than that the empirical. For the cells with context-modulated variance, the mean model cannot capture the differences in firing rate variance, and is thus highly uncertain for all trials, while the empirical can predict with certainty for the firing rate distribution extrema. For the cells with context-dependent 95th percentiles, the mean model has an increased mean for the context in which the elevated trials occur. Consequently, it correctly predicts the elevated trials, yet incorrectly predicts with certainty all of the higher firing rates from the non-elevated context. The empirical model correctly predicts the elevated trials with great certainty and the rest with confidence near chance. Future studies may be able to use these comparative decode certainties to automatically determine cell-type in large-scale populations.

We have applied the decoding methods to recorded neural spike data from CA1 and dcMEC. For each region, the population of individually uninformative cells had increased in predictive power when combined. There are some notable differences in the nature of the population decodes for each region. For both models, the trial-by-trial certainty matrices from CA1 (Fig. 4.5b, c) are much cleaner than those of dcMEC (Fig. 4.6b, c). That is, for the most part, the trials in CA1 are either certain and correct, or uncertain, whereas the single cell decodes in dcMEC contain more certain and incorrect trials. However, when we look at the population trial-by-trial certainty, the dcMEC population is more confident on average in its population prediction than the CA1 population, for both models. This suggests that while some individual dcMEC cells may provide strong incorrect information on a given trial, the population has enough cells responding with correct information for that same trial, thereby being robust to noise. Such a scenario falls in line with the other finding (Jackson and Redish 2007), which suggests that the activation of neurons within these populations does not occur at random.

In addition, there is a larger difference between the mean and empirical population decodes for dcMEC than for CA1. A potential reason for this is that cells from dcMEC have been shown to have strong turn-direction selectivity, whereas cells from CA1 are more spatially selective (Lipton et al. 2007). This can be observed in the firing rate traces. The cells in CA1 (Fig. 4.5a, e.g., cells 3 and 4) have more distinct receptive fields than the cells in dcMEC (Fig. 4.5a, e.g., cells 5 and 6). In addition, cells in dcMEC are more likely to be ICD, and are typically more complex in the way in which the ICD is manifested in the firing rate structure

(Prerau et al. 2014). Therefore, relative to CA1, there will be more cells in dcMEC with ICD firing rate distributions for which the empirical firing rate model will outperform the mean firing rate model.

It should be noted that decoding alone cannot determine causality between the neural activity and behavior. In fact, it has been shown that continuous spatial alternation is not a hippocampal-dependent task (Ainge et al. 2007), although the hippocampus is vital when a delay between trials is introduced. However, the accuracy with which it is possible to decode behavioral context serves as an excellent measure of the information content within a single neuron or population. Thus, while the contextual information within CA1 and dcMEC is not required for continuous spatial alternation, it is certainly present there, and perhaps used in other ways.

Within the development of this analysis, however, we found certain cells and populations for which both methods decoded poorly or even significantly worse than chance. On the single-cell level, these models will fail for neurons that have non-stationary representations of context. The decoding methods assume that the models of firing activity for each context are independent of time, and thus would perform poorly from these cells due to model misspecification. Non-stationarity can be modeled using state-space analyses (Frank et al. 2002; Czanner et al. 2008), and future context-decoding algorithms can build on these methods to create models of context that adapt over time. Another reason why single cells may decode poorly is that they may have some robust multiple-state representations of an external stimulus that is independent of context. Thus, by improperly specifying the collections of data from which to build the discrete state models, the algorithms will decode poorly. Future work may use the prediction rate of a decoding algorithm to automatically select the two most separable trial groups. Decoding from populations can be improved by designing algorithms that incorporate multiple models based on various cell types. A simple solution would be to create a mixture model that includes both individual trial and mean firing rates. Such models would be able to accurately capture both splitter and ICD cells.

Ultimately, a large-scale analysis of populations from various regions of the hippocampus and EC must be performed to fully understand the purpose of intermittent context-dependent representations of behavioral context. One possible hypothesis is that a form of multiplexing may occur, such that splitter cells in CA1 may arise from the convergence of connections from many ICD cells in dcMEC. Or perhaps an inverse multiplexing scenario may occur, in which the information from a splitter cell is divided amongst other cells that manifests as ICD in the target cells. Here, we present strong evidence that information relating to behavioral context is distributed across trials throughout individually uninformative cells within the hippocampus and EC, which provides new insight into processing and representing episodic memories in the brain.

References

Ainge, J. A., van der Meer, M. A. A., Langston, R. F., & Wood, E. R. (2007). Exploring the role of context-dependent hippocampal activity in spatial alternation behavior. *Hippocampus, 17*, 988–1002.

Barbieri, R., Frank, L. M., Nguyen, D. P., Quirk, M. C., Solo, V., Wilson, M. A., et al. (2004). Dynamic analyses of information encoding in neural ensembles. *Neural Computation, 16*(2), 277–307.

Barbieri, R., Quirk, M. C., Frank, L. M., Wilson, M. A., & Brown, E. N. (2001). Construction and analysis of non-Poisson stimulus-response models of neural spiking activity. *Journal of Neuroscience Methods, 105*, 25–37.

Brillinger, D. R. (1988). Maximum-likelihood analysis of spike trains of interacting nerve-cells. *Biological Cybernetics, 59*, 189–200.

Brillinger, D. R. (1992). Nerve-cell spike train data-analysis—A progression of technique. *Journal of the American Statistical Association, 87*, 260–271.

Brown, E. N., Barbieri, R., Eden, U. T., & Frank, L. M. (2003). Likelihood methods for neural data analysis. In J. Feng (Ed.), *Computational neuroscience: A comprehensive approach* (pp. 253–286). London: Chapman and Hall/CRC Press.

Brown, E. N., Frank, L. M., Tang, D., Quirk, M. C., & Wilson, M. A. (1998). A statistical paradigm for neural spike train decoding applied to position prediction from ensemble firing patterns of rat hippocampal place cells. *Journal of Neuroscience, 18*(18), 7411–7425.

Brown, E. N., Kass, R. E., & Mitra, P. P. (2004). Multiple neural spike train data analysis: State-of-the-art and future challenges. *Nature Neuroscience, 7*, 456–461.

Brown, E. N., Ngyuen, D. P., Frank, L. M., Wilson, M. A., & Solo, V. (2001). An analysis of neural receptive field plasticity by point process adaptive filtering. *Proceedings of National Academy of Sciences USA, 98*, 12261–12266.

Churchland, M. M., Yu, B. M., Cunningham, J. P., Sugrue, L. P., Cohen, M. R., Corrado, G. S., et al. (2010). Stimulus onset quenches neural variability: A widespread cortical phenomenon. Nature Neuroscience, 13(3), 369–378. http://doi.org/10.1038/nn.2501.

Cohen, N. J., & Squire, L. R. (1980). Preserved learning and retention of pattern-analyzing skill in amnesia: Dissociation of knowing how and knowing that. *Science, 210*, 207–210.

Czanner, G., Eden, U. T., Wirth, S., Yanike, M., Suzuki, W. A., & Brown, E. N. (2008). Analysis of between-trial and within-trial neural spiking dynamics. *Journal of Neurophysiology, 99*(5), 2672–2693.

Dayan, P., & Abbott, L. F. (2001). *Theoretical neuroscience: Computational and mathematical modeling of neural systems*. Cambridge, MA: MIT Press.

Eden, U. T., Frank, L. M., Barbieri, R., Solo, V., & Brown, E. N. (2004). Dynamic analysis of neural encoding by point process adaptive filtering. *Neural Computation, 16*(5), 971–998.

Efron, B., & Gong, G. (1983). A leisurely look at the Bootstrap, the Jackknife, and cross-validation. *American Statistician, 37*, 36–48.

Eichenbaum, H., Dudchenko, P., Wood, E., Shapiro, M., & Tanila, H. (1999). The hippocampus, memory, and place cells: Is it spatial memory or a memory space? *Neuron, 23*, 209–226.

Eichenbaum, H., Otto, T., & Cohen, N. J. (1994). Functional components of the hippocampal memory system. *Behavioral and Brain Sciences, 17*, 449–472.

Fenton, A. A., & Muller, R. U. (1998). Place cell discharge is extremely variable during individual passes of the rat through the firing field. *Proceedings of National Academy of Sciences USA, 95*(6), 3182–3187.

Ferbinteanu, J., & Shapiro, M. L. (2003). Prospective and retrospective memory coding in the hippocampus. *Neuron, 40*, 1227–1239.

Frank, L. M., Brown, E. N., & Wilson, M. A. (2000). Trajectory encoding in the hippocampus and entorhinal cortex. *Neuron, 27*, 169–178.

Frank, L. M., Eden, U. T., Solo, V., Wilson, M. A., & Brown, E. N. (2002). Contrasting patterns of receptive field plasticity in the hippocampus and the entorhinal cortex: An adaptive filtering approach. *Journal of Neuroscience, 22*, 3817–3830.

Fyhn, M., Molden, S., Witter, M. P., Moser, E. I., & Moser, M. B. (2004). Spatial representation in the entorhinal cortex. *Science, 305*, 1258–1264.

Griffin, A. L., Eichenbaum, H., & Hasselmo, M. E. (2007). Spatial representations of hippocampal CA1 neurons are modulated by behavioral context in a hippocampus-dependent memory task. *Journal Neuroscience, 27*, 2416–2423.

Huang, Y., Brandon, M. P., Griffin, A. L., Hasselmo, M. E., & Eden, U. T. (2009). Decoding movement trajectories through a T-maze using point process filters applied to place field data from rat hippocampal region CA1. *Neural Computation, 21*(12), 3305–3334.

Jackson, J., & Redish, A. D. (2007). Network dynamics of hippocampal cell-assemblies resemble multiple spatial maps within single tasks. *Hippocampus, 17*, 1209–1229.

Johnson, A., & Redish, A. D. (2007). Neural ensembles in CA3 transiently encode paths forward of the animal at a decision point. *Journal of Neuroscience, 27*, 12176–12189.

Kass, R. E., & Ventura, V. (2001). A spike-train probability model. *Neural Computation, 13*(8), 1713–1720.

Kulkarni, J. E., & Paninski, L. (2008). State-space decoding of goal-directed movements. *IEEE Signal Processing Magazine, 25*, 78–86.

Lansky, P., Fenton, A. A., & Vaillant, J. (2001). The overdispersion in activity of place cells. *Neurocomputing, 38*, 1393–1399.

Latimer, K. W., Yates, J. L., Meister, M. L. R., Huk, A. C., & Pillow, J. W. (2015). Single-trial spike trains in parietal cortex reveal discrete steps during decision-making. *Science, 349*, 184–187.

Lee, I., Griffin, A. L., Zilli, E. A., Eichenbaum, H., & Hasselmo, M. E. (2006). Gradual translocation of spatial correlates of neuronal firing in the hippocampus toward prospective reward locations. *Neuron, 51*, 639–650.

Lin, L., Osan, R., Shoham, S., Jin, W., Zuo, W., & Tsien, J. Z. (2005). Identification of network-level coding units for real-time representation of episodic experiences in the hippocampus. *Proceedings of National Academy of Sciences USA, 102*, 6125–6130.

Lipton, P. A., White, J. A., & Eichenbaum, H. (2007). Disambiguation of overlapping experiences by neurons in the medial entorhinal cortex. *Journal of Neuroscience, 27*, 5785–5789.

McCullagh, P. (1984). Generalized linear-models. *European Journal of Operation Research, 16*, 285–292.

Mizumori, S. J. Y., Ward, K. E., & Lavoie, A. M. (1992). Medial septal modulation of entorhinal single unit-activity in anesthetized and freely moving rats. *Brain Research, 570*, 188–197.

Moeliker, C. (2001). The first case of homosexual necrophilia in the mallard Anas platyrhynchos (Aves: Anatidae). *Deinsea, 8*, 243–247.

Muller, R. (1996). A quarter of a century of place cells. *Neuron, 17*, 813–822.

O'Keefe, J. (1979). A review of the hippocampal place cells. *Progress in Neurobiology, 13*, 419–439.

O'Keefe, J., & Dostrovsky, J. (1971). The hippocampus as a spatial map. Preliminary evidence from unit activity in the freely-moving rat. *Brain Research, 34*(1), 171–175.

O'Keefe, J., & Nadel, L. (1978). *The hippocampus as a cognitive map*. London: Oxford University Press.

Parzen, E. (1962). On estimation of a probability density function and mode. *Annals of Mathematical Statistics, 33*, 1065–1076.

Prerau, M. J., & Eden, U. T. (2011). A general likelihood framework for characterizing the time course of neural activity. *Neural Computation, 23*(10), 2537–2566.

Prerau, M. J., Lipton, P. A., Eichenbaum, H., & Eden, U. T. (2014). Characterizing context-dependent differential firing activity in the hippocampus and entorhinal cortex. *Hippocampus, 24*(4), 476–492.

Prerau, M. J., Smith, A. C., Eden, U. T., Kubota, Y., Yanike, M., Suzuki, W., et al. (2009). Characterizing learning by simultaneous analysis of continuous and binary measures of performance. *Journal of Neurophysiology, 102*(5), 3060–3072.

Prerau, M. J., Smith, A. C., Eden, U. T., Yanike, M., Suzuki, W., & Brown, E. N. (2008). A mixed filter algorithm for cognitive state estimation from simultaneously recorded continuous and binary measures of performance. *Biological Cybernetics, 99*, 1–14.

Quirk, G. J., Muller, R. U., Kubie, J. L., & Ranck, J. B. (1992). The positional firing properties of medial entorhinal neurons—Description and comparison with hippocampal place cells. *Journal of Neuroscience, 12*, 1945–1963.

Redish, A. D., & Touretzky, D. S. (1997). Cognitive maps beyond the hippocampus. *Hippocampus, 7*, 15–35.

Rieke, F., Warland, D., de Ruyter van Steveninck, R. R., & Bialek, W. (1997). *Spikes: Exploring the neural code*. Cambridge, MA: MIT Press.

Scoville, W. B., & Milner, B. (1957). Loss of recent memory after bilateral hippocampal lesions. *Journal of Neurology, Neurosurgery, and Psychiatry, 20*, 11–21.

Serruya, M. D., Hatsopoulos, N. G., Paninski, L., Fellows, M. R., & Donoghue, J. P. (2002). Instant neural control of a movement signal. *Nature, 416*, 141–142.

Smith, A. C., Frank, L. M., Wirth, S., Yanike, M., Hu, D., Kubota, Y., et al. (2004). Dynamic analysis of learning in behavioral experiments. *Journal of Neuroscience, 24*(2), 447–461.

Smith, A. C., Stefani, M. R., Moghaddam, B., & Brown, E. N. (2005). Analysis and design of behavioral experiments to characterize population learning. *Journal of Neurophysiology, 93*, 1776–1792.

Smith, A. C., Wirth, S., Suzuki, W. A., & Brown, E. N. (2007). Bayesian analysis of interleaved learning and response bias in behavioral experiments. *Journal of Neurophysiology, 97*, 2516–2524.

Smith, D. M., & Mizumori, S. J. (2006). Hippocampal place cells, context, and episodic memory. *Hippocampus, 16*, 716–729.

Snyder, D. L., & Miller, M. I. (1991). *Random point processes in time and space*. New York: Springer.

Tolman, E. C. (1948). Cognitive maps in rats and men. *Psychological Review, 55*, 189–208.

Touretzky, D., & Muller, R. (2006). Place field dissociation and multiple maps in hippocampus. *Neurocomputing, 69*, 1260–1263.

Touretzky, D., & Redish, A. D. (1996). Theory of rodent navigation based on interacting representations of space. *Hippocampus, 6*, 247–270.

Truccolo, W., Eden, U. T., Fellows, M. R., Donoghue, J. P., & Brown, E. N. (2005). A point process framework for relating neural spiking activity to spiking history, neural ensemble, and extrinsic covariate effects. *Journal of Neurophysiology, 93*(2), 1074–1089.

Turlach, B. A. (1993). *Bandwidth Selection in Kernel Density Estimation: A Review*. Technical report. Institut de Statistique, Louvain-la-Neuve, Belgium. Discussion Paper 9317.

Ventura, V., Cai, C., & Kass, R. E. (2005). Trial-to-trial variability and its effect on time-varying dependency between two neurons. *Journal of Neurophysiology, 94*, 2928–2939.

Wessberg, J., Stambaugh, C. R., Kralik, J. D., Beck, P. D., Laubach, M., Chapin, J. K., et al. (2000). Real-time prediction of hand trajectory by ensembles of cortical neurons in primates. *Nature, 408*, 361–365.

Wiener, M. C., & Richmond, B. J. (2003). Decoding spike trains instant by instant using order statistics and the mixture-of-Poissons model. *Journal of Neuroscience, 23*, 2394–2406.

Wilson, M. A., & McNaughton, B. L. (1993). Dynamics of the hippocampal ensemble code for space. *Science, 261*, 1055–1058.

Wood, E. R., Dudchenko, P. A., Robitsek, R. J., & Eichenbaum, H. (2000). Hippocampal neurons encode information about different types of memory episodes occurring in the same location. *Neuron, 27*, 623–633.

Zhang, K., Ginzburg, I., McNaughton, B. L., & Sejnowski, T. J. (1998). Interpreting neuronal population activity by reconstruction: Unified framework with application to hippocampal place cells. *Journal of Neurophysiology, 79*(2), 1017–1044.

Chapter 5
Sparsity Meets Dynamics: Robust Solutions to Neuronal Identification and Inverse Problems

Behtash Babadi

5.1 Introduction

One of the first steps in neural signal processing is the construction of forward models. Forward models relate observed neural activity to intrinsic and extrinsic stimuli as well as to neural states and sources. In the context of spike recordings from neuronal networks, numerous forward models have been proposed such as the conductance-based models (Dayan and Abbott 2001), integrate-and-fire models (Tuckwell 1988), rate-based models (McCulloch and Pitts 1943; Rosenblatt 1958; Werbos 1974), and the more recent statistical models based on generalized linear models (GLMs) and point processes (Brown et al. 2004, 2001; Paninski 2004; Paninski et al. 2007; Pillow et al. 2011; Truccolo et al. 2005) for modeling emerging phenomena such as receptive fields. Forward models for neuroimaging modalities of electroencephalography (EEG) and magnetoencephalography (MEG) are based on solutions to the Maxwell's equations over head models obtained by MRI images, which relate the post-synaptic neuronal population activity to the out-of-scalp electromagnetic recordings (Hämäläinen et al. 1993; Hämäläinen and Sarvas 1989; Marin et al. 1998; Mosher et al. 1999).

From a signal processing viewpoint, forward models have two main applications. First, when combined with models accounting for the dynamics of neuronal, dipolar, or voxel-based networks, they can be used to infer information regarding the function of the underlying neural systems given the observed activity and extrinsic stimuli. We refer to these signal processing problems as *Neural Identification Problems*. For instance, estimating the interaction parameters of a network of

B. Babadi (✉)
University of Maryland, College Park, MD, USA
e-mail: behtash@umd.edu

Z. Chen, S.V. Sarma (eds.), *Dynamic Neuroscience*,
https://doi.org/10.1007/978-3-319-71976-4_5

neurons from multi-unit recordings can be used to obtain measures of synchrony (Kass et al. 2011), functional connectivity (Okatan et al. 2005), or Granger causality (Kim et al. 2011).

Second, these forward models can be used as a basis for solving the so-called *Neural Inverse Problems*. Neural inverse problems can be thought of as duals to neural identification problems, where the parameters of the underlying neuronal models are assumed to be fixed, and the observations are used to estimate extrinsic stimuli, or intrinsic processes such as perception, decision-making, intention, and attention. Examples of neural inverse problems include EEG/MEG source localization (Babadi et al. 2014; Daunizeau and Friston 2007; Gramfort et al. 2012; Hämäläinen and Ilmoniemi 1994; Sato et al. 2004), trajectory decoding from hippocampal place cells (Brown et al. 1998; Eden et al. 2004; Huang et al. 2009), visual stimulus decoding from fMRI (Kay et al. 2008; Nishimoto et al. 2011), or motor intention decoding from neural implants (Hochberg et al. 2006; Velliste et al. 2008).

In modern neural data applications, networks of size $\sim 10^4$–10^5 neurons, dipoles, or voxels often need to be considered in solving neural identification and inverse problems. Considering a simple scalar interaction parameter between pairs of neurons, dipoles, or voxels, this amounts to model parameters of the order of $\sim 10^8$–10^{10} to be estimated per time sample. Given the millisecond sampling resolution typical of EEG/MEG or spike recordings, parameter estimation for even 1 min of data becomes computationally infeasible. Moreover, the neural identification and inverse problems are highly ill-posed, as there are usually far fewer number of sensors available than the number of unknown parameters. Various sophisticated solutions have been proposed to overcome the ill-posed nature of inverse problems (e.g., Friston et al. (2008), Gramfort et al. (2012), Lamus et al. (2012) for EEG/MEG inverse problems), which are successful for low data dimensions. However, they do not scale well with the dimensions of modern-day neural data, and are particularly not well-suited for real-time applications. With the emergence of neural prostheses and brain-computer interface (BCI) systems, there is a growing demand for scalable signal processing solutions, which largely remains unaddressed by existing methods.

Analyses of neural data recorded through various modalities have revealed three main features of these data: first, neural activity is stochastic and exhibits significant variability across trials; second, the underlying statistics of a neural system often undergo rapid changes in order to adapt to changing stimulus salience and behavioral context; and third, neural signals and systems exhibit a degree of sparsity that is manifested in different forms: place cells in the hippocampus (Frank et al. 2004) and spectrotemporally tuned cells in the primary auditory cortex (Depireux et al. 2001) exhibit sparsity in their tuning characteristics; brain rhythms manifested in EEG/MEG have sparse spectrotemporal structures (Buzsaki 2006); and spike trains can be considered as sequences of sparse events in time. Hence, in order to gain insight into the functional mechanism of the underlying neural system, it is crucial to develop inference algorithms that simultaneously capture the stochasticity, dynamicity, and sparsity of neural activity. The main objective of this chapter is to exploit the aforementioned features of neural signals and systems in order to construct scalable solutions to neural identification and inverse problems, with provable performance guarantees.

5.2 State-Space Models from Robust Statistics

In what follows, we denote by $\mathbf{y}_t$ the generic vector of neural data at time t, for $t = 1, 2, \ldots, T$, where T denotes the observation length. In order to define the neural identification and inverse problems more formally, let Θ_t denote the set of parameters modeling the structural or functional properties of the system at time t and let Ψ_t denote the stimulus and other neural covariates at time t.

We model the stochastic nature of the observed data using a probability distribution $p_{\text{obs}}(\cdot)$, which in general relates the observations to the system parameters, stimuli, and other covariates. The most widely used observation model in neural data analysis is the Gaussian model, with a corresponding quadratic log-likelihood. In particular, in common applications to binary spiking data, the spikes are first smoothed out through windowing to form continuous covariates to be used in least squares procedures. Although fitting a quadratic model to data can be carried out efficiently via least squares, it often fails to capture the likely non-Gaussian structure of the data and the underlying dynamics.

Inspired by the seminal contributions of Brown et al. (1998, 2001, 2002, 2004), we take the approach of constructing observation models that are informed by the measurement mechanism as well as the underlying stochasticity and biophysical dynamics of the neural signals and systems.

Table 5.1 shows a few examples of such observation models. The first model $p_{\text{obs},1}$ is commonly employed to model binary neurons (Brown et al. 1998). The second model $p_{\text{obs},2}$ captures the dynamics of a binary neuron, with a forgetting factor mechanism that favors recent observations, as a means to account for the temporal variability of the underlying parameters. The last model $p_{\text{obs},3}$ corresponds to two-photon fluorescence recordings $\mathbf{y}_t$ observed in Gaussian noise, exhibiting autoregressive dynamics. In Sects. 5.4 through 5.6, we will use these observation models to analyze real data and highlight the achievable performance gains obtained by using models informed by the underlying dynamics.

Table 5.1 Examples of dynamic observation models

Observation models	Biophysical motivation
$\log p_{\text{obs},1}\left(\{y_t\}_{t=1}^T\right) \propto \sum_{t=1}^T \{y_t \log(\lambda_t \Delta) - \lambda_t \Delta\}$	Poisson statistics with conditional intensity λ_t, e.g. hippocampal place cells
$\log p_{\text{obs},2}\left(\{y_t\}_{t=1}^T\right) \propto$ $\sum_{t=1}^T \beta^{T-t} \{y_t \log(\lambda_t \Delta) + (1-y_t)\log(1-\lambda_t \Delta)\}$	Bernoulli statistics with conditional intensity λ_t, weighted with forgetting factor β, e.g. single neuron undergoing plasticity
$\log p_{\text{obs},3}\left(\{\mathbf{y}_t\}_{t=1}^T\right) \propto -\sum_{t=1}^T \|\mathbf{y}_t - a\mathbf{y}_{t-1}\|_2^2$	Autoregressive dynamics, e.g. two-photon fluorescence traces of neuronal activity

As summarized in Table 5.2, general identification and inverse neural problems can be posed as maximum *a posteriori* (MAP) estimation problems, where p_{idn} and p_{inv} are the prior probability densities used for identification and inverse parameters, respectively.

In order to simultaneously capture the sparsity and dynamicity of data, we construct biophysically inspired priors p_{inv} and p_{idn} through dynamic extensions of priors from robust statistics (Huber 2011). Robust priors are commonly used in regression and are known to be effective in outlier rejection and denoising data due to their heavy-tail nature (Rousseeuw and Leroy 2005). The Laplace distribution used is a widely used robust prior in regression, and is central to compressed sensing. A large family of robust priors correspond to Normal/Independent distributions with desirable analytical properties (Dempster et al. 1980) including the *multivariate Student's t, Power-Exponential*, and *Stable* distributions (Lange and Sinsheimer 1993).

The dynamic extension of these priors can be carried out by fusing temporal Markovian dynamics with robust priors. Table 5.3 lists a few examples of such extensions. $p_1(\cdot)$ is an extension of the multivariate Power-Exponential distribution to the spectrotemporal domain, promoting spectral sparsity and temporal smoothness of a spectrogram $\mathbf{s}_t$, which can be used to model the spectrotemporal dynamics of brain rhythms. $p_2(\cdot)$ is a mixture of Laplace and Gaussian priors, promoting

Table 5.2 Duality of identification and inverse problems under MAP estimation

Problem	Known	Unknown	MAP estimation
Identification	$\{\mathbf{y}_t, \Psi_t\}_{t=1}^T$	$\{\Theta_t\}_{t=1}^T$	$\underset{\{\Theta_t\}_{t=1}^T}{\operatorname{argmax}} \left\{\log p_{\mathsf{obs}}\left(\{\mathbf{y}_t\}_{t=1}^T \middle| \{\Theta_t, \Psi_t\}_{t=1}^T\right) + \log p_{\mathsf{idn}}\left(\{\Theta_t\}_{t=1}^T\right)\right\}$
Inverse	$\{\mathbf{y}_t, \Theta_t\}_{t=1}^T$	$\{\Psi_t\}_{t=1}^T$	$\underset{\{\Psi_t\}_{t=1}^T}{\operatorname{argmax}} \left\{\log p_{\mathsf{obs}}\left(\{\mathbf{y}_t\}_{t=1}^T \middle| \{\Theta_t, \Psi_t\}_{t=1}^T\right) + \log p_{\mathsf{inv}}\left(\{\Psi_t\}_{t=1}^T\right)\right\}$

Table 5.3 Examples of dynamic extensions of robust priors

Dynamic extensions of robust priors	Biophysical motivation
$\log p_1\left(\{\mathbf{s}_t\}_{t=1}^T\right) \propto -\gamma \sum_{n=1}^{N} \sqrt{\sum_{t=1}^T ((\mathbf{s}_t)_n - (\mathbf{s}_{t-1})_n)^2}$	Sparse and smoothly varying spectrogram, e.g. brain rhythms
$\log p_2\left(\{\boldsymbol{\alpha}_t\}_t^T\right) \propto -\sum_{t=1}^{T} \gamma_1 \left\|\boldsymbol{\Phi}\boldsymbol{\alpha}_t\right\|_1 + \gamma_2 \left\|\boldsymbol{\alpha}_t - \boldsymbol{\alpha}_{t-1}\right\|_2^2$	Sparse in transform domain spanned by $\boldsymbol{\Phi}$, smoothly varying in time, e.g. neuronal plasticity over Gabor domain
$\log p_3\left(\{\mathbf{x}_t\}_{t=1}^T\right) \propto -\gamma \sum_{t=1}^{T} \|\mathbf{x}_t - a\mathbf{x}_{t-1}\|_1$	Sparse in space and dynamic with sparse innovations, e.g. two-photon fluorescence traces of ensemble neuronal spiking
$\log p_4\left(\{\theta_t^{(m,l)}\}_{t,m,l=1}^{T,M,M}\right) \propto -\sum_{i=1}^G \gamma_i \sqrt{\sum_{m,l \in g_i} \sum_{t=1}^T \left(\theta_t^{(m,l)} - \theta_{t-1}^{(m,l)}\right)^2}$	Group-sparse interaction induced by the partition $\{g_i\}_{i=1}^G$, smoothly varying in time, e.g. functional connectivity

sparsity in a transform domain spanned by $\boldsymbol{\Phi}$ and smoothness in time, which can be used to model neuronal plasticity or spectrotemporal modulations. p_3 is a dynamic extension of the Laplace prior and captures sparsity in space as well as the temporal innovations of the signal, and can be used to model the statistics of two-photon fluorescence traces from ensemble neuronal activity. Finally, $p_4(\cdot)$ is a dynamic extension of the Power-Exponential density induced by group-sparse structure. Given M neurons and the partitioning of $\{1, 2, \cdots, M\}$ into $\{g_i\}_{i=1}^{G}$, $p_4(\cdot)$ can account for the sparse dynamic nature of the functional connectivity parameters $\{\theta_t^{(m,l)}\}_{m,l=1}^{M,M}$ in ensemble neural activity.

The hyper-parameters appearing in the robust state-space models (e.g., γ in $p_1(\cdot)$) can be chosen based on several mechanisms. First, cross-validation techniques can be used to choose these parameters in the absence of any prior information (Hastie et al. 2009). Second, when prior biophysical information is available, a fully Bayesian approach can be employed to construct hierarchical priors on the hyper-parameters (Gelman et al. 2013). Third, analytical results can identify some of the hyper-parameters and their scaling properties with respect to the problem dimension for simplified models to gain insight for data applications (see Sect. 5.4.2).

Extensions of the ℓ_1 norm to time domain and 2D plane have previously appeared in image processing/denoising literature such as the fused LASSO (Tibshirani et al. 2005) or total variation denoising (Rudin et al. 1992). Our approach is distinct in that we explicitly model the biophysical features of the underlying neural signals and systems using the dynamic extension of robust priors.

In what follows, we will examine three specific problems in neural data analysis in which the sparsity and dynamics are simultaneously modeled and captured using the preceding models: analysis of spectrotemporal receptive field plasticity (identification problem), spike deconvolution from two-photon calcium imaging (inverse problem), and spectrotemporal decomposition of oscillatory neural signals (inverse problem).

5.3 Notation and Preliminaries

We denote vectors and matrices by boldface lowercase and uppercase letters, respectively. For a vector $\mathbf{x} \in \mathbb{R}^M$, we denote by $(\mathbf{x})_i$ the ith component of $\mathbf{x}$. Similarly, for a matrix $\mathbf{A} \in \mathbb{R}^{N \times M}$, we denote by $(\mathbf{A})_i$ and $(\mathbf{A})_{i,j}$ the ith column and the (i, j)th element, respectively.

For a sparsity level $L < M$, we denote by $\mathscr{L} \subset \{1, 2, \ldots, M\}$ the support of the L highest elements of $\mathbf{x}$ in absolute value, and by $\mathbf{x}_L$ the best L-term approximation to $\mathbf{x}$. We also define

$$\sigma_L(\mathbf{x}) := \|\mathbf{x} - \mathbf{x}_L\|_1 \tag{5.1}$$

to capture the compressibility of the vector $\mathbf{x}$. Recall that for $\mathbf{x} \in \mathbb{R}^M$, the ℓ_1-norm is defined as $\|\mathbf{x}\|_1 := \sum_{i=1}^{M} |x_i|$. When $\sigma_L(\mathbf{x}) = 0$, the vector $\mathbf{x}$ is called L-sparse.

If $\sigma_L(\mathbf{x}) = \mathcal{O}(L^{1-\frac{1}{\xi}})$ for some $\xi \in (0, 1)$, the vector is called (ξ, L)-compressible (Needell and Tropp 2009).

5.4 Sparsity Meets Dynamics for Spectrotemporal Receptive Field Plasticity Analysis

In this section, we focus on the analysis of spectrotemporal receptive field (STRF) plasticity as an identification problem under the foregoing MAP estimation framework. The responses of a group of neurons in the primary auditory cortex (A1) can be characterized by their STRFs, where each neuron is tuned to a specific region in the time-frequency plane, and only significantly spikes when the acoustic stimulus contains spectrotemporal contents matching its tuning region (Depireux et al. 2001). In addition, several experimental studies have revealed that receptive fields undergo rapid changes in their characteristics during attentive behavior in order to capture salient stimulus modulations (Fritz et al. 2003, 2005; Mesgarani et al. 2010). Our goal is therefore to capture both the adaptivity and sparsity of these receptive fields using scalable and robust algorithms. It is worth mentioning that the approach taken here integrates that pioneered by Emery N. Brown in Brown et al. (2001) with the theory of compressed sensing (CS).

5.4.1 Problem Definition

We first give a brief introduction to point process models (Daley and Vere-Jones 2007). Consider a stochastic process defined by a sequence of discrete events at random points in time, noted by $\tau_1^J = [\tau_1, \tau_2, \ldots, \tau_J]^\top$, and a counting measure given by

$$dN(\tau) = \sum_{k=1}^{J} \delta(\tau - \tau_k), \quad \text{and} \quad N(\tau) = \int_0^{\tau} dN(u), \tag{5.2}$$

where $\delta(.)$ is the Dirac's measure. The conditional intensity function (CIF) for this process, denoted by $\lambda(\tau|H_\tau)$, is defined as

$$\lambda(\tau|H_\tau) := \lim_{\varepsilon\to 0} \frac{\mathbb{P}\left(N(\tau+\varepsilon) - N(\tau) = 1|H_\tau\right)}{\varepsilon}, \tag{5.3}$$

where H_τ denotes the history of the process as well as the covariates up to time τ. The CIF can be interpreted as the *instantaneous rate* given the history of the process and the covariates. A point process model is fully characterized by its CIF. For instance, $\lambda(\tau|H_\tau) = \lambda$ corresponds to the homogenous Poisson process with rate λ. A discretized version of this process can be obtained by binning $N(\tau)$ within an observation interval of $[0, \mathscr{T}]$ by bins of length Δ, that is

$$y_t := N(t\Delta) - N((t-1)\Delta), \quad t = 1, 2, \ldots, T, \tag{5.4}$$

where $T := \lceil \mathscr{T}/\Delta \rceil$ and $N(0) := 0$. In what follows, $\{y_t\}_{t=1}^T$ will be considered as the observed spiking sequence, which will be used for estimation purposes. Also, by approximating Eq. (5.3) for small $\Delta \ll 1$, and defining $\lambda_t := \lambda(t\Delta|H_{t\Delta})$, we have

$$\begin{aligned} \mathbb{P}(y_t = 0) &= 1 - \lambda_t\Delta + o(\Delta), \\ \mathbb{P}(y_t = 1) &= \lambda_t\Delta + o(\Delta), \\ \mathbb{P}(y_t \geq 2) &= o(\Delta). \end{aligned} \tag{5.5}$$

In discrete time, the orderliness of the process is equivalent to the requirement that with high probability not more than one event fall into any given bin. In practice, this can always be achieved by choosing Δ small enough. An immediate consequence of Eq. (5.5) is that $\{y_t\}_{t=1}^T$ can be approximated by a sequence of Bernoulli random variables with success probabilities $\{\lambda_t\Delta\}_{t=1}^T$.

A popular class of models for the CIF is given by generalized linear models (GLMs). In its general form, a GLM consists of two main components: an observation model (which is given by Eq. (5.5) in this paper) and an equation expressing some (possibly nonlinear) function of the observation mean as a *linear* combination of the covariates. In neuronal systems, the covariates consist of extrinsic covariates (e.g., neural stimuli) as well as intrinsic covariates (e.g., the history of the process). In this paper, we only consider GLMs with purely extrinsic covariates, although most of our results can be generalized to incorporate intrinsic covariates as well.

At time t, let s_t denote the stimulus, $[\omega_{0,t}, \omega_{1,t}, \ldots, \omega_{M-2,t}]^\top$ denote the vector of stimulus modulation parameters, and μ_t denote the baseline firing rate. The stimulus modulation parameters and the baseline firing rate are typically assumed to be constant. But, in order to capture the dynamics of these parameters, we consider the foregoing general time-varying form. We adopt a logistic regression model for the CIF as follows:

$$\text{logit}(\lambda_t\Delta) := \log\left(\frac{\lambda_t\Delta}{1-\lambda_t\Delta}\right) = \mu_t + \sum_{i=0}^{M-2} \omega_{i,t} s_{t-i}. \tag{5.6}$$

By defining $\boldsymbol{\theta}_t := [\mu_t, \omega_{0,t}, \omega_{1,t}, \ldots, \omega_{M-2,t}]^\top$ and $\mathbf{x}_t := [1, s_t, \ldots, s_{t-M+2}]^\top$, we can equivalently write:

$$\lambda_t\Delta = \text{logit}^{-1}(\boldsymbol{\theta}_t^\top \mathbf{x}_t) := \frac{\exp(\boldsymbol{\theta}_t^\top \mathbf{x}_t)}{1 + \exp(\boldsymbol{\theta}_t^\top \mathbf{x}_t)}. \tag{5.7}$$

The model above is also known as the logistic-link CIF model. The significance of this model is that $\text{logit}^{-1}(.)$ maps the real line $(-\infty, +\infty)$ to the unit probability interval $(0, 1)$, making it a feasible model for describing statistics of binary events independent of the scaling of the covariates and modulation parameters. We refer to $\mathbf{x}_t$ and $\boldsymbol{\theta}_t$ as the covariate vector and the modulation parameter vector at time t, respectively.

In our applications of interest in analyzing spectrotemporal receptive field plasticity, the modulation parameter vector exhibits a degree of sparsity (Truccolo et al. 2005; Chen et al. 2011). That is, only certain components in the stimulus modulation have significant contribution in determining the statistics of the process. These components can be thought of as the preferred or intrinsic tuning features of the underlying neuron.

The identification problem of this section can be stated as follows: *given binary observations* $\{y_t\}_{t=1}^T$ *and covariates* $\{\mathbf{x}_t\}_{t=-M+1}^T$ *from a point process with a CIF given by Eq.* (5.7), *the goal is to estimate the M-dimensional parameter vectors* $\{\boldsymbol{\theta}_t\}_{t=1}^T$ *in an online and stable fashion.*

5.4.2 ℓ_1-Regularized Exponentially Weighted Maximum Likelihood Estimation

In order to allow the identification problem to operate at possibly a different time-scale than the sampling interval, we consider piece-wise constant dynamics for the modulation parameter vector. That is, we assume that $\boldsymbol{\theta}_t$ remains constant over windows of arbitrary length $W \geq 1$ samples, for some integer W, such that $K := \frac{T}{W}$ is also an integer (without loss of generality). By segmenting the corresponding spiking data $\{y_t\}_{t=1}^T$ into K windows of length W samples each, the CIF for each time point $(k-1)W + 1 \leq t \leq kW$ is governed by $\boldsymbol{\theta}_t = \boldsymbol{\theta}_k$, for $k = 1, 2, \ldots, K$.

Invoking the Bernoulli approximation to the spiking statistics for $\Delta \ll 1$, and assuming conditional independence of the spiking events, the joint log-likelihood of the observations within window i evaluated at a generic $\boldsymbol{\theta}$ can be expressed as:

$$\mathcal{L}_i(\boldsymbol{\theta}) := \sum_{j=1}^{W} \left\{ n_{(i-1)W+j} \mathbf{x}_{(i-1)W+j}^{\top} \boldsymbol{\theta} - \log\left(1 + \exp\left(\mathbf{x}_{(i-1)W+j}^{\top} \boldsymbol{\theta}\right)\right)\right\}. \tag{5.8}$$

In order to enforce adaptivity in the log-likelihood function, we adopt the forgetting factor mechanism of the recursive least squares (RLS) algorithm, where the log-likelihood of each window is exponentially weighted regressively in time, with a forgetting factor $0 < \beta \leq 1$. At window k, we define:

$$\mathcal{L}^{\beta}(\boldsymbol{\theta}_k) := \sum_{i=1}^{k} \beta^{k-i} \mathcal{L}_i(\boldsymbol{\theta}_k). \tag{5.9}$$

This model corresponds to $p_{\text{obs},2}$ in Table 5.1. Note that for $\beta = 1$, $\mathcal{L}^1(\boldsymbol{\theta}_k)$ coincides with the natural data log-likelihood. Next, in order to promote sparsity, we regularize the exponentially weighted log-likelihood in order to estimate $\boldsymbol{\theta}_k$ as:

$$\widehat{\boldsymbol{\theta}}_k = \underset{\boldsymbol{\theta}_k}{\text{argmax}} \quad \left\{\mathcal{L}^{\beta}(\boldsymbol{\theta}_k) - \gamma \|\boldsymbol{\theta}_k\|_1\right\}, \tag{5.10}$$

where γ is a regularization parameter controlling the trade-off between the log-likelihood fit and the sparsity of estimated parameters.

The following theorem from Sheikhattar et al. (2016) quantifies the benefits of the objective function in Eq. (5.10):

Theorem 1 (Theorem 1 of Sheikhattar et al. (2016)) *Suppose that binary observations from a point process with a CIF given by Eq. (5.7) are given over K windows of length W each. Suppose that the stimulus sequence* $\{s_t\}_{t=-M+1}^{T}$ *consists of independent (but not necessarily identically distributed) random variables with a variance of* σ^2 *which are uniformly bounded by a constant* $B > 0$ *in absolute value. Consider the setting where* $\boldsymbol{\theta}_k = \boldsymbol{\theta}$ *for all k. Then, under mild technical assumptions, for an arbitrarily chosen positive constant* $d > 0$*, there exist constants* C, C', *and* C'' *such that for* $M > 10L$, $1 - \frac{C'}{L^2 \log M} \leq \beta < 1$, $K \geq \frac{\log 2}{\log(\frac{1}{\beta})}$, *and a choice of* $\gamma = C'' \sqrt{\frac{\log M}{1-\beta}}$, *any solution* $\widehat{\boldsymbol{\theta}}$ *to Eq. (5.10) satisfies the bound*

$$\left\|\widehat{\boldsymbol{\theta}} - \boldsymbol{\theta}\right\|_2 \leq C\sqrt{(1-\beta)L\log M} + \sqrt{C\sigma_L(\boldsymbol{\theta})}\sqrt[4]{(1-\beta)L\log M},$$

with probability at least $1 - \frac{5}{M^d}$.

Proof The proof uses techniques from compressed sensing as well as the concentration of dependent random variables. See Sheikhattar et al. (2016) for details.

The result of Theorem 1 has several implications. First, assuming that $\sigma_L(\boldsymbol{\theta}) = 0$, the error bound scales with $\sqrt{(1-\beta)L\log M}$, the sparsity level, as opposed to $\sqrt{(1-\beta)M}$ for the maximum likelihood (ML) estimate, implying a putative performance gain of order $\mathcal{O}\left(\frac{M}{L\log M}\right)$ in terms of estimation error. Nevertheless, the bound holds for general non-sparse $\boldsymbol{\theta}$, but is sharpest when $\sigma_L(\boldsymbol{\theta})$ is negligible. Second, the theorem prescribes a lower bound on the forgetting factor which results in significant performance improvement. Third, the theorem reveals the scaling of the regularization parameter in terms of M and β. In particular, this scaling is significant as it reveals another role for the forgetting factor mechanism: not only the forgetting factor mechanism allows for adaptivity of the estimates, it also controls the scaling of the ℓ_1-regularization term with respect to the log-likelihood term. Fourth, unlike conventional results in the analysis of adaptive filters which concern the expectation of the error in the asymptotic regime, our result holds for a single realization with probability polynomially approaching 1, in the non-asymptotic regime.

5.4.3 Adaptive Parameter Identification

Several standard optimization techniques, such as interior point methods, can be used to find the maximizer of Eq. (5.10). However, most of these techniques operate offline and do not meet the real-time requirements of the adaptive filtering setting where the observations arrive in a streaming fashion. In order to avoid the increasing runtime complexity and memory requirements of the batch-mode computation, we seek a recursive approach which can perform low-complexity updates in an online fashion upon the arrival of new data in order to form the estimates. To this end, we adopt the proximal gradient approach. Each iteration of the algorithm moves the previous iterate along the gradient of the log-likelihood function, which will then pass through a shrinkage operator.

Let $\mathbf{y}_k := [y_{(k-1)W+1}, y_{(k-1)W+2}, \ldots, y_{kW}]^\top$ denote the vector of observed spikes within window k, for $k = 1, 2, \ldots, K$. Similarly, let $\boldsymbol{\lambda}_k := \left[\lambda_{(k-1)W+1}, \lambda_{(k-1)W+2}, \ldots, \lambda_{kW}\right]^\top$ denote the vector of CIFs within window k. By extending the domain of the $\text{logit}^{-1}(\cdot)$ to vectors in a component-wise fashion, we define $\boldsymbol{\lambda}_k(\boldsymbol{\theta})$ for any window k and any parameter $\boldsymbol{\theta}$ to be:

$$\boldsymbol{\lambda}_k(\boldsymbol{\theta}) := \frac{1}{\Delta} \text{logit}^{-1}\left(\mathbf{X}_k \boldsymbol{\theta}\right), \tag{5.11}$$

where $\mathbf{X}_k := \left[\mathbf{x}_{(k-1)W+1}, \mathbf{x}_{(k-1)W+2}, \ldots, \mathbf{x}_{kW}\right]^\top$ is the data matrix of size $W \times M$ with rows corresponding to the covariate vectors in window k. Suppose that at window k, we have an iterate denoted by $\widehat{\boldsymbol{\theta}}_k^{(\ell)}$, for $\ell = 0, 1, \ldots, R$, with R being an integer denoting the total number of iterations. The gradient of $\mathcal{L}^\beta(\cdot)$ evaluated at $\widehat{\boldsymbol{\theta}}_k^{(\ell)}$ can be written as:

$$\nabla_{\boldsymbol{\theta}} \mathcal{L}^\beta\left(\widehat{\boldsymbol{\theta}}_k^{(\ell)}\right) = \sum_{i=1}^{k} \beta^{k-i} \mathbf{X}_i^\top \boldsymbol{\varepsilon}_i\left(\widehat{\boldsymbol{\theta}}_k^{(\ell)}\right) =: \mathbf{g}_k\left(\widehat{\boldsymbol{\theta}}_k^{(\ell)}\right), \tag{5.12}$$

where $\boldsymbol{\varepsilon}_i(\cdot) := \mathbf{y}_i - \boldsymbol{\lambda}_i(\cdot)\Delta$ represents the innovation vector of the point process at window i. The proximal gradient iteration for the ℓ_1-regularization can be written in the compact form as:

$$\widehat{\boldsymbol{\theta}}_k^{(\ell+1)} = \mathscr{S}_{\gamma\alpha}\left(\widehat{\boldsymbol{\theta}}_k^{(\ell)} + \alpha \mathbf{g}_k\left(\widehat{\boldsymbol{\theta}}_k^{(\ell)}\right)\right) \tag{5.13}$$

where $\mathscr{S}_\tau(\cdot)$ is the element-wise soft thresholding operator at a level of τ defined as:

$$(\mathscr{S}_\tau(x))_i := \text{sgn}(x_i)(|x_i| - \tau)_+,$$

for $i = 1, 2, \ldots, M$, with sgn denoting the standard signum function, and $(a)_+ := \max\{a, 0\}$. The final estimate at window k is obtained following the Rth iteration, and is denoted by $\widehat{\boldsymbol{\theta}}_k := \widehat{\boldsymbol{\theta}}_k^{(R)}$. In order to achieve a recursive updating rule for $\mathbf{g}_k$, we can rewrite Eq. (5.12) as:

$$\mathbf{g}_k\left(\widehat{\boldsymbol{\theta}}_k^{(\ell)}\right) = \beta\, \mathbf{g}_{k-1}\left(\widehat{\boldsymbol{\theta}}_k^{(\ell)}\right) + \mathbf{X}_k^\top \boldsymbol{\varepsilon}_k\left(\widehat{\boldsymbol{\theta}}_k^{(\ell)}\right). \tag{5.14}$$

However, in an adaptive setting, we only have access to values of $\mathbf{g}_{k-1}$ evaluated at $\widehat{\boldsymbol{\omega}}_{k-1}^{(1:R)}$. In order to turn Eq. (5.14) into a fully recursive updating rule, we exploit the smoothness of the logistic function and employ the Taylor series expansion of the CIF to approximate the required recursive update. To this end, by retaining the first two terms in the Taylor expansion of $\boldsymbol{\lambda}_i(\cdot)$, we get:

$$\boldsymbol{\lambda}_i\left(\widehat{\boldsymbol{\theta}}_k^{(\ell)}\right)\Delta \approx \boldsymbol{\lambda}_i(\widehat{\boldsymbol{\theta}}_i)\Delta + \boldsymbol{\Lambda}_i(\widehat{\boldsymbol{\theta}}_i)\mathbf{X}_i\left(\widehat{\boldsymbol{\theta}}_k^{(\ell)} - \widehat{\boldsymbol{\theta}}_i\right), \tag{5.15}$$

where $\boldsymbol{\Lambda}_i(\widehat{\boldsymbol{\theta}}_i)$ is a diagonal $W \times W$ matrix with the (m, m)-th diagonal element given by $\lambda_{(i-1)W+m}\Delta(1 - \lambda_{(i-1)W+m}\Delta)$. Using the first-order approximation above, we can approximate the gradient $\mathbf{g}_k$ by $\mathbf{g}_k^1$, as:

$$\mathbf{g}_k^1\left(\widehat{\boldsymbol{\theta}}_k^{(\ell)}\right) = \sum_{i=1}^{k} \beta^{k-i}\mathbf{X}_i^\top\left(\boldsymbol{\varepsilon}_i(\widehat{\boldsymbol{\theta}}_i) - \boldsymbol{\Lambda}_i(\widehat{\boldsymbol{\theta}}_i)\mathbf{X}_i(\widehat{\boldsymbol{\theta}}_k^{(\ell)} - \widehat{\boldsymbol{\theta}}_i)\right). \tag{5.16}$$

By defining:

$$\mathbf{u}_k := \sum_{i=1}^{k} \beta^{k-i}\mathbf{X}_i^\top\left(\boldsymbol{\varepsilon}_i(\widehat{\boldsymbol{\theta}}_i) + \boldsymbol{\Lambda}_i(\widehat{\boldsymbol{\theta}}_i)\mathbf{X}_i\widehat{\boldsymbol{\theta}}_i\right), \quad \text{and} \quad \mathbf{B}_k := \sum_{i=1}^{k} \beta^{k-i}\mathbf{X}_i^\top\boldsymbol{\Lambda}_i(\widehat{\boldsymbol{\theta}}_i)\mathbf{X}_i,$$

we can express $\mathbf{g}_k^1\left(\widehat{\boldsymbol{\theta}}_k^{(\ell)}\right)$ as:

$$\mathbf{g}_k^1\left(\widehat{\boldsymbol{\theta}}_k^{(\ell)}\right) = \mathbf{u}_k - \mathbf{B}_k\widehat{\boldsymbol{\theta}}_k^{(\ell)} = \beta\,\mathbf{g}_{k-1}^1\left(\widehat{\boldsymbol{\theta}}_k^{(\ell)}\right) + \mathbf{X}_k^\top\boldsymbol{\varepsilon}_k\left(\widehat{\boldsymbol{\theta}}_k^{(\ell)}\right).$$

which replaces the gradient $\mathbf{g}_k\left(\widehat{\boldsymbol{\theta}}_k^{(\ell)}\right)$ in the shrinkage step given by Eq. (5.13). It is then straightforward to check that both $\mathbf{u}_k$ and $\mathbf{B}_k$ can be updated recursively (Babadi et al. 2010) as:

$$\mathbf{u}_k = \beta\,\mathbf{u}_{k-1} + \mathbf{X}_k^\top\left(\boldsymbol{\varepsilon}_k\left(\widehat{\boldsymbol{\theta}}_k^{(R)}\right) + \boldsymbol{\Lambda}_k\left(\widehat{\boldsymbol{\theta}}_k^{(R)}\right)\mathbf{X}_k\widehat{\boldsymbol{\theta}}_k^{(R)}\right),$$

$$\mathbf{B}_k = \beta\,\mathbf{B}_{k-1} + \mathbf{X}_k^\top\boldsymbol{\Lambda}_k\left(\widehat{\boldsymbol{\theta}}_k^{(R)}\right)\mathbf{X}_k.$$

Note that the update rules for both $\mathbf{B}_k$ and $\mathbf{u}_k$ involve simple rank-W operations. We refer to the resulting filter as the ℓ_1-regularized Point Process Filter of the First Order (ℓ_1-PPF$_1$).

Next, we will briefly describe how to characterize the statistical confidence bounds for the ℓ_1-PPF$_1$ estimates. Confidence bounds are crucial for interpreting the results of our analysis as they allow to test the validity of hypotheses. Although construction of confidence bounds for linear models in the absence of regularization is well understood and widely applied, regularized ML estimates are usually deemed as point estimates for which the construction of statistical confidence regions is not straightforward. A series of recent results in high-dimensional statistics (Javanmard and Montanari 2014; Van de Geer et al. 2014; Zhang and Zhang 2014) have addressed this issue by providing techniques to construct confidence intervals for ℓ_1-regularized ML estimates of GLMs. These approaches are based on a careful inspection of the *Karush-Kuhn-Tucker* (KKT) conditions for the regularized estimates. To this end, they provide a procedure to decompose the estimates into a bias term plus an asymptotically Gaussian term (referred to as "de-sparsifying" in Van de Geer et al. (2014)), which can be computed using a nodewise regression (Meinshausen and Bühlmann 2006) of the covariates.

In what follows, we give a brief description of how the methods of Van de Geer et al. (2014) apply to our setting, and refer the reader to Sheikhattar et al. (2016) for details. Following the techniques in Van de Geer et al. (2014), the estimate $\widehat{\boldsymbol{\theta}}_k$ as the maximizer of Eq. (5.10) can be decomposed as:

$$\widehat{\boldsymbol{\theta}}_k = \widehat{\boldsymbol{\Theta}}_k \mathbf{g}_k(\widehat{\boldsymbol{\theta}}_k) + \widehat{\mathbf{w}}_k, \tag{5.17}$$

where $\widehat{\boldsymbol{\Theta}}_k$ is an approximate inverse to the Hessian of $\mathcal{L}^{\beta}(\boldsymbol{\theta})$ evaluated at $\widehat{\boldsymbol{\theta}}_k$, $\mathbf{g}_k$ is the gradient of $\mathcal{L}^{\beta}(\boldsymbol{\theta})$ previously defined in Eq. (5.12), and $\widehat{\mathbf{w}}_k$ is an unbiased and asymptotically Gaussian random vector with a covariance matrix of $\mathsf{cov}(\widehat{\mathbf{w}}_k) = \widehat{\boldsymbol{\Theta}}_k \mathbf{G}_k(\widehat{\boldsymbol{\theta}}_k)\widehat{\boldsymbol{\Theta}}_k^{\top}$, with

$$\mathbf{G}_k(\widehat{\boldsymbol{\theta}}_k) := \sum_{i=1}^{k} \beta^{2(k-i)} \mathbf{X}_i^{\top} \boldsymbol{\varepsilon}_i(\widehat{\boldsymbol{\theta}}_k)\boldsymbol{\varepsilon}_i(\widehat{\boldsymbol{\theta}}_k)^{\top} \mathbf{X}_i. \tag{5.18}$$

The first term in Eq. (5.17) is a bias term which can be directly computed given $\widehat{\boldsymbol{\Theta}}_k$. Given $\mathsf{cov}(\widehat{\mathbf{w}}_k)$, statistical confidence bounds for the second term at desired levels can be constructed in a standard way. The main technical issue in the aforementioned procedure in our setting is the computation of $\widehat{\boldsymbol{\Theta}}_k$ in a recursive fashion. Since the rows of $\widehat{\boldsymbol{\Theta}}_k$ are computed using ℓ_1-regularized least squares, we use the SPARLS algorithm (Babadi et al. 2010) as an efficient method to carry out the computation in a recursive fashion.

5.4.4 *Application: Spectrotemporal Receptive Field Plasticity Analysis*

In Fritz et al. (2003), it is suggested that this rapid plasticity has a significant role in the functional processes underlying active listening. However, most of the widely used estimation techniques (e.g., normalized reverse correlation) provide static estimates of the receptive field with a temporal resolution of the order of minutes. Moreover, they do not systematically capture the inherent sparsity manifested in the receptive field characteristics.

We model the STRF as an $(I \times J)$-dimensional matrix, where I and J denote the number of time lags and frequency bands, respectively. By vectorizing this matrix, we obtain an $(M-1)$-dimensional vector $\boldsymbol{\omega}_k$ at window k, where $M = I \times J + 1$. Augmenting the baseline rate parameter μ_k, we can model the activity of the A1 neurons using the logistic CIF with a parameter $\boldsymbol{\theta}_k := [\mu_k, \boldsymbol{\omega}_k]^\top$. The stimulus vector at time t, $\mathbf{s}_t$ is given by the vectorized version of the spectrogram of the acoustic stimulus with J frequency bands and I lags. In order to capture the sparsity of the STRF in the time-frequency plane, we further represent $\boldsymbol{\omega}_k$ over a Gabor time-frequency dictionary consisting of Gaussian windows centered around a regular subset of the $I \times J$ time-frequency plane. That is, for $\boldsymbol{\omega}_k = \boldsymbol{\Phi}\boldsymbol{\xi}_k$, where $\boldsymbol{\Phi}$ is the dictionary matrix and $\boldsymbol{\xi}_k$ is the sparse representation of the STRF. Note that the resulting prior on $\boldsymbol{\xi}_k$ is a special case of the prior p_2 in Table 5.3. The estimation procedures of this paper can be applied to $\boldsymbol{\xi}_k$, by absorbing the dictionary matrix into the data matrix $\mathbf{X}_k$ at window k.

We apply the ℓ_1-PPF$_1$ filter to multi-unit spike recordings from the ferrets A1 during a series of passive listening conditions and active auditory task conditions (data from the Neural Systems Laboratory, University of Maryland, College Park). During each active task, ferrets attended to the temporal dynamics of the sounds, and discriminated the rate of acoustic clicks (Fritz et al. 2005). The STRFs were estimated from the passive condition, where the quiescent animal listened to a series of broadband noise-like acoustic stimuli known as *Temporally Orthogonal Ripple Combinations* (TORC). The experiment consisted of 2 active and 11 passive blocks. Within each passive block, 30 TORCs were randomly repeated a total of 4–5 times each. In our analysis, we pool the spiking data corresponding to the same repeated TORC within each block. Therefore, the time axis corresponds to the experiment time modulo repetitions within each block. We discretize the resulting duration of $\mathcal{T} = 990$ s to time bins of size $\Delta = 1$ ms, and segment data to windows of size $W = 10$ samples (10 ms). The STRF dimensions are 50×50, regularly spanning lags of 1–50 ms and frequency bands of 0.5–16 kHz (in logarithmic scale). The dictionary $\boldsymbol{\Phi}$ consists of 13×13 Gabor atoms, evenly spaced within the STRF domain. Each atom is a two-dimensional Gaussian kernel with a variance of $D^2/4$ per dimension, where D denotes the spacing between the atoms. We selected a forgetting factor of $\beta = 0.9998$, a step size of $\alpha = \frac{4(1-\beta)}{MW\bar{\sigma}^2}$, where $\bar{\sigma}^2$ is the average variance of the spectrogram components, $R = 1$ iteration per sample, and a regularization parameter of $\gamma = 40$ via twofold even-odd cross validation.

Figure 5.1a shows a schematic depiction the experimental setup. The sequence of passive (green) and active (red) tasks is shown in Fig. 5.1b. Figure 5.1c depicts five snapshots taken at $\{180, 360, 540, 630, 990\}$ s corresponding to the end-points

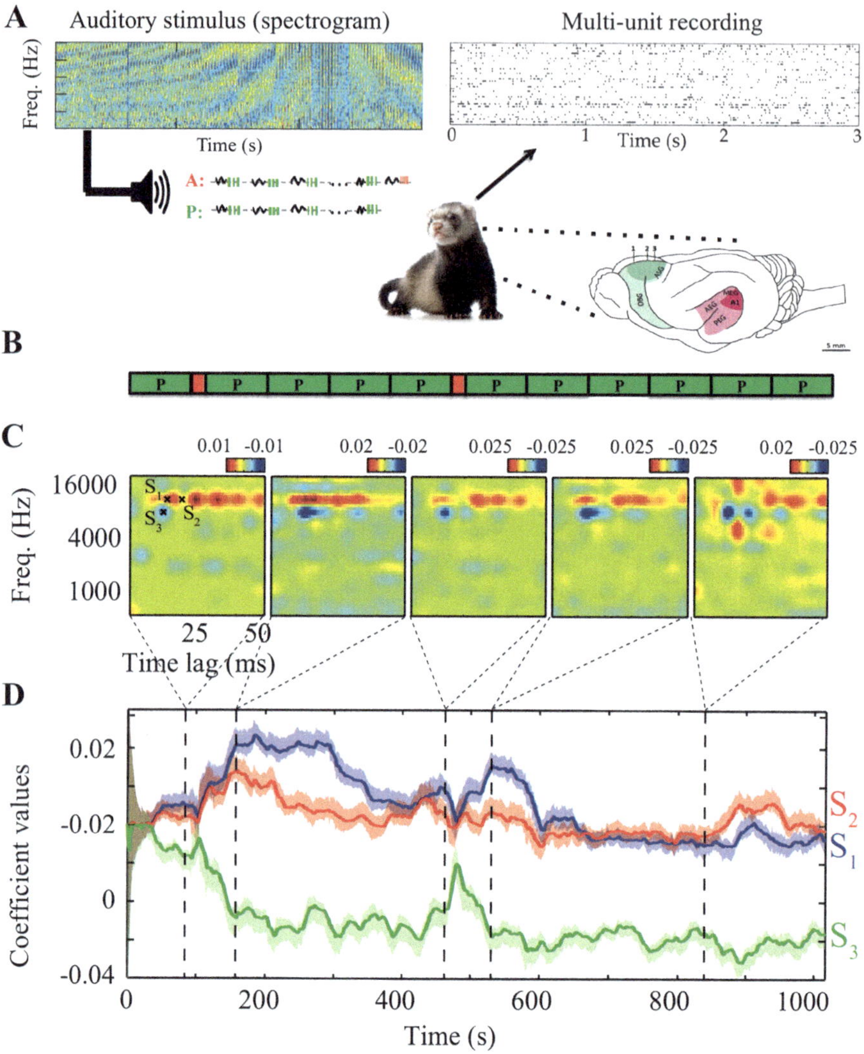

Fig. 5.1 Analysis of ferret STRF plasticity. (**a**) The response of ferret A1 neurons to TORC stimuli is captured by multi-unit recordings during a series of passive and active auditory tasks. (**b**) Snapshots of the STRF at five selected points in time, marked by the dashed vertical lines. (**c**) The time-course of three selected points (S_1, S_2, and S_3) in the STRF marked on the leftmost panel. The colored hulls show 95% confidence intervals. The ℓ_1-PPF$_1$ filter is capable of detecting rapid changes in the STRF, while capturing the sparsity of spectrotemporal tuning. Figure modified from Sheikhattar et al. (2016)

of the $\{2, 4, 6, 7, 11\}$th passive tasks. The bottom row shows the time-course of three selected points (marked as S_1, S_2, and S_3 in the leftmost panel) of the STRF during the experiment. The STRF snapshots at times 180 and 540 s correspond to 90 s after the two active tasks, respectively, and verify the sharpening effect of the excitatory region ($\sim$30 ms, 8 kHz) due to the animal's attentive behavior following the active task reported in Fritz et al. (2003). Moreover, the STRF snapshots at times 360 and 630 s reveal the weakening of the excitatory region long after the active task and returning to the pre-active state, highlighting the plasticity of A1 neurons. Previous studies have revealed the STRF dynamics with a resolution of the order of minutes (Mesgarani et al. 2010). The result in Fig. 5.1 provides a temporal resolution of the order of seconds, while capturing the STRF sparsity in a robust fashion (Sheikhattar et al. 2016).

5.5 Sparsity Meets Dynamics for Signal Deconvolution

In this section, we consider signal deconvolution from two-photon calcium imaging data as an inverse problem (Kazemipour et al. 2017). In many signal processing applications such as estimation of brain activity from MEG time-series (Phillips et al. 1997), estimation of time-varying networks (Kolar et al. 2010), EEG analysis (Nunez and Cutillo 1995), calcium imaging (Vogelstein et al. 2010), functional magnetic resonance imaging (fMRI) (Chang and Glover 2010), and video compression (Jung and Ye 2010), the signals often exhibit abrupt changes that are blurred through convolution with unknown kernels due to intrinsic measurement constraints. Extracting the underlying signals from blurred and noisy measurements is often referred to as signal deconvolution.

Traditionally, state-space models have been used for such signal deconvolution problems, where the states correspond to the unobservable signals. Gaussian state-space models in particular are widely used to model smooth state transitions. When applied to observations from abruptly changing states, however, Gaussian state-space models exhibit poor performance in recovering sharp transitions of the states due to their underlying smoothing property. Our goal is therefore to construct state-space models, along with fast and robust estimation algorithms, to capture abrupt state transitions in calcium imaging data arising from spiking activity, from noisy and undersampled observations. The approach taken here builds up on joint work of the author and his colleagues including Emery N. Brown (Ba et al. 2012), in modeling the sparsity of state innovations under the CS framework (Kazemipour et al. 2017).

5.5.1 *Problem Formulation*

Consider a linear state-space model given by

$$\mathbf{x}_t = \boldsymbol{\Theta}\mathbf{x}_{t-1} + \mathbf{w}_t, \qquad \mathbf{y}_t = \mathbf{A}_t\mathbf{x}_t + \mathbf{v}_t, \tag{5.19}$$

where $\{\mathbf{x}_t\}_{t=1}^T \in \mathbb{R}^M$ denote the sequence of unobservable states, $\boldsymbol{\Theta}$ is the state transition matrix satisfying $\|\boldsymbol{\Theta}\| < 1$, $\mathbf{w}_t \in \mathbb{R}^M$ is the state innovation sequence, $\{\mathbf{y}_t\}_{t=1}^T \in \mathbb{R}^{N_t}$ are the linear observations, $\mathbf{A}_t \in \mathbb{R}^{N_t \times M}$ denotes the measurement matrix, and $\mathbf{v}_t \in \mathbb{R}^{N_t}$ denotes the measurement noise. The main problem is to estimate the unobserved sequence $\{\mathbf{x}_t\}_{t=1}^T$ (and possibly $\boldsymbol{\Theta}$), given the sequence of observations $\{\mathbf{y}_t\}_{t=1}^T$. This problem is in general ill-posed, when $N_t < M$, for some t. We therefore need to make additional assumptions in order to seek a stable solution.

As in the previous section, we assume that the state innovations are compressible, i.e. $\mathbf{w}_t = \mathbf{x}_t - \boldsymbol{\Theta}\mathbf{x}_{t-1}$ is (L_t, ξ)-compressible with $L_1 \gg L_t$ for $t \in [T]\backslash\{1\}$. We thus denote the model of Eq. (5.19) by a *compressible* state-space model. We further assume that $1 \ll L_t < N_t \ll M$.

For simplicity of notation, we define $\mathbf{x}_0$ to be the all-zero vector in $\mathbb{R}^M$. For a matrix $\mathbf{A}$, we denote restriction of $\mathbf{A}$ to its first n rows by $(\mathbf{A})_n$. We say that the matrix $\mathbf{A} \in \mathbb{R}^{N \times M}$ satisfies the restricted isometry property (RIP) (Candès 2006) of order L, if for all L-sparse $\mathbf{x} \in \mathbb{R}^M$, we have

$$(1 - \delta_L)\|\mathbf{x}\|_2^2 \le \|\mathbf{A}\mathbf{x}\|_2^2 \le (1 + \delta_L)\|\mathbf{x}\|_2^2, \tag{5.20}$$

where $\delta_L \in (0, 1)$ is the smallest constant for which Eq. (5.20) holds (Candès and Wakin 2008). We assume that the rows of $\mathbf{A}_t$ are a subset of the rows of $\mathbf{A}_1$, i.e. $\mathbf{A}_t = (\mathbf{A}_1)_{N_t}$, and define $\widetilde{\mathbf{A}}_t = \sqrt{\frac{N_1}{N_t}}\mathbf{A}_t$. Other than its technical usefulness, the latter assumption helps avoid prohibitive storage of all the measurement matrices.

The inverse problem of this section can be stated as follows: *given under-determined and noisy observations* $\{\mathbf{y}_t\}_{t=1}^T$ *from an ensemble of neurons and a compressible signal evolution model given by Eq.* (5.19), *the goal is to estimate the underlying signal sequence* $\{\mathbf{x}_t\}_{t=1}^T$ *in a stable fashion.*

5.5.2 A MAP Formulation for Sparse Signal Deconvolution

In order to promote sparsity of the state innovations, we consider the dynamic ℓ_1-regularization problem defined as

$$\min_{\{\mathbf{x}_t\}_{t=1}^T, \boldsymbol{\Theta}} \sum_{t=1}^{T} \frac{\|\mathbf{x}_t - \boldsymbol{\Theta}\mathbf{x}_{t-1}\|_1}{\sqrt{L_t}} \quad \text{s.t.} \ \|\mathbf{y}_t - \mathbf{A}_t\mathbf{x}_t\|_2 \le \sqrt{\frac{N_t}{N_1}}\eta. \tag{5.21}$$

where η is an upper bound on the observation noise, i.e., $\|v_t\|_2 \le \eta$ for all t. Note that this problem is a variant of the dynamic CS problem introduced in Ba et al. (2012).

In order to cast this problem in the MAP framework as an inverse problem, we consider the modified Lagrangian form of Eq. (5.21) given by

$$\min_{\{\mathbf{x}_t\}_{t=1}^T, \boldsymbol{\Theta}} \quad \gamma \sum_{t=1}^{T} \frac{\|\mathbf{x}_t - \boldsymbol{\Theta}\mathbf{x}_{t-1}\|_1}{\sqrt{L_t}} + \frac{1}{N_t} \frac{\|\mathbf{y}_t - \mathbf{A}_t\mathbf{x}_t\|_2^2}{2\sigma^2}. \tag{5.22}$$

for some constants σ^2 and $\gamma \geq 0$. Note that if $\mathbf{v}_t \sim \mathcal{N}(\mathbf{0}, n_t\sigma^2\mathbf{I})$, then Eq. (5.22) is algebraically similar to the MAP estimator of the states in Eq. (5.19), assuming that the *state* innovations were independent Laplace random variables with respective parameters $\gamma/\sqrt{s_t}$. Note that this prior coincides with p_3 in Table 5.3. We will later use this analogy to derive fast solutions to the optimization problem in Eq. (5.22). The following theorem establishes the properties of the minimizer in Eq. (5.21):

Theorem 2 (Theorem 1 in Kazemipour et al. (2017)) *Let $\{\mathbf{x}_t\}_{t=1}^T \in \mathbb{R}^M$ be a sequence of states with a known transition matrix $\boldsymbol{\Theta} = \theta\mathbf{I}$, where $|\theta| < 1$ and $\widetilde{\mathbf{A}}_t$, $t \geq 1$ satisfies RIP of order $4L$ with $\delta_{4L} < 1/3$. Suppose that $N_1 > N_2 = N_3 = \cdots = N_T$. Then, the solution $\{\widehat{\mathbf{x}}_t\}_{t=1}^T$ to the dynamic CS problem (5.21) satisfies*

$$\frac{1}{T}\sum_{t=1}^T \|\mathbf{x}_t - \widehat{\mathbf{x}}_t\|_2 \leq \frac{1-\theta^T}{1-\theta}\left(12.6\left(1 + \frac{1}{T}\sqrt{\frac{N_1}{N_2}} - \frac{1}{T}\right)\eta + \frac{3}{T}\sum_{t=1}^T \frac{\sigma_{L_t}(\mathbf{x}_t - \boldsymbol{\Theta}\mathbf{x}_{t-1})}{\sqrt{L_t}}\right).$$

Proof The proof is given in Kazemipour et al. (2017).

The first term on the right-hand side of the statement of Theorem 2 implies that the average reconstruction error of the sequence $\{\mathbf{x}_t\}_{t=1}^T$ is upper bounded proportional to the noise level η, which implies the stability of the estimate. The second term is a measure of compressibility of the innovation sequence and vanishes when the sparsity condition is exactly met.

5.5.3 *Fast Iterative Solution via the EM Algorithm*

Due to the high dimensional nature of the state estimation problem, algorithms with polynomial complexity exhibit poor scalability. Moreover, when the state transition matrix is not known, the dynamic CS optimization problem (5.22) is not convex in $\left(\{\mathbf{x}_t\}_{t=1}^T, \boldsymbol{\Theta}\right)$. Therefore standard convex optimization solvers cannot be directly applied. This problem can be addressed by employing the expectation-maximization (EM) algorithm (Shumway and Stoffer 1982). A related existing result considers weighted ℓ_1-regularization to adaptively capture the state dynamics (Charles and Rozell 2013). Our approach is distinct in that we derive a fast solution to (5.22) via two nested EM algorithms, in order to jointly estimate the states and their transition matrix. The outer EM algorithm converts the estimation problem to a form suitable for the usage of the traditional fixed-interval smoothing (FIS) by invoking the EM interpretation of the iterative re-weighted least squares (IRLS) algorithms (Ba et al. 2012). The inner EM algorithm performs state and parameter estimation efficiently using the FIS. We refer to our estimated as the *fast compressible state-space* (FCSS) estimator.

In order to employ the EM theory, we first note that the problem of Eq. (5.22) can be interpreted as a MAP problem: the first term corresponds to the state-space prior $-\log p(\mathbf{x}_t|\mathbf{x}_{t-1}, \boldsymbol{\Theta}) = -\log p_{L_t}(\mathbf{x}_t - \boldsymbol{\Theta}\mathbf{x}_{t-1})$, where $p_{L_t}(\mathbf{x}) \sim \exp\left(-\gamma\|\mathbf{x}\|_1/\sqrt{L_t}\right)$ denoting the Laplace distribution; the second term is the negative log-likelihood of the data given the state, assuming a zero-mean Gaussian observation noise with covariance $\sigma^2\mathbf{I}$.

It is more convenient to work with the ϵ-perturbed ℓ_1-norm defined by

$$\|\mathbf{x}\|_{1,\epsilon} := \sqrt{x_1^2 + \epsilon^2} + \sqrt{x_2^2 + \epsilon^2} + \cdots + \sqrt{x_p^2 + \epsilon^2}. \tag{5.23}$$

Note that for $\epsilon = 0$, $\|\mathbf{x}\|_{1,\epsilon}$ coincides with the usual ℓ_1-norm. We define the ϵ-perturbed version of the dual problem (5.22) by

$$\min_{\{\mathbf{x}_t\}_{t=1}^T, \boldsymbol{\Theta}} \quad \gamma \sum_{t=1}^{T} \frac{\|\mathbf{x}_t - \boldsymbol{\Theta}\mathbf{x}_{t-1}\|_{1,\epsilon}}{\sqrt{L_t}} + \frac{1}{N_t} \frac{\|\mathbf{y}_t - \mathbf{A}_t\mathbf{x}_t\|_2^2}{2\sigma^2}. \tag{5.24}$$

As it will become evident shortly, this slight modification is carried out for the sake of numerical stability. The ϵ-perturbation only adds a term of the order $\mathcal{O}(\epsilon p)$ to the estimation error bound of Theorem 2, which is negligible for small enough ϵ (Ba et al. 2012).

If instead of the $\ell_{1,\epsilon}$-norm, we had the square ℓ_2 norm, then the above problem could be efficiently solved using the FIS. The outer EM algorithm transforms the problem of Eq. (5.24) into a quadratic form, by invoking the equivalence of the IRLS algorithm as an instance of the EM algorithm for solving $\ell_{1,\epsilon}$-minimization problems via the Normal/Independent (N/I) characterization of the ϵ-perturbed Laplace distribution (Ba et al. 2012). That is, given the estimates $\{\widehat{\mathbf{x}}_t^{(\ell)}\}_{t=1}^T, \widehat{\boldsymbol{\Theta}}^{(\ell)}$ at the end of the ℓ-th iteration, the outer EM algorithm transforms the optimization problem to:

$$\min_{\{\mathbf{x}_t\}_{t=1}^T, \boldsymbol{\Theta}} \quad \frac{\gamma}{2} \sum_{j=1}^{M} \sum_{t=1}^{T} \frac{(\mathbf{x}_t - \boldsymbol{\Theta}\mathbf{x}_{t-1})_j^2 + \epsilon^2}{\sqrt{L_t}\sqrt{\left(\widehat{\mathbf{x}}_t^{(\ell)} - \widehat{\boldsymbol{\Theta}}^{(\ell)}\widehat{\mathbf{x}}_{t-1}^{(\ell)}\right)_j^2 + \epsilon^2}} + \sum_{t=1}^{T} \frac{1}{N_t} \frac{\|\mathbf{y}_t - \mathbf{A}_t\mathbf{x}_t\|_2^2}{2\sigma^2}, \tag{5.25}$$

in order to find $\{\widehat{\mathbf{x}}_t^{(\ell+1)}\}_{t=1}^T$ and $\widehat{\boldsymbol{\Theta}}^{(\ell+1)}$. Under mild conditions, convergence of the solution of Eq. (5.25) to that of Eq. (5.22) was established in Ba et al. (2012). The objective function (5.25) is still not jointly convex in $\left(\{\mathbf{x}_t\}_{t=1}^T, \boldsymbol{\Theta}\right)$. Therefore, to carry out the optimization, i.e. the outer M-step, we will employ another instance of the EM algorithm, which we call the inner EM algorithm, to alternate between estimating of $\{\mathbf{x}_t\}_{t=1}^T$ and $\boldsymbol{\Theta}$. To this end, let $\mathbf{W}_t^{(\ell)}$ be a diagonal matrix such that

$$\left(\mathbf{W}_t^{(\ell)}\right)_{j,j} = L_t^{-1/2}\left\{\left(\widehat{\mathbf{x}}_t^{(\ell)} - \widehat{\boldsymbol{\Theta}}^{(\ell)}\widehat{\mathbf{x}}_{t-1}^{(\ell)}\right)_j^2 + \epsilon^2\right\}^{-1/2}.$$

Consider an estimate $\widehat{\boldsymbol{\Theta}}^{(\ell,m)}$, corresponding to the m-th iteration of the inner EM algorithm within the ℓ-th M-step of the outer EM. In this case, Eq. (5.25) can be thought of the MAP estimate of the Gaussian state-space model given by:

$$\begin{aligned}\mathbf{x}_t &= \widehat{\boldsymbol{\Theta}}^{(\ell,m)}\mathbf{x}_{t-1} + \mathbf{w}_t, \quad \mathbf{w}_t \sim \mathcal{N}\left(\mathbf{0}, \tfrac{1}{\gamma}\mathbf{W}_t^{(\ell)^{-1}}\right),\\ \mathbf{y}_t &= \mathbf{A}_t\mathbf{x}_t + \mathbf{v}_t, \quad \mathbf{v}_t \sim \mathcal{N}(\mathbf{0}, N_t\sigma^2\mathbf{I}).\end{aligned} \tag{5.26}$$

In order to obtain the inner E-step, one needs to find the density of $\{\mathbf{x}_t\}_{t=1}^T$ given $\{\mathbf{y}_t\}_{t=1}^T$ and $\widehat{\boldsymbol{\Theta}}^{(\ell,m)}$. Given the Gaussian nature of the state-space in Eq. (5.26), this density is a multivariate Gaussian density, whose means and covariances can be efficiently computed using the FIS. For all $t \in [T]$, the FIS performs a forward Kalman filter and a backward smoother to generate (Rauch et al. 1965; Anderson and Moore 1979):

$$\mathbf{x}_{t|T}^{(\ell,m+1)} := \mathbb{E}\left\{\mathbf{x}_t \middle| \{\mathbf{y}_t\}_{t=1}^T, \widehat{\boldsymbol{\Theta}}^{(\ell,m)}\right\}, \quad \boldsymbol{\Sigma}_{t|T}^{(\ell,m+1)} := \mathbb{E}\left\{\mathbf{x}_t\mathbf{x}_t^\top \middle| \{\mathbf{y}_t\}_{t=1}^T, \widehat{\boldsymbol{\Theta}}^{(\ell,m)}\right\}, \quad \text{and}$$

$$\boldsymbol{\Sigma}_{t-1,t|T}^{(\ell,m+1)} = \boldsymbol{\Sigma}_{t,t-1|T}^{(\ell,m+1)} = \mathbb{E}\left\{\mathbf{x}_{t-1}\mathbf{x}_t^\top \middle| \{\mathbf{y}_t\}_{t=1}^T, \widehat{\boldsymbol{\Theta}}^{(\ell,m)}\right\}.$$

Note that due to the quadratic nature of all the terms involving $\{\mathbf{x}_t\}_{t=1}^T$, the outputs of the FIS suffice to compute the expectation of the objective function in Eq. (5.25), i.e., the inner E-step, which results in:

$$\begin{aligned}\max_{\boldsymbol{\Theta}} \; &-\frac{\gamma}{2}\left(\boldsymbol{\Theta}\left(\sum_{t=1}^T \mathbf{W}_t^{(\ell)}\left(\mathbf{x}_{t-1|T}^{(\ell,m+1)}\mathbf{x}_{t-1|T}^{(\ell,m+1)\top} + \boldsymbol{\Sigma}_{t-1|T}^{(\ell,m+1)}\right)\right)\boldsymbol{\Theta}^\top\right)\\ &+\frac{\gamma}{2}\operatorname{Tr}\left(\boldsymbol{\Theta}\left(\sum_{t=1}^T \mathbf{W}_t^{(\ell)}\left(\mathbf{x}_{t-1|T}^{(\ell,m+1)}\mathbf{x}_{t|T}^{(\ell,m+1)\top} + \mathbf{x}_{t|T}^{(\ell,m+1)}\mathbf{x}_{t-1|T}^{(\ell,m+1)\top} + 2\boldsymbol{\Sigma}_{t-1,t|T}^{(\ell,m+1)}\right)\right)\right),\end{aligned} \tag{5.27}$$

to obtain $\widehat{\boldsymbol{\Theta}}^{(\ell,m+1)}$. The solution has a closed-form given by:

$$\begin{aligned}\widehat{\boldsymbol{\Theta}}^{(\ell,m+1)} = &\left(\sum_{t=1}^T 2\mathbf{W}_t^{(\ell)}\left(\mathbf{x}_{t-1|T}^{(\ell,m+1)}\mathbf{x}_{t-1|T}^{(\ell,m+1)\top} + \boldsymbol{\Sigma}_{t-1|T}^{(\ell,m+1)}\right)\right)^{-1}\\ &\left(\sum_{t=1}^T \mathbf{W}_t^{(\ell)}\left(\mathbf{x}_{t-1|T}^{(\ell,m+1)}\mathbf{x}_{t|T}^{(\ell,m+1)\top} + \mathbf{x}_{t|T}^{(\ell,m+1)}\mathbf{x}_{t-1|T}^{(\ell,m+1)\top} + 2\boldsymbol{\Sigma}_{t-1,t|T}^{(\ell,m+1)}\right)\right).\end{aligned} \tag{5.28}$$

This process is repeated for I_1 iterations for the inner EM and I_2 iterations for the outer EM, until a convergence criterion is met.

5.5.4 *Application: Spike Deconvolution from Two-Photon Calcium Imaging Data*

Calcium imaging takes advantage of intracellular calcium flux to directly visualize calcium signaling in living neurons. This is done by using calcium indicators, which are fluorescent molecules that can respond to the binding of calcium ions by changing their fluorescence properties and using a fluorescence or two-photon microscope and a CCD camera to capture the visual patterns (Smetters et al. 1999; Stosiek et al. 2003). Since spikes are believed to be the units of neuronal computation, inferring spiking activity from calcium recordings, referred to as calcium deconvolution, is an important problem in neural data analysis. Several approaches to calcium deconvolution have been proposed in the neuroscience literature, including model-free approaches such as sequential Monte Carlo methods (Vogelstein et al. 2009) and model-based approaches such as nonnegative deconvolution methods (Vogelstein et al. 2010; Pnevmatikakis et al. 2016). These approaches require solving convex optimization problems, which do not scale well with the temporal dimension of the data. In addition, they lack theoretical performance guarantees and do not provide clear measures for assessing the statistical significance of the detected spikes.

In order to construct confidence bounds for our estimates, we employ recent results from high-dimensional statistics (Van de Geer et al. 2014). We first compute the confidence intervals around the outputs of the FCSS estimates using the node-wise regression procedure of Van de Geer et al. (2014), at a confidence level of $1-\frac{\alpha}{2}$. We perform the node-wise regression separately for each time t. For an estimate $\widehat{\mathbf{x}}_t$, we obtain $\widehat{\mathbf{x}}_t^{\mathsf{u}}$ and $\widehat{\mathbf{x}}_t^{\mathsf{l}}$ as the upper and lower confidence bounds, respectively. Next, we partition the estimates into small segments, starting with a local minimum (trough) and ending in a local maximum (peak). For the i-th component of the estimate, let t_{min} and t_{max} denote the time index corresponding to two such consecutive troughs and peaks. If the difference $(\widehat{\mathbf{x}}_{t_{\mathsf{max}}}^{\mathsf{l}})_i - (\widehat{\mathbf{x}}_{t_{\mathsf{min}}}^{\mathsf{u}})_i$ is positive, the detected innovation component is declared significant (i.e., spike) at a confidence level of $1-\alpha$, otherwise it is discarded (i.e., no spike). We refer to this procedure as Pruned-FCSS (PFCSS).

We apply the FCSS algorithm for calcium deconvolution in a scenario where the ground-truth spiking is recorded *in vitro* through simultaneous electrophysiology (cell-attached patch clamp) and two-photon calcium imaging (See Kazemipour et al. (2017) for experimental procedures). The calcium trace and the ground-truth spikes are shown for a sample neuron in Fig. 5.2a. The FCSS denoised estimate of the states (black) and the detected spikes (blue) using 95% confidence intervals (orange hulls) and the corresponding quantities for the constrained f-oopsi algorithm (Pnevmatikakis et al. 2016) are shown in Fig. 5.2b and c, respectively.

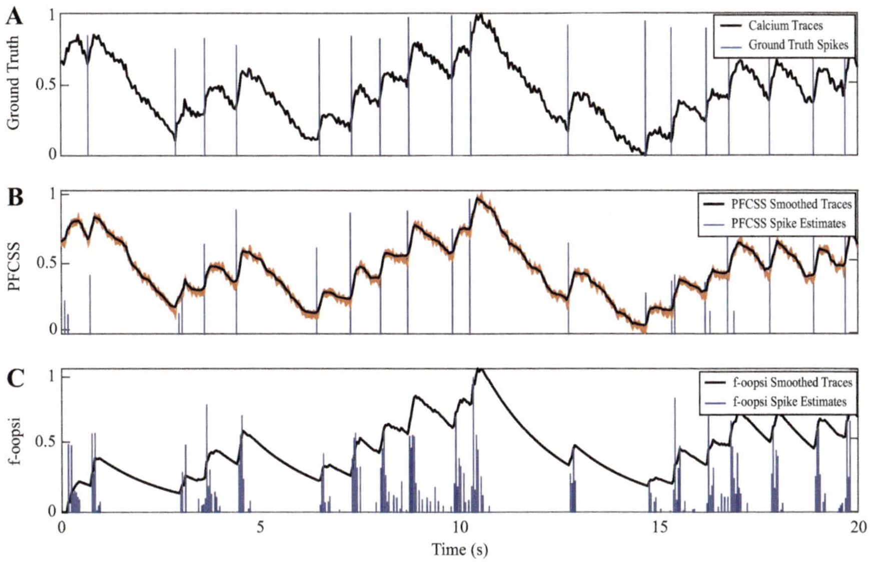

Fig. 5.2 Ground-truth performance comparison between PFCSS and constrained f-oopsi. (**a**) the observed calcium traces (black) and ground-truth electrophysiology data (blue). (**b**) PFCSS state estimates (black) with 95% confidence intervals (orange) and the detected spikes (blue). (**c**) The constrained f-oopsi state estimates (black) and the detected spikes (blue). The FCSS spike estimates closely match the ground-truth spikes with only a few false detections, while the constrained f-oopsi estimates contain significant clustered false detections. Figure modified from Kazemipour et al. (2017)

Both algorithms detect the large dynamic changes in the data, corresponding to the spikes, which can also be visually captured in this case. However, in doing so, the f-oopsi algorithm incurs a high rate of false positive errors, manifested as clustered spikes around the ground truth events. Similar to f-oopsi, most state-of-the-art model-based methods suffer from high false positive rate, which makes the inferred spike estimates unreliable. Thanks to the aforementioned pruning process based on the confidence bounds, the PFCSS is capable of rejecting the insignificant innovations, and hence achieve a lower false positive rate. One factor responsible for this performance gap can be attributed to the underestimation of the calcium decay rate in the transition matrix estimation step of f-oopsi. However, we believe the performance gain achieved by FCSS is mainly due to the explicit modeling of the sparse nature of the spiking activity by going beyond the Gaussian state-space modeling paradigm.

5.6 Sparsity Meets Dynamics for Spectral Decomposition

In this section, we consider spectral decomposition of time series observed in noise as an inverse problem and thereby cast it as a MAP problem. Across nearly all fields of science and engineering, non-stationary behavior in time series data is a ubiquitous phenomenon. Common examples include speech (Quatieri 2008), image and video (Lim 1990) signals; neural spike trains (Truccolo et al. 2005) and EEG (Mitra and Bokil 2007) measurements; seismic and oceanographic recordings (Emery and Thomson 2001) and radar emissions (Haykin and Steinhardt 1992). Due to the exploratory nature of these applications and the complexity of the underlying spectrotemporal features, nonparametric spectral techniques, rather than parametric approaches (Kitagawa and Gersch 1996), are among the most widely used in the analysis of these data.

Nonparametric spectral techniques based on Fourier methods (Thomson 1982; Percival 1993; Thomson and Vernon 1998), wavelets (Daubechies 1990; Daubechies et al. 2011), and data-dependent approaches, such as the empirical mode decomposition (EMD) (Huang et al. 1998; Wu and Huang 2009), use sliding windows to take account of the non-stationarity. Although analysis with sliding windows is universally accepted, this approach has several drawbacks including low spectral resolution due to short window lengths and lack of a mechanism to integrate data from adjacent windows in order to capture the inherent temporal smoothness as well as spectral sparsity of these data. Our goal is therefore to capture the spectrotemporal dynamics of noisy time series whose non-stationary mean is the superposition of a small number of smooth harmonic components. This section is based on a joint work of the author and his colleagues, including Emery N. Brown (Ba et al. 2014b).

5.6.1 Problem Formulation

Consider a discrete-time signal $y_t, t = 1, 2, \ldots, T$ obtained by sampling of an underlying noisy continuous-time signal at a rate F_s above the Nyquist rate. Given an arbitrary window of length W, let $\mathbf{y}_k := [y_{(k-1)W+1}, y_{(k-1)W+2}, \ldots, y_{kW}]^\top$ for $k = 1, 2, \ldots, K$ with $K := \frac{T}{W}$ being an integer without loss of generality. For some integer M, consider the following harmonic representation of $\mathbf{y}_k$ as

$$\mathbf{y}_k = \mathbf{F}_k \mathbf{x}_k + \mathbf{v}_k \tag{5.29}$$

where $(\mathbf{F}_k)_{l,m} := \cos\left(2\pi\left((k-1)W + l\right)\frac{m}{M}\right)$ and $(\mathbf{F}_k)_{l,m+M/2} := \sin\left(2\pi\left((k-1)W + l\right)\frac{m+M/2}{M}\right)$ for $l = 1, 2, \ldots, W$ and $m = 0, 1, \ldots, \frac{M}{2} - 1$, $\mathbf{x}_k \in \mathbb{R}^M$ is the vector of harmonic coefficients and $\mathbf{v}_k$ is independent, identically distributed, additive zero-mean Gaussian noise. By defining $\mathbf{F}$ as a $T \times MK$ block-diagonal matrix with $\mathbf{F}_k$ on the diagonal blocks:

$$\mathbf{F} := \begin{pmatrix} \mathbf{F}_1 & & & \\ & \mathbf{F}_2 & & \\ & & \ddots & \\ & & & \mathbf{F}_K \end{pmatrix}, \tag{5.30}$$

we may rewrite Eq. (5.29) in a compact form as

$$\mathbf{y} = \mathbf{F}\mathbf{x} + \mathbf{v}, \tag{5.31}$$

with new notations $\mathbf{y} = [y_1, y_2, \ldots, y_T]^\top \in \mathbb{R}^T$, $\mathbf{x} = [\mathbf{x}_1^\top, \mathbf{x}_2^\top, \ldots, \mathbf{x}_K^\top]^\top \in \mathbb{R}^{MK}$, and $\mathbf{v} = [\mathbf{v}_1^\top, \mathbf{v}_2^\top, \ldots, \mathbf{v}_K^\top]^\top \in \mathbb{R}^T$. The vector $\mathbf{x}$ can be viewed as a time-frequency representation of the non-stationary signal $\mathbf{y}$.

Our goal is to compute an estimate $\widehat{\mathbf{x}}$ of $\mathbf{x}$ given the data $\mathbf{y}$. Classical spectral estimation techniques use sliding windows with overlap to *implicitly* enforce temporal smoothness of the harmonic components, but they do not consider sparsity in the frequency domain. In contrast, we take a direct approach which treats $\{\mathbf{x}_k\}_{k=1}^K$ as a sequence of random variables and *explicitly* imposes a stochastic continuity constraint on its elements across time, as well as a sparsity constraint across frequency.

To this end, starting with an initial condition $\mathbf{x}_0 = \mathbf{0}$, we can express the stochastic continuity constraint in the form of the first-order difference equation

$$\mathbf{x}_k = \mathbf{x}_{k-1} + \mathbf{w}_k, \tag{5.32}$$

where $\mathbf{w}_k$ is a random innovation vector. To promote a desired spectrotemporal structure, we consider an ϵ-perturbed variant of the prior p_1 from Table 5.3:

$$\log p_{\mathsf{inv}}(\mathbf{w}_1, \mathbf{w}_2, \ldots, \mathbf{w}_K) \propto -\gamma \sum_{m=1}^{M} \sqrt{\sum_{k=1}^{K} \sqrt{(\mathbf{w}_k)_m^2 + \epsilon^2}}, \tag{5.33}$$

where $\gamma > 0$ and $\epsilon > 0$ is a small constant.

The inverse problem of this section can be stated as follows: *given the noisy time-series* $\{y_t\}_{t=1}^T$ *and an evolution model of the underlying harmonic components given by Eq.* (5.32), *the goal is to estimate the underlying harmonic sequence* $\{\mathbf{x}_t\}_{t=1}^T$ *in a robust and scalable fashion.*

5.6.2 A MAP Formulation for Spectral Decomposition

Given the observation model of Eq. (5.29) and the prior model (5.33), we can find $\widehat{\mathbf{x}}$ by solving the following MAP problem:

$$\max_{\{\mathbf{x}_k\}_{k=1}^K} -\sum_{k=1}^K \frac{1}{2\sigma^2}\|\mathbf{y}_k - \mathbf{F}_k\mathbf{x}_k\|_2^2 - \gamma \sum_{m=1}^M \sqrt{\sum_{k=1}^K \sqrt{((\mathbf{x}_k)_m - (\mathbf{x}_{k-1})_m)^2 + \epsilon^2}}. \tag{5.34}$$

We call the MAP estimation problem of Eq. (5.34) the *spectrotemporal pursuit* (SP) problem, and its solution the SP estimate. Note that we can absorb the constant σ in γ and henceforth assume that $\sigma = 1$. Similar to the previous sections, we seek solutions which are iterative and therefore scale favorably with growing data dimensions.

Using the equivalence of IRLS and EM algorithms for N/I densities (Ba et al. 2014a,b), an iterative algorithm can be obtained as follows. Suppose that at ℓ-th iteration an estimate $\widehat{\mathbf{x}}^{(\ell)}$ is given. Then, we solve:

$$\max_{\{\mathbf{x}_k\}_{k=1}^K} -\sum_{k=1}^K \frac{1}{2\sigma^2}\|\mathbf{y}_k - \mathbf{F}_k\mathbf{x}_k\|_2^2 - \sum_{m=1}^M\sum_{k=1}^K \frac{((\mathbf{x}_k)_m - (\mathbf{x}_{k-1})_m)^2}{2\left(\mathbf{Q}_k^{(\ell)}\right)_{m,m}}. \tag{5.35}$$

to find the estimate at iteration $(\ell + 1)$, where $\mathbf{Q}_k^{(\ell)}$ is an $M \times M$ diagonal matrix given by:

$$\left(\mathbf{Q}_k^{(\ell)}\right)_{m,m} = \frac{2\sqrt{\left(\left(\widehat{\mathbf{x}}_k^{(\ell)}\right)_m - \left(\widehat{\mathbf{x}}_{k-1}^{(\ell)}\right)_m\right)^2 + \epsilon^2}\sqrt{\sum_{k'=1}^K \sqrt{\left(\left(\widehat{\mathbf{x}}_{k'}^{(\ell)}\right)_m - \left(\widehat{\mathbf{x}}_{k'-1}^{(\ell-1)}\right)_m\right)^2 + \epsilon^2}}}{\gamma}. \tag{5.36}$$

The solution to Eq. (5.35) is given by the FIS (Rauch et al. 1965), which exploits tridiagonal structure of the quadratic cost function to obtain a recursive solution via forward-backward substitution. In Ba et al. (2014b), it has been shown that as $\ell \to \infty$, the solution to Eq. (5.35) converges to a fixed point of Eq. (5.34). We repeat the iterative process for a total of I iterations or until some convergence criterion is met.

5.6.3 Application: Robust Spectrotemporal Decomposition of EEG

We illustrate the application of SP by computing the spectrogram of frontal EEG data recorded from a patient during propofol-induced general anesthesia for a surgical procedure (Ba et al. 2014b). The patient received a bolus intravenous injection of propofol at approximately 3.5 min, followed by a propofol infusion which was maintained until minute 27, when the case ended.

When administered to initiate general anesthesia, propofol produces profound slow (<1 Hz) and delta (1–4 Hz) oscillations (Fig. 5.3, minute 5) (Purdon et al. 2013; Lewis et al. 2012). With maintenance of general anesthesia, using propofol we observe an alpha oscillation (8–12 Hz) in addition to the slow and delta oscillations. The presence of the alpha oscillations along with the slow and delta oscillations is a marker of unconsciousness (Purdon et al. 2013; Lewis et al. 2012). Developing a precise characterization of the spectrotemporal dynamics of neural activity under propofol general anesthesia is important in understanding the neural circuit mechanisms of this anesthetic.

We computed the spectrogram for $T = 35$ min of EEG data, sampled at a rate $F_s = 250$ Hz, using the multitaper method (Thomson 1982; Babadi and Brown 2014) with 1 s temporal resolution (Fig. 5.3a), multitaper method with 0.5 Hz frequency resolution (Fig. 5.3b) and the SP estimator (Fig. 5.3c). The right panels show a zoomed-in views of the spectrogram from minute 15 to minute 18. For the SP analysis, $W = M = 500$, $K = 1050$ and we select γ by splitting the data into two sequences consisting of its even and odd times, respectively, and performing a form of twofold cross validation (Friedman et al. 2007). For each 2 s window of

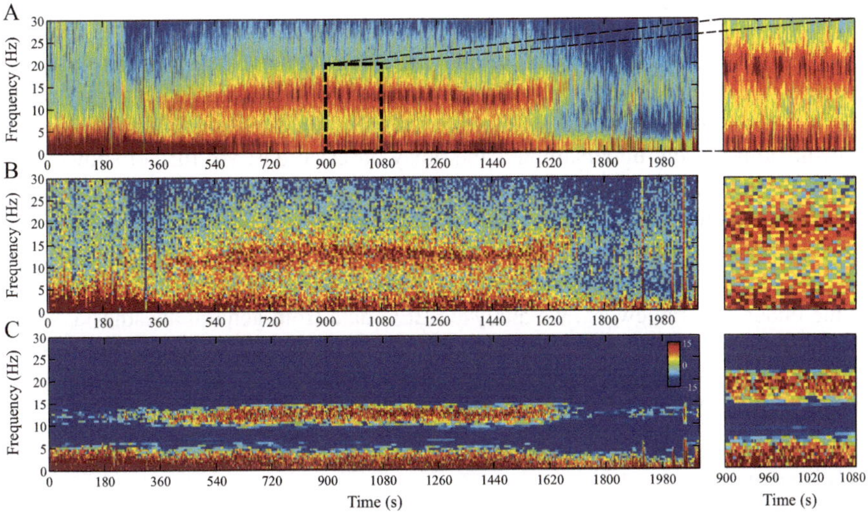

Fig. 5.3 Spectral decomposition of frontal EEG from a subject undergoing propofol-induced general anesthesia. (**a**) Multitaper with 2 s temporal resolution, (**b**) multitaper with 0.5 Hz frequency resolution, and (**c**) SP estimate. The right panels show the respective zoomed-in view from $t = 15$ min to $t = 18$ min. The color scale is in dB. The SP estimate significantly denoises the spectrogram, and captures the spectrotemporal dynamics at high resolution. Figure modified from Ba et al. (2014b)

data, $\mathbf{F}_k$ is the 500×500 matrix which is the Fourier basis for the discrete-time interval $[(k-1)W+1, kW]$ for $k = 1, 2, \ldots, K$.

By the choice of the window length and time-bandwidth product in the multitaper method (Babadi and Brown 2014), it is possible to achieve either high frequency or high temporal resolution. In contrast, SP achieves high temporal resolution, high spatial resolution, and significantly denoises the spectrogram. As a consequence, in the SP analysis the slow and delta oscillations are clearly delineated during the induction of anesthesia (minute 3.5), whereas during the maintenance period (minutes 5–27), the oscillations are strongly localized in the slow, delta and alpha bands. Furthermore, the denoising achieved by SP creates a $\sim$30 dB contrast between these spectral bands and the other frequencies in the spectrum.

5.7 Concluding Remarks

In this chapter, we considered neural identification and inverse problems cast in a Bayesian MAP estimation framework. We exploited two salient features of neural data, namely dynamicity and sparsity, to construct biophysically inspired forward models and priors. We further showed that it is possible to design inference algorithms for solving these problems in a scalable and provably robust fashion.

As for a case study for neural identification under this framework, we analyzed the STRF plasticity of neurons in the ferret primary auditory cortex. Our theoretical analysis as well as application to real data revealed substantial gains in terms of increasing the temporal resolution and capturing the sparsity in spectrotemporal tuning. We also considered two inverse neural problems under this framework. First, we employed state-space models with compressible innovations for signal deconvolution from undersampled and noisy observations. We further showed that it is possible to achieve robust and scalable spike deconvolution from two-photon calcium imaging of ensemble neuronal activity using these models. Second, we considered the problem of spectral decomposition of noisy non-stationary data as an inverse problem. By invoking the sparsity in frequency and smoothness in time under our framework, we analyzed EEG data from general anesthesia, which highlighted the utility of our techniques in delineating spectrotemporal features of EEG at high resolution.

References

Anderson, B., & Moore, J. B. (1979). *Optimal filtering*. Englewood Cliffs: Prentice-Hall.

Ba, D., Babadi, B., Purdon, P., & Brown, E. (2012). Exact and stable recovery of sequences of signals with sparse increments via differential ℓ_1-minimization. In *Proceedings of the 2012 Advances in Neural Information Processing Systems (NIPS)*, 3–8 December 2012, Lake Tahoe, NV (pp. 2627–2635).

Ba, D., Babadi, B., Purdon, P. L., & Brown, E. N. (2014a). Convergence and stability of iteratively re-weighted least squares algorithms. *IEEE Transactions on Signal Processing, 62*(1–4), 183–195.

Ba, D., Babadi, B., Purdon, P. L., & Brown, E. N. (2014b). Robust spectrotemporal decomposition by iteratively reweighted least squares. *Proceedings of the National Academy of Sciences USA, 111*(50), E5336–E5345.

Babadi, B., & Brown, E. N. (2014). A review of multitaper spectral analysis. *IEEE Transactions on Biomedical Engineering, 61*(5), 1555–1564.

Babadi, B., Kalouptsidis, N., & Tarokh, V. (2010). SPARLS: The sparse RLS algorithm. *IEEE Transactions on Signal Processing, 58*(8), 4013–4025.

Babadi, B., Obregon-Henao, G., Lamus, C., Hämäläinen, M. S., Brown, E. N., & Purdon, P. L. (2014). A subspace pursuit-based iterative greedy hierarchical solution to the neuromagnetic inverse problem. *Neuroimage, 87*, 427–443.

Brown, E. N., Barbieri, R., Ventura, V., Kass, R. E., & Frank, L. M. (2002). The time-rescaling theorem and its application to neural spike train data analysis. *Neural Computation, 14*(2), 325–346.

Brown, E. N., Frank, L. M., Tang, D., Quirk, M. C., & Wilson, M. A. (1998). A statistical paradigm for neural spike train decoding applied to position prediction from ensemble firing patterns of rat hippocampal place cells. *Journal of Neuroscience, 18*(18), 7411–7425.

Brown, E. N., Kass, R. E., & Mitra, P. P. (2004). Multiple neural spike train data analysis: State-of-the-art and future challenges. *Nature Neuroscience, 7*(5), 456–461.

Brown, E. N., Nguyen, D. P., Frank, L. M., Wilson, M. A., & Solo, V. (2001). An analysis of neural receptive field plasticity by point process adaptive filtering. *Proceedings of the National Academy of Sciences, 98*(21), 12261–12266.

Buzsaki, G. (2006). *Rhythms of the brain*. New York: Oxford University Press.

Candès, E. J. (2006). Compressive sampling. In *Proceedings of the International Congress of Mathematicians, Madrid, August 22–30* (pp. 1433–1452).

Candès, E. J., & Wakin, M. B. (2008). An introduction to compressive sampling. *IEEE Signal Processing Magazine, 25*(2), 21–30.

Chang, C., & Glover, G. H. (2010). Time-frequency dynamics of resting-state brain connectivity measured with fMRI. *Neuroimage, 50*(1), 81–98.

Charles, A. S., & Rozell, C. J. (2013). Dynamic filtering of sparse signals using reweighted ℓ_1. In *Proceedings of IEEE ICASSP* (pp. 6451–6455).

Chen, Z., Putrino, D. F., Ghosh, S., Barbieri, R., & Brown, E. N. (2011). Statistical inference for assessing functional connectivity of neuronal ensembles with sparse spiking data. *IEEE Transactions on Neural Systems and Rehabilitation Engineering, 19*(2), 121–135.

Daley, D., & Vere-Jones, D. (2007). *An introduction to the theory of point processes: Volume II: General theory and structure*. Berlin: Springer Science & Business Media.

Daubechies, I. (1990). The wavelet transform, time-frequency localization and signal analysis. *IEEE Transactions on Information Theory, 36*, 961–1005.

Daubechies, I., Lu, J., & Wu, H.-T. (2011). Synchrosqueezed wavelet transforms: An empirical mode decomposition-like tool. *Applied and Computational Harmonic Analysis, 30*(2), 243–261.

Daunizeau, J., & Friston, K. J. (2007). A mesostate-space model for EEG and MEG. *Neuroimage, 38*(1), 67–81.

Dayan, P., & Abbott, L. F. (2001). *Theoretical neuroscience*, Cambridge, MA: MIT Press.

Dempster, A. P., Laird, N. M., & Rubin, D. B. (1980). Iteratively reweighted least squares for linear regression when errors are normal/independent distributed. In *Multivariate analysis* (pp. 35–57). New York: North-Holland.

Depireux, D. A., Simon, J. Z., Klein, D. J., & Shamma, S. A. (2001). Spectro-temporal response field characterization with dynamic ripples in ferret primary auditory cortex. *Journal of Neurophysiology, 85*(3), 1220–1234.

Eden, U., Frank, L., Barbieri, R., Solo, V., & Brown, E. (2004). Dynamic analysis of neural encoding by point process adaptive filtering. *Neural Computation, 16*(5), 971–998.

Emery, W. J., & Thomson, R. E. (2001). *Data analysis methods in physical oceanography*. New York: Elsevier Science.

Frank, L. M., Stanley, G. B., & Brown, E. N. (2004). Hippocampal plasticity across multiple days of exposure to novel environments. *Journal of Neuroscience, 24*(35), 7681–7689.
Friedman, J., Hastie, T., Höfling, H., & Tibshirani, R. (2007). Pathwise coordinate optimization. *Annals of Applied Statistics, 1*(2), 302–332.
Friston, K., Harrison, L., Daunizeau, J., Kiebel, S., Phillips, C., Trujillo-Barreto, N., et al. (2008). Multiple sparse priors for the M/EEG inverse problem. *NeuroImage, 39*(3), 1104–1120.
Fritz, J., Elhilali, M., & Shamma, S. (2005). Active listening: Task-dependent plasticity of spectrotemporal receptive fields in primary auditory cortex. *Hearing Research, 206*(1), 159–176.
Fritz, J., Shamma, S., Elhilali, M., & Klein, D. (2003). Rapid task-related plasticity of spectrotemporal receptive fields in primary auditory cortex. *Nature Neuroscience, 6*(11), 1216–1223.
Gelman, A., Carlin, J. B., Stern, H. S., Dunson, D. B., Vehtari, A., & Rubin, D. B. (2013). *Bayesian data analysis*. Boca Raton: CRC press.
Gramfort, A., Kowalski, M., & Hämäläinen, M. (2012). Mixed-norm estimates for the M/EEG inverse problem using accelerated gradient methods. *Physics in Medicine and Biology, 57*(7), 1937.
Hämäläinen, M. S., Hari, R., Ilmoniemi, R. J., Knuutila, J., & Lounasmaa, O. V. (1993). Magnetoencephalography—Theory, instrumentation, and applications to noninvasive studies of the working human brain. *Reviews of Modern Physics, 65*(2), 413–497.
Hämäläinen, M. S., & Ilmoniemi, R. J. (1994). Interpreting magnetic fields of the brain: Minimum norm estimates. *Medical & Biological Engineering & Computing, 32*(1), 35–42.
Hämäläinen, M. S., & Sarvas, J. (1989). Realistic conductivity geometry model of the human head for interpretation of neuromagnetic data. *IEEE Transactions on Biomedical Engineering, 36*(2), 165–171.
Hastie, T., Tibshirani, R., & Friedman, J. (2009). *The elements of statistical learning*. New York: Springer.
Haykin, S., & Steinhardt, A. O. (1992). *Adaptive radar detection and estimation* (Vol. 11). Hoboken: Wiley-Interscience.
Hochberg, L. R., Serruya, M. D., Friehs, G. M., Mukand, J. A., Saleh, M., Caplan, A. H., et al. (2006). Neuronal ensemble control of prosthetic devices by a human with tetraplegia. *Nature, 442*(7099), 164–171.
Huang, N. E., Shen, Z., Long, S. R., Wu, M. C., Shih, H. H., Zheng, Q., et al. (1998). The empirical mode decomposition and the Hilbert spectrum for nonlinear and non-stationary time series analysis. *Proceedings of the Royal Society of London. Series A: Mathematical, Physical and Engineering Sciences, 454*(1971), 903–995.
Huang, Y., Brandon, M. P., Griffin, A. L., Hasselmo, M. E., & Eden, U. T. (2009). Decoding movement trajectories through a T-maze using point process filters applied to place field data from rat hippocampal region CA1. *Neural Computation, 21*(12), 3305–3334.
Huber, P. J. (2011). *Robust statistics*. New York: Springer.
Javanmard, A., & Montanari, A. (2014). Confidence intervals and hypothesis testing for high-dimensional regression. *Journal of Machine Learning Research, 15*(1), 2869–2909.
Jung, H., & Ye, J. C. (2010). Motion estimated and compensated compressed sensing dynamic magnetic resonance imaging: What we can learn from video compression techniques. *International Journal of Imaging Systems and Technology, 20*(2), 81–98.
Kass, R. E., Kelly, R. C., & Loh, W.-L. (2011). Assessment of synchrony in multiple neural spike trains using loglinear point process models. *Annals of Applied Statistics, 5*(2B), 1262–1292.
Kay, K. N., Naselaris, T., Prenger, R. J., & Gallant, J. L. (2008). Identifying natural images from human brain activity. *Nature, 452*(7185), 352–355.
Kazemipour, A., Liu, J., Solarana, K., Nagode, D., Kanold, P., Wu, M., et al. (2017, December 15). Fast and stable signal deconvolution via compressible state-space models. *IEEE Transactions on Biomedical Engineering* (in press). https://doi.org/10.1109/TBME.2017.2694339
Kim, S., Putrino, D., Ghosh, S., & Brown, E. N. (2011). A granger causality measure for point process models of ensemble neural spiking activity. *PLoS Computational Biology, 7*(3), e1001110.

Kitagawa, G., & Gersch, W. (1996). *Smoothness priors analysis of time series* (Vol. 116). New York: Springer.
Kolar, M., Song, L., Ahmed, A., & Xing, E. P. (2010). Estimating time-varying networks. *Annals of Applied Statistics, 4*(1), 94–123.
Lamus, C., Hämäläinen, M. S., Temereanca, S., Brown, E. N., & Purdon, P. L. (2012). A spatiotemporal dynamic distributed solution to the MEG inverse problem. *NeuroImage, 63*(2), 894–909.
Lange, K., & Sinsheimer, J. S. (1993). Normal/independent distributions and their applications in robust regression. *Journal of Computational and Graphical Statistics, 2*(2), 175–198.
Lewis, L., Weiner, V., Mukamel, E., Donoghue, J., Eskandar, E., Madsen, J., et al. (2012). Rapid fragmentation of neuronal networks at the onset of propofol-induced unconsciousness. *Proceedings of the National Academy of Sciences, 109*(49), E3377–E3386.
Lim, J. S. (1990). *Two-dimensional signal and image processing*. Englewood Cliffs, NJ: Prentice Hall.
Marin, G., Guerin, C., Baillet, S., Garnero, L., & Meunier, G. (1998). Influence of skull anisotropy for the forward and inverse problem in EEG: Simulation studies using FEM on realistic head models. *Human Brain Mapping, 6*(4), 250–269.
McCulloch, W. S., & Pitts, W. (1943). A logical calculus of the ideas immanent in nervous activity. *The Bulletin of Mathematical Biophysics, 5*(4), 115–133.
Meinshausen, N., & Bühlmann, P. (2006). High-dimensional graphs and variable selection with the lasso. *The Annals of Statistics, 34*(3), 1436–1462.
Mesgarani, N., Fritz, J., & Shamma, S. (2010). A computational model of rapid task-related plasticity of auditory cortical receptive fields. *Journal of Computational Neuroscience, 28*(1), 19–27.
Mitra, P., & Bokil, H. (2007). *Observed brain dynamics*. New York: Oxford University Press.
Mosher, J. C., Leahy, R. M., & Lewis, P. S. (1999). EEG and MEG: Forward solutions for inverse methods. *IEEE Transactions on Biomedical Engineering, 46*(3), 245–259.
Needell, D., & Tropp, J. A. (2009). CoSaMP: Iterative signal recovery from incomplete and inaccurate samples. *Applied and Computational Harmonic Analysis, 26*(3), 301–321.
Nishimoto, S., Vu, A. T., Naselaris, T., Benjamini, Y., Yu, B., & Gallant, J. L. (2011). Reconstructing visual experiences from brain activity evoked by natural movies. *Current Biology, 21*(19), 1641–1646.
Nunez, P. L., & Cutillo, B. A. (1995). *Neocortical dynamics and human EEG rhythms*. New York: Oxford University Press.
Okatan, M., Wilson, M. A., & Brown, E. N. (2005). Analyzing functional connectivity using a network likelihood model of ensemble neural spiking activity. *Neural Computation, 17*(9), 1927–1961.
Paninski, L. (2004). Maximum likelihood estimation of cascade point-process neural encoding models. *Network: Computation in neural systems, 15*(4), 243–262.
Paninski, L., Pillow, J., & Lewi, J. (2007). Statistical models for neural encoding, decoding, and optimal stimulus design. *Progress in Brain Research, 165*, 493–507.
Percival, D. B. (1993). *Spectral analysis for physical applications*. Cambridge: Cambridge University Press.
Phillips, J. W., Leahy, R. M., & Mosher, J. C. (1997). MEG-based imaging of focal neuronal current sources. *IEEE Transactions on Medical Imaging, 16*(3), 338–348.
Pillow, J. W., Ahmadian, Y., & Paninski, L. (2011). Model-based decoding, information estimation, and change-point detection techniques for multineuron spike trains. *Neural Computation, 23*(1), 1–45.
Pnevmatikakis, E. A., Soudry, D., Gao, Y., Machado, T. A., Merel, J., Pfau, D., et al. (2016). Simultaneous denoising, deconvolution, and demixing of calcium imaging data. *Neuron, 89*(2), 285–299.
Purdon, P. L., Pierce, E. T., Mukamel, E. A., Prerau, M. J., Walsh, J. L., Wong, K. F. K., et al. (2013). Electroencephalogram signatures of loss and recovery of consciousness from propofol. *Proceedings of the National Academy of Sciences USA, 110*(12), E1142–E1151.
Quatieri, T. F. (2008). *Discrete-time speech signal processing: Principles and practice*. Upper Saddle River: Prentice Hall.

Rauch, H. E., Striebel, C., & Tung, F. (1965). Maximum likelihood estimates of linear dynamic systems. *AIAA Journal, 3*(8), 1445–1450.
Rosenblatt, F. (1958). The perceptron: A probabilistic model for information storage and organization in the brain. *Psychological Review, 65*(6), 386.
Rousseeuw, P. J. & Leroy, A. M. (2005). *Robust regression and outlier detection*. Hoboken: Wiley.
Rudin, L. I., Osher, S., & Fatemi, E. (1992). Nonlinear total variation based noise removal algorithms. *Physica D: Nonlinear Phenomena, 60*(1), 259–268.
Sato, M., Yoshioka, T., Kajihara, S., Toyama, K., Goda, N., Doya, K., et al. (2004). Hierarchical Bayesian estimation for MEG inverse problem. *Neuroimage, 23*(3), 806–826.
Sheikhattar, A., Fritz, J. B., Shamma, S. A., & Babadi, B. (2016). Recursive sparse point process regression with application to spectrotemporal receptive field plasticity analysis. *IEEE Transactions on Signal Processing, 64*(8), 2026–2039.
Shumway, R. H. & Stoffer, D. S. (1982). An approach to time series smoothing and forecasting using the em algorithm. *Journal of Time Series Analysis, 3*(4), 253–264.
Smetters, D., Majewska, A., & Yuste, R. (1999). Detecting action potentials in neuronal populations with calcium imaging. *Methods, 18*(2), 215–221.
Stosiek, C., Garaschuk, O., Holthoff, K., & Konnerth, A. (2003). In vivo two-photon calcium imaging of neuronal networks. *Proceedings of the National Academy of Sciences USA, 100*(12), 7319–7324.
Thomson, D. J. (1982). Spectrum estimation and harmonic analysis. *Proceedings of the IEEE, 70*(9), 1055–1096.
Thomson, D. J. & Vernon, III, F. L. (1998). Signal extraction via multitaper spectra of nonstationary data. Technical Report BL0112170–981218-37TM, Bell Laboratories, Lucent Technologies.
Tibshirani, R., Saunders, M., Rosset, S., Zhu, J., & Knight, K. (2005). Sparsity and smoothness via the fused lasso. *Journal of the Royal Statistical Society: Series B (Statistical Methodology), 67*(1), 91–108.
Truccolo, W., Eden, U. T., Fellows, M. R., Donoghue, J. P., & Brown, E. N. (2005). A point process framework for relating neural spiking activity to spiking history, neural ensemble, and extrinsic covariate effects. *Journal of Neurophysiology, 93*(2), 1074–1089.
Tuckwell, H. (1988). *Introduction to Theoretical Neurobiology*. Cambridge: Cambridge University Press.
Van de Geer, S., Bühlmann, P., Ritov, Y., & Dezeure, R. (2014). On asymptotically optimal confidence regions and tests for high-dimensional models. *Annals of Statistics, 42*(3), 1166–1202.
Velliste, M., Perel, S., Spalding, M. C., Whitford, A. S., & Schwartz, A. B. (2008). Cortical control of a prosthetic arm for self-feeding. *Nature, 453*(7198), 1098–1101.
Vogelstein, J. T., Packer, A. M., Machado, T. A., Sippy, T., Babadi, B., Yuste, R., et al. (2010). Fast nonnegative deconvolution for spike train inference from population calcium imaging. *Journal of Neurophysiology, 104*(6), 3691–3704.
Vogelstein, J. T., Watson, B. O., Packer, A. M., Yuste, R., Jedynak, B., & Paninski, L. (2009). Spike inference from calcium imaging using sequential Monte Carlo methods. *Biophysical Journal, 97*(2), 636–655.
Werbos, P. (1974). *Beyond Regression: New Tools for Prediction and Analysis in the Behavioral Sciences*. PhD thesis, Harvard University.
Wu, Z., & Huang, N. E. (2009). Ensemble empirical mode decomposition: A noise-assisted data analysis method. *Advances in Adaptive Data Analysis, 1*(1), 1–41.
Zhang, C.-H., & Zhang, S. S. (2014). Confidence intervals for low dimensional parameters in high dimensional linear models. *Journal of the Royal Statistical Society: Series B (Statistical Methodology), 76*(1), 217–242.

Chapter 6
RETRACTED CHAPTER: Artifact Rejection for Concurrent TMS-EEG Data

Wei Wu, Corey Keller, and Amit Etkin

The chapter "Artifact Rejection for Concurrent TMS-EEG Data" by "Wei Wu, Corey Keller, and Amit Etkin" published in the book "Dynamic Neuroscience", pages 141–173, https://doi.org/10.1007/978-3-319-71976-4_6, has been retracted at the request of the authors, as this work will be appearing in another publication.

Z. Chen, S.V. Sarma (eds.), *Dynamic Neuroscience*,
https://doi.org/10.1007/978-3-319-71976-4_6

RETRACTED CHAPTER

RETRACTED CHAPTER

Part II
Modeling and Control Theory

Chapter 7
Characterizing Complex Human Behaviors and Neural Responses Using Dynamic Models

Sridevi V. Sarma and Pierre Sacré

7.1 Introduction

Many experiments conducted in neuroscience entail applications of stimuli and recordings of behavioral responses and neural activity. Traditional approaches to understanding how the brain encodes stimuli often compute correlations between stimuli and neural activity time-locked to behavioral events. For example, when studying motor control, investigators train a participant to move the arm in different directions while activity from premotor and primary motor regions are measured (Carpenter et al. 1999; Schieber 2004). Then, to understand how neurons encode movement direction, firing rates of neurons are modeled as functions of behavior right after the onset of movement (Agarwal et al. 2015). In this example, both direction of movement and neural activity are measured outputs, and behavior is primarily driven by a target cue.

Now, let's consider experiments wherein behavior is not only driven by stimuli provided by the experimentalist, but also by internal factors within the participant that are not easily measurable. A first example is when participants are performing a gambling task, wherein they are betting virtual money and then perhaps get emotional if they are winning or losing (Sacré et al. 2016a,b,c). Although objective measures of emotion have been proposed such as skin conductance response and heart rate variability, these measures are typically delayed or only accurate over several minutes, while emotions can fluctuate at a faster time scale during gambling (Mauss and Robinson 2009). A second example is when participants are performing a Stroop-like task, wherein distractors are present to confuse participants while they attempt to make correct associations between presented stimuli and

S.V. Sarma (✉) · P. Sacré
Johns Hopkins University, Baltimore, MD, USA
e-mail: ssarma2@jhu.edu; psacre1@jhu.edu

Z. Chen, S.V. Sarma (eds.), *Dynamic Neuroscience*,
https://doi.org/10.1007/978-3-319-71976-4_7

appropriate responses (Shoham et al. 2003; Smith et al. 2015). During this task, participants make errors and their motivational and attentional states vary over the session. Motivation and attention are not directly measurable, yet they may influence behavior in a profound way (Shoham et al. 2003).

In both of the gambling and Stroop-like tasks described above, behavioral responses and neural activity are influenced by external stimuli *and* internal states of participants. Thus, when looking for neural correlates, how behavior changes with stimuli and underlying dynamic state variables must first be characterized. In this chapter, we present a systematic approach based on existing methodologies to (1) estimate internal dynamic states of participants from measured data to explain behavior variability within and across participants, and (2) identify neural substrates of behavior and internal states. The proposed approach is a two-step procedure wherein one first constructs participant-specific state-space models that capture the dynamics of internal states and how they evolve with administered stimuli, and how measured behavior depends on these states and stimuli; and then, one relates stimuli, responses, and states back to neural activity. We discuss the challenges that arise in each step of the process and provide suggestions on how to successfully complete these two steps. We present examples from two data sets involving a gambling and Stroop-like task.

7.2 Methods

In this section, we first describe a general dynamic state-space modeling framework and the maximum likelihood procedure used to estimate parameters of participant-specific models of behavior. Then, we discuss how to map model variables and estimated internal states back to neural data using nonparametric statistical tests and point process models (PPMs).

7.2.1 Dynamic State-Space Modeling

The first step of the proposed approach is to build a mathematical model with inputs $\boldsymbol{u}$ and outputs $\boldsymbol{y}$ that explains the variability that we observe in the data. In this context, we can distinguish between two types of models. A model is *static* (or without memory) if the value of the output signal at a particular time depends only on the value of the input signal at the same time. Otherwise, it is *dynamic* (or with memory).

The general *dynamic* state-space model for a discrete-time system can be written as follows

$$\boldsymbol{x}_{k+1} \sim f_\theta(\boldsymbol{x}_{k+1} \mid \boldsymbol{x}_k, \boldsymbol{u}_k), \tag{7.1a}$$

$$\boldsymbol{y}_k \sim h_\theta(\boldsymbol{y}_k \mid \boldsymbol{x}_k, \boldsymbol{u}_k), \tag{7.1b}$$

where $\boldsymbol{x}_k \in \mathbb{R}^n$ is the n-dimensional state-vector, $\boldsymbol{u}_k \in \mathbb{R}^m$ is the m-dimensional input-vector, and $\boldsymbol{y}_k \in \mathbb{R}^p$ is the p-dimensional output-vector. The state-transition map $f_{\boldsymbol{\theta}}$ and the measurement map $h_{\boldsymbol{\theta}}$ are conditional probability distributions of $\boldsymbol{x}_{k+1}$ given $(\boldsymbol{x}_k, \boldsymbol{u}_k)$, and of $\boldsymbol{y}_k$ given $(\boldsymbol{x}_k, \boldsymbol{u}_k)$, respectively. The initial state vector $\boldsymbol{x}^0$ is distributed according to $p_{\boldsymbol{\theta}}(\boldsymbol{x}^0)$. The model parameters are denoted by $\boldsymbol{\theta}$. It is often convenient to define its equivalent *static* model, that is, $\boldsymbol{y}_k \sim \tilde{h}_{\boldsymbol{\theta}}(\boldsymbol{y}_k \mid \boldsymbol{u}_k) = h_{\boldsymbol{\theta}}(\boldsymbol{y}_k \mid 0, \boldsymbol{u}_k)$, where the state-vector is fixed to zero for all k.

For our neuroscience applications, the inputs $\boldsymbol{u}_k$ represent stimuli (or functions of stimuli) of the task on trial k, the outputs $\boldsymbol{y}_k$ represent measured behavioral responses (or functions of responses) in the task (e.g., reaction time, correct/incorrect answer) on trial k, and the states $\boldsymbol{x}_k$ represent internal states on trial k that influence behavior (e.g., attentional state, motivation, emotion).

As a first step to model behavioral data, it often is sufficient to begin with a time-invariant state-space model with a linear state equation and a generalized linear output equation, which reduces to

$$\boldsymbol{x}_{k+1} = \boldsymbol{A}\,\boldsymbol{x}_k + \boldsymbol{B}\,\boldsymbol{u}_k + \boldsymbol{w}_k, \tag{7.2a}$$

$$\boldsymbol{y}_k \sim h(\boldsymbol{y}_k \mid \boldsymbol{C}\boldsymbol{x}_k + \boldsymbol{D}\,\boldsymbol{u}_k, \boldsymbol{\tau}), \tag{7.2b}$$

where $\boldsymbol{w}_k \in \mathbb{R}^n$ is the n-dimensional zero-mean Gaussian noise-vector with unknown covariance matrix $\boldsymbol{\Sigma}_w$ and h is a probability distribution from the exponential family, that is conditioned on an affine combination of states and inputs and the dispersion parameter $\boldsymbol{\tau}$. The initial state vector $\boldsymbol{x}^0$ is assumed to follow a Gaussian distribution with mean $\bar{\boldsymbol{x}}^0$ and covariance matrix $\boldsymbol{\Sigma}^0$. The model parameters are then $\boldsymbol{\theta} = \{\boldsymbol{A}, \boldsymbol{B}, \boldsymbol{\Sigma}_w, \bar{\boldsymbol{x}}^0, \boldsymbol{\Sigma}^0, \boldsymbol{C}, \boldsymbol{D}, \boldsymbol{\tau}\}$.

The model estimation problem then boils down to: given N input-output measurements $\boldsymbol{u}_{1:N} = \{\boldsymbol{u}_1, \ldots, \boldsymbol{u}_N\}$ and $\boldsymbol{y}_{1:N} = \{\boldsymbol{y}_1, \ldots, \boldsymbol{y}_N\}$, estimate the model parameters $\boldsymbol{\theta}$ and the state $\boldsymbol{x}_{1:N} = \{\boldsymbol{x}_1, \ldots, \boldsymbol{x}_N\}$. One approach is to estimate $\boldsymbol{\theta}$ and $p_{\boldsymbol{\theta}}(\boldsymbol{x}_{1:N} \mid \boldsymbol{u}_{1:N}, \boldsymbol{y}_{1:N})$ from data to maximize the likelihood function (Van Trees 1968; Louis 1991; Moon 1996). The likelihood function is the family of probability distributions considered as a function of $\boldsymbol{\theta}$, for fixed $\boldsymbol{y}_{1:N}$ and $\boldsymbol{u}_{1:N}$. It is often more convenient to work with its logarithm, which is called the log-likelihood function, and denoted as ℓ:

$$\ell(\boldsymbol{\theta}) = \log p_{\boldsymbol{\theta}}(\boldsymbol{y}_{1:N} \mid \boldsymbol{u}_{1:N}). \tag{7.3}$$

Now, the problem is to estimate the value of the parameters $\boldsymbol{\theta}$. A widely used method, called maximum likelihood estimation, is to estimate $\boldsymbol{\theta}$ as

$$\hat{\boldsymbol{\theta}}_{\mathrm{ml}} = \arg\max_{\boldsymbol{\theta} \in \boldsymbol{\Theta}} p_{\boldsymbol{\theta}}(\boldsymbol{y}_{1:N} \mid \boldsymbol{u}_{1:N}) = \arg\max_{\boldsymbol{\theta} \in \boldsymbol{\Theta}} \ell(\boldsymbol{\theta}), \tag{7.4}$$

where $\boldsymbol{\theta} \in \boldsymbol{\Theta}$ gives the prior information or other constraints on the parameter vector $\boldsymbol{\theta}$. In the context of the estimation of a dynamic model, the state is not observed and we can write the likelihood as follows

$$p_{\boldsymbol{\theta}}(\boldsymbol{y}_{1:N} \mid \boldsymbol{u}_{1:N}) = \int_{\mathcal{X}} p_{\boldsymbol{\theta}}(\boldsymbol{x}_{1:N}, \boldsymbol{y}_{1:N} \mid \boldsymbol{u}_{1:N})\, d\boldsymbol{x}. \tag{7.5}$$

One way to solve this problem is to use the expectation-maximization (EM) algorithm. The EM algorithm is an iterative algorithm that is composed of two steps at each iteration: an expectation step and a maximization step.

E-step The idea of the E-step is to take the expectation with respect to the unknown underlying states, using the current estimate of the parameters $\boldsymbol{\theta}^*$ and conditioned upon the observation, that is,

$$Q(\boldsymbol{\theta} \mid \boldsymbol{\theta}^*) = \mathbf{E}\left[\log\left(\frac{p_{\boldsymbol{\theta}}(\boldsymbol{x}_{1:N}, \boldsymbol{y}_{1:N} \mid \boldsymbol{u}_{1:N})}{p_{\boldsymbol{\theta}^*}(\boldsymbol{x}_{1:N}, \boldsymbol{y}_{1:N} \mid \boldsymbol{u}_{1:N})}\right) \mid \boldsymbol{U}_{1:N} = \boldsymbol{u}_{1:N}, \boldsymbol{Y}_{1:N} = \boldsymbol{y}_{1:N}, \boldsymbol{\theta}^*\right], \tag{7.6}$$

$$= \int_{\mathscr{X}} \log\left(\frac{p_{\boldsymbol{\theta}}(\boldsymbol{x}_{1:N}, \boldsymbol{y}_{1:N} \mid \boldsymbol{u}_{1:N})}{p_{\boldsymbol{\theta}^*}(\boldsymbol{x}_{1:N}, \boldsymbol{y}_{1:N} \mid \boldsymbol{u}_{1:N})}\right) p_{\boldsymbol{\theta}^*}(\boldsymbol{x}_{1:N} \mid \boldsymbol{u}_{1:N}, \boldsymbol{y}_{1:N})\, d\boldsymbol{x}. \tag{7.7}$$

M-step The idea of the M-step is to provide a new estimate $\boldsymbol{\theta}^{**}$ of the parameters, that is,

$$\boldsymbol{\theta}^{**} = \arg\max_{\boldsymbol{\theta} \in \boldsymbol{\theta}} Q(\boldsymbol{\theta} \mid \boldsymbol{\theta}^*). \tag{7.8}$$

Finally, there are several ways to establish the degree of agreement between the model and observed data. In particular, they are mainly two families of statistics that we can compute: the first family measures the *goodness-of-fit* of the model with the data and the second family measures the *improvement of goodness-of-fit* from a static model to a dynamic model. In both families, we can use different test statistics such as log-likelihood, deviance and Pearson residuals, predictive power, etc. The first family is interested in the absolute value of this statistic; while the second family is interested in the relative difference between the statistics for the dynamic and static models. The statistical significance of these test statistics can be evaluated using a nonparametric permutation test.

7.2.2 Neural Correlates Informed by State-Space Model

The second step of the proposed approach is to relate the model variables (inputs, outputs, and estimated states) back to the neural data. Below we describe this second step if one has recorded continuous neural activity or spike train observations.

7.2.2.1 Continuous Neural Data

If neural activity measured is continuous (e.g., local field potential, electroencephalogram, electrocorticography), then a common approach to analyzing the data is to move to the spectral domain. In particular, select a time window of interest (e.g., 500-ms window time-centered to an applied stimulus). Then, compute a spectrogram including frequencies of interest (e.g., 2–150 Hz) for that time window for each trial. This will generate a stack of spectrograms, one for each trial in the recorded session.

Once the stack of spectrograms is computed, identify time-frequency clusters within the stack that distinguish between two conditions of interest (e.g., high bet vs low bet, or moving up vs moving down). In particular, take a model variable of interest (e.g., player's card) and split trials into those when the variable takes on low values and those when the variable takes on high values. Low-value trials may be defined, for example, as the bottom third of the variable distribution over all trials, and high-value trials may be defined as the top third. Then apply a cluster-based nonparametric statistical test to leverage the dependency between adjacent time-frequency windows in order to avoid over-penalizing with multiple comparison corrections (Maris and Oostenveld 2007).

For each window in the spectrogram, create a null distribution by shuffling the condition labels 1000–5000 times between trials within each participant. Within each shuffle, compute a t-statistic and a p-value for each window of the newly labeled spectrograms (independent two-sample t-test with both tails, unequal sample sizes, and unequal variances). Clusters are formed by grouping windows with significant p-values (e.g., $p < 0.05$) that are adjacent in either time or frequency. The cluster-level test statistic is then calculated by taking the sum of absolute values of the t-statistics for each window in the cluster. This prioritizes clusters that have both strong differences and large sizes. A null distribution of cluster statistics is created using the same process but with the 1000–5000 spectrograms obtained from the originally shuffled labels. The observed cluster statistic is then compared against this null distribution of cluster statistics in order to obtain the final p-value of the test.

Data from all patients can be pooled together but the labels are permuted within each participant only. This process of finding time-frequency cluster correlated to a model or task variable can be repeated for each variable and the estimated state trajectories across participants.

7.2.2.2 Spike Train Data

If neural activity is measured as spike trains, then one can use point process models to identify how behavior influences neuronal spiking activity. Several examples of how PPMs are estimated used for different experimental setups are given in Coleman and Sarma (2007), Coleman and Sarma (2010), Santaniello et al. (2010), Santaniello et al. (2012), Sarma et al. (2010), Sarma et al. (2012).

A point process is a series of 0/1 random events that occur in continuous time. For a neural spike train, the 1's are individual spike times and the 0's are the times at which no spikes occur. To define a point process model of neural spiking activity, in this analysis, one can consider an observation interval $(0, T]$, and let $N_k(t)$ be the number of spikes counted in the interval $(0, t]$ for $t \in (0, T]$ for a trial k.

A point process model of a neural spike train can be completely characterized by its conditional intensity function (CIF) $\lambda_k(t \mid H_t)$ defined as follows:

$$\lambda_k(t|H_t, \boldsymbol{u}_k, \boldsymbol{x}_k) = \lim_{\Delta \to 0} \frac{\Pr(N_k(t+\Delta) - N_k(t) = 1 \mid H_t, \boldsymbol{u}_k, \boldsymbol{x}_k)}{\Delta}, \tag{7.9}$$

where H_t denotes the history of spikes and covariates up to time t. It follows from (7.9) that the probability of a single spike in a small interval $(t, t+\Delta]$ is approximately

$$\Pr(\text{spike in } (t, t+\Delta] \text{ on trial } k \mid H_t, \boldsymbol{u}_k, \boldsymbol{x}_k) \;= \lambda_k(t \mid H_t, \boldsymbol{u}_k, \boldsymbol{x}_k)\, \Delta. \tag{7.10}$$

Details can be found in Cox and Isham (1980), Snyder and Miller (1991).

The CIF generalizes the rate function of a Poisson process to a rate function that is history dependent. Because the CIF completely characterizes a spike train, defining a model for the CIF defines a model for the spike train (Brown et al. 2003).

For neural correlate analyses, use a generalized linear model (GLM) to define CIF models by expressing for each neuron, the log of its CIF in terms of the neurons spike history H_t, relevant model inputs $\boldsymbol{u}_k$, and the state trajectory $\boldsymbol{x}_k$. The GLM is an extension of the multiple linear regression model, in which the variable being predicted (e.g., in this case spike times) need not be Gaussian (McCullagh and Nelder 1989). GLM also provides an efficient computational scheme for model parameter estimation and a likelihood framework for conducting statistical inferences (McCullagh and Nelder 1989).

One can express the CIF for each neuron at each time step (e.g., millisecond) as a function of task stimuli which can turn on and/or off over time, the state variable value which typically is constant over a trial, and the neuron's spiking history. Instead of estimating the CIF continuously throughout the entire trial, one can estimate it over time windows around key epochs and at discrete time intervals each 1 ms in duration.

In particular, one can express the CIF as follows:

$$\lambda_k(t \mid H_t, \boldsymbol{u}_k, \boldsymbol{x}_k, \boldsymbol{\theta}) = \lambda^S(\boldsymbol{u}_k \mid \boldsymbol{\theta})\, \lambda^X(\boldsymbol{x}_k \mid \boldsymbol{\theta})\, \lambda^H(t \mid H_t, \boldsymbol{\theta}) \tag{7.11}$$

where $\lambda^S(\boldsymbol{u}_k \mid \boldsymbol{\theta})$ describes the effect of the stimulus on the neural response, $\lambda^X(\boldsymbol{x}_k \mid \boldsymbol{\theta})$ describes the effect of the state variable on the neural response, and $\lambda^H(t \mid H_t, \boldsymbol{\theta})$ describes the effect of spiking history on the neural response. $\boldsymbol{\theta}$ is a parameter vector to be estimated from data. The units of $\lambda^S(t \mid \boldsymbol{\theta})$ is spikes per second and $\lambda^H(t \mid H_t, \boldsymbol{\theta})$ is dimensionless. Finally, one can compute maximum-likelihood (ML) estimates for $\boldsymbol{\theta}$ and 95% confidence intervals of $\boldsymbol{\theta}$ for each neuron using `glmfit` in MATLAB.

It is important to establish the degree of agreement between a PPM and observations of the spike train and associated experimental variables is a prerequisite for using the point process analysis to make scientific inferences. One can use Kolmogorov-Smirov (KS) plots based on the time-rescaling theorem to assess the model goodness-of-fit. The time-rescaling theorem is a well-known result in probability theory, which states that any point process with an integrable CIF may be transformed into a Poisson process with unit rate (Johnson and Kotz 1970). A KS plot, which plots the empirical cumulative distribution function of the transformed spike times versus the cumulative distribution function of a unit rate exponential, is used to visualize the goodness-of-fit for each model. The model is better if its corresponding KS plot lies near the 45° line. One can compute the 95% confidence bounds for the degree of agreement using the distribution of the KS statistic (Johnson and Kotz 1970). If a model's KS plot was within the 95% confidence bounds, then it can be included it in the analyses.

7.3 Results

In this section, we present two applications where we applied our approach to reveal new insights on the neural mechanisms involved in a Stroop-like task where spike train observations were made and a gambling task where local field potentials were measured.

7.3.1 *Multi-Source Interference Task*

This example is taken from Sklar et al. (2017). Two participants being treated at the Columbia University Medical Center performed the behavioral task in their hospital rooms using methods previously described in Johnson et al. (2014). Behavioral data were simultaneously acquired on the same time base as the electrophysiology data. Participants performed the multi-source interference task (MSIT) (Shoham et al. 2003). The MSIT is a Stroop-like task in which the participant is presented with three integers ranging from 0 to 3. Two of the three integers presented are the same integer. The goal of the MSIT is to indicate the identity of the different integer on the number pad (e.g., cue: `0 2 0`; correct response: button 2; Fig. 7.1b).

Conflict is introduced in this task by changing the position of the target number (e.g., `0 0 1`; correct response: button 1; Simon or spatial interference) or by changing the identity of the distracting integers to potential responses (e.g., `1 2 1`; correct response: button 2; Eriksen or flanker interference). Additionally, both types of interference can occur (e.g., `3 1 3`; correct response: button 1). These four groups of trials were presented randomly, with a uniform frequency distribution.

7.3.1.1 Dynamic State-Space Modeling

In this study, we hypothesized that the "cognitive state" of each participant influences behavior and modulates neuronal activity in the dorsal anterior cingulate cortex (dACC). In particular, we hypothesized that when participants require more

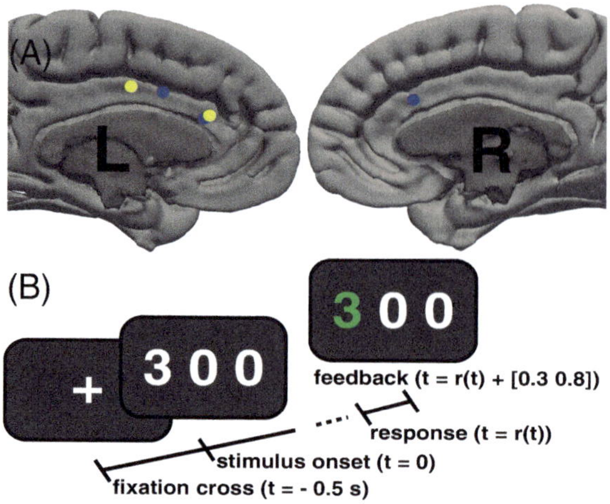

Fig. 7.1 MSIT task and microwire recording locations. (**a**) Microwire recording locations in the dACC. Colors represent recording locations corresponding to each participant on each hemisphere (L and R). (**b**) MSIT task diagram showing an example trial structure. In each trial, a fixation cross appears on the screen for 0.5 s prior to the stimulus presentation. The stimulus remains on the screen until the participant indicates her response on the button pad. Feedback is delivered between 0.3 and 0.8 s after the participant indicates her response. Figure reproduced with permission from Sklar et al. (2017)

cognitive control, (1) they are more likely to react to the stimulus slowly and (2) their cingulate neurons are modulated. Since such a cognitive state is not directly measurable, we compute it from measurable data.

Before constructing the state-space model of behavior, we first looked to see whether behavior varied for different stimuli, and for the same stimuli over the session. To examine behavioral variability, we plotted a moving average reaction time for each stimulus type (easy, hard, Simon, Flanker). As shown in Fig. 7.2, the reaction times for each stimulus type change over time, suggesting dynamics in the behavior that may be explained by a latent state variable.

Therefore, we constructed a cognitive state variable x_k that updates for each trial k as follows:

$$x_{k+1} = a\, x_k + \sum_{i=1}^{5} b_i\, u_{i,k} = a\, x_k + \boldsymbol{B}\, \boldsymbol{u}_k \tag{7.12}$$

where $\boldsymbol{u}_k = [u_{1,k}, u_{2,k}, u_{3,k}, u_{4,k}, u_{5,k}]^\top$ is an input column vector dependent on the trial conditions:

- $u_{1,k} = 1$ if no interference on trial k, and 0 otherwise;
- $u_{2,k} = 1$ if both interferences on trial k, and 0 otherwise;
- $u_{3,k} = 1$ if spatial interference on trial k, and 0 otherwise;
- $u_{4,k} = 1$ if flanker interference on trial k, and 0 otherwise;
- $u_{5,k} = 1$ if trial type on trial k changed from previous trial, and 0 otherwise.

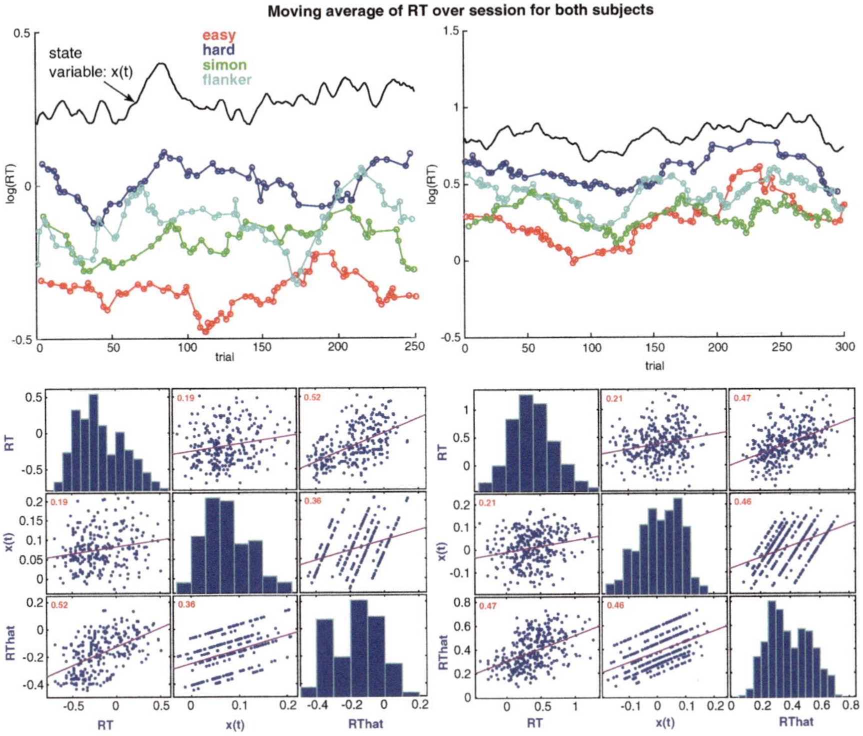

Fig. 7.2 Relationship between reaction time variability and cognitive state. (*Top*) Moving average of reaction times for each trial type for participant 1 (*left*) and participant 2 (*right*). The estimated cognitive states are overlaid in black. (*Bottom*) Correlation plots between actual reaction times and x_k and predicted reaction times for participant 1 (*left*) and participant 2 (*right*). Figure reproduced with permission from Sklar et al. (2017)

The parameter a represents the decaying influence of previous trials on the cognitive state, and $\boldsymbol{B} = [b_1, b_2, b_3, b_4, b_5]$ dictate the effects that the trial conditions have on the state x_{k+1}. The solution to the state-space equation is

$$x_k = a^k x_1 + \sum_{i=1}^{5} \sum_{s=1}^{k-1} a^{(k-s-1)}\, b_i\, u_{i,s}, \tag{7.13}$$

which can be used to determine the parameters $\{a, \boldsymbol{B}\}$ by inserting the solution x_k as a covariate into a GLM. The output of the GLM is y_k, defined as the log of the reaction time modeled as

$$y_k = \log(r_k) = x_k + \boldsymbol{D}\,\boldsymbol{u}_k + d_0 + \epsilon_k, \tag{7.14}$$

where r_k is the reaction time of the trial, the ϵ_k are independent zero-mean Gaussian random inputs with variance σ_ϵ^2, and $\boldsymbol{D}$ is a vector of the form of $\boldsymbol{B}$, that represents the direct influence of the current input on the reaction time of a trial.

The state-space model above includes a state that is completely deterministic, and the output is stochastic. If the state is deterministic, then the EM algorithm is unnecessary. It is worth beginning with a deterministic state variable to help identify model structure (what inputs to include into the state-space model) that best explains the observed data.

To estimate the parameters a, $\boldsymbol{B}$, $\boldsymbol{D}$, and d_0 of the state-space model, we gridded the parameter space a and for each parameter value, (1) we computed each term of the sum of the state trajectory (7.13) for $b_i = 1$ over the session, then (2) we substituted x_k by each term of the sum in a GLM and estimated $\boldsymbol{B}$ and $\boldsymbol{D}$ that maximize the data likelihood function. We then selected $\boldsymbol{\theta} = \{a, \boldsymbol{B}, \boldsymbol{D}, d_0\}$ that produce the maximum of all likelihoods over the entire grid.

Figure 7.2 overlays the estimate state variables (black trajectories) for the two participants. The state trajectories follow the dynamics of mean reaction times over the session for one or more stimuli. For participant 1 (left panel), the estimated x_k attempts to capture the variability of reaction times over the session for all four task types, but is not able to characterize behavior for all stimuli. In this case, a second state variable may better explain the behavior. On the other hand, the reaction time dynamics for participant 2 (right panel) are very similar across all stimuli suggesting that a scalar state variable is sufficient to explain participant 2's variability in behavior.

The bottom panels in Fig. 7.2 show the correlation plots between actual reaction times and r_k and predicted reaction times, $\hat{r}_k$ for participant 1 (left) and participant 2 (right), where $\log(\hat{r}_k) = x_k + \hat{\boldsymbol{D}}\,\boldsymbol{u}_k + \hat{d}_0$. The state-space models for both participants suggest that the inclusion of the state helps explain the variability in reaction times over the session that cannot be entirely explained with task stimuli that changes over the session.

7.3.1.2 Neural Correlates Informed by Dynamic State-Space Model

Now that behavior is sufficiently explained by the state-space model described above, we search to explain neuronal responses to both task stimuli and cognitive state estimates. We thus formulated a PPM to relate the spiking of each dACC neuron for each participant to factors associated with the neuron's spiking history and the cognitive state variable. We use these model parameters to analyze temporal dynamics in neuronal activity due to the cognitive state variable after the stimulus is shown.

As described in Sect. 7.2.2.2, we use the GLM framework to define the CIFs of our PPMs by expressing, for each neuron, the log of its CIF in terms of the neuron's spike history and relevant covariates (Truccolo et al. 2005). We express the CIF for each neuron as a function of the neuron's spiking history, λ_k^H, in the preceding 240 ms and our derived cognitive state variable, λ^X. Specifically, for trial k and time bin t:

$$\lambda_k(t \mid H_t, \boldsymbol{\theta}) = \lambda^X(x_k \mid \boldsymbol{\theta})\, \lambda^H(t \mid H_t, \boldsymbol{\theta}), \tag{7.15}$$

such that

$$\lambda^X(x_k \mid \boldsymbol{\theta}) = \alpha\, x_k \tag{7.16}$$

and

$$\log(\lambda^H(t \mid H_t, \boldsymbol{\theta})) = \gamma_0 + \sum_{j=1}^{8} \gamma_j\, n_{t-5j:t-5(j-1)} + \sum_{j=1}^{8} \beta_j\, n_{t-40-25j:t-40-25(j-1)}, \tag{7.17}$$

where $n_{A:B}$ is the number of spikes observed in the time interval $[A, B)$ during the epoch analyzed. The $\{\gamma_j\}$ coefficients capture short-term history effects going back to 40 ms in the past in 5-ms bins. The $\{\beta_j\}$ coefficients capture long-term history effects going back to 240 ms in the past in 25-ms bins, and α captures the effect of the cognitive state. We computed ML estimates for all coefficients and their associated 95% confidence intervals for each neuron model using `glmfit` in MATLAB.

We examined the activity of 12 units (10 in patient 1 and 2 in patient 2). Figure 7.3a shows the spiking frequency in two units from both participants during the first second after stimulus presentation in each trial. These spike counts are overlaid with the cognitive state variables x_k for each participant. The neurons' spiking frequencies appear to have a negative correlation with x_k dropping markedly when x_k rises.

In participant 1, we found 10 units whose activity was predicted using the cognitive state variable (the GLM fit coefficient for the x_k covariate was significantly non-zero, with $p < 0.05$). Some units had increased activity as x_k increased, while some displayed decreased activity. The covariate coefficients for participant 1's PPM are shown in Fig. 7.3b. The dependence on short- and long-term spiking history is displayed in the lower two plots and shows refractoriness in the first 15 ms after a spike, and an increased likelihood to fire in the 25–100 ms interval. The upper right plot displays the PPM coefficient for x_k with 2 standard deviation error bars. For this unit, x_k was a strong predictor of the spiking behavior, with spiking probability decreasing for higher x_k. The upper left plot shows the goodness of fit of the model using a KS plot, with 95% confidence bars. In participant 2, two units' spiking could be significantly predicted by x_k (see Sklar et al. (2017) for details). The PPM coefficients and goodness of fit for participant 2 for one neuron are also shown in Fig. 7.3b. This unit had significantly longer inter-spike-intervals, so the coefficient values for the short-term history bins have higher uncertainty.

These preliminary results suggest that neurons in the dACC slowly track subjects' overall need for cognitive control, while simultaneously maintaining faster task-related dynamics. A latent cognitive state variable correlates with both reaction times and neuronal activity in two patients. These results provide support for an additional representation of task state or attentional motivation in the dACC.

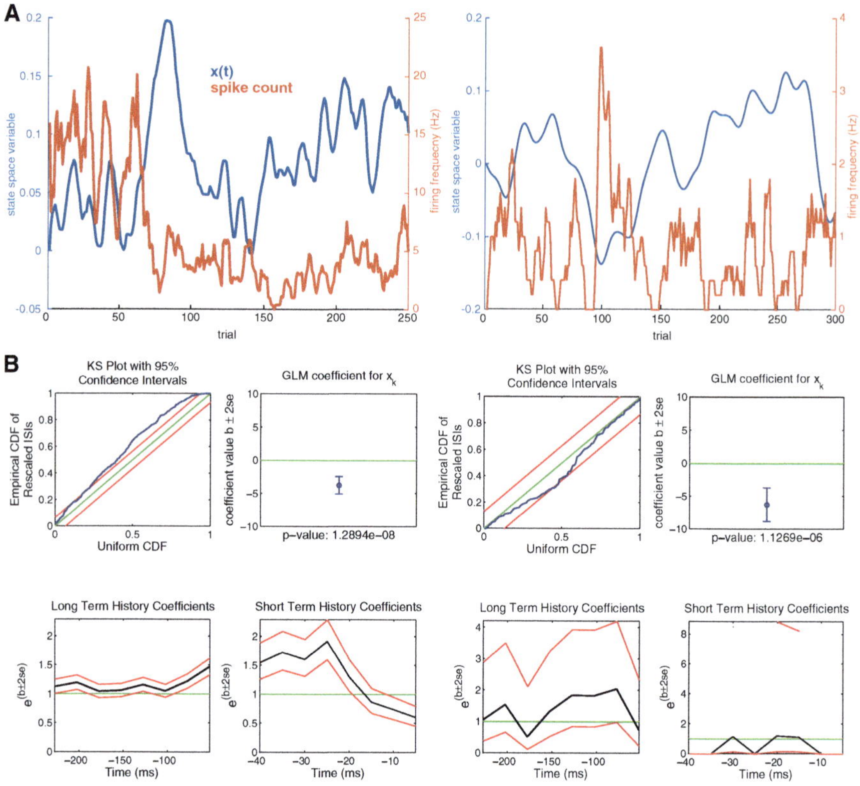

Fig. 7.3 Neural correlates of cognitive states. (**a**) Cognitive states over sessions overlaid with spike counts for one unit for participant 1 (*left*), and one unit for participant 2 (*right*). (**b**) Point process model for same unit as above from participant 1 (the 1st and 2nd columns) for first second after stimulus onset, and same unit as above from participant 2 (the 3rd and 4th columns). *Top left*: KS plot. *Top right*: coefficient for x_k. *Bottom left*: long-term history coefficients with 95% confidence bounds. *Bottom right:* short-term history coefficients with 95% confidence bounds. Figure reproduced with permission from Sklar et al. (2017)

7.3.2 Gambling Task

This example is taken from Sacré et al. (2016a). Five participants being treated at the Cleveland Clinic Epilepsy Center performed the behavioral task in their hospital rooms using methods previously described in Johnson et al. (2014). The gambling task (Fig. 7.4 top left) is based on a simple game of high card where participants would win virtual money if their card beat the computer's card. Specifically, in the beginning of each trial, the participant controls a cursor via a planar manipulandum to a fixation target. During fixation, participants must center the cursor in less than 8 s. Once centered, the participant is shown his card (only 2, 4, 6, 8, or 10 are in the deck) for a duration of 2 s. The card is randomly chosen with equal distribution.

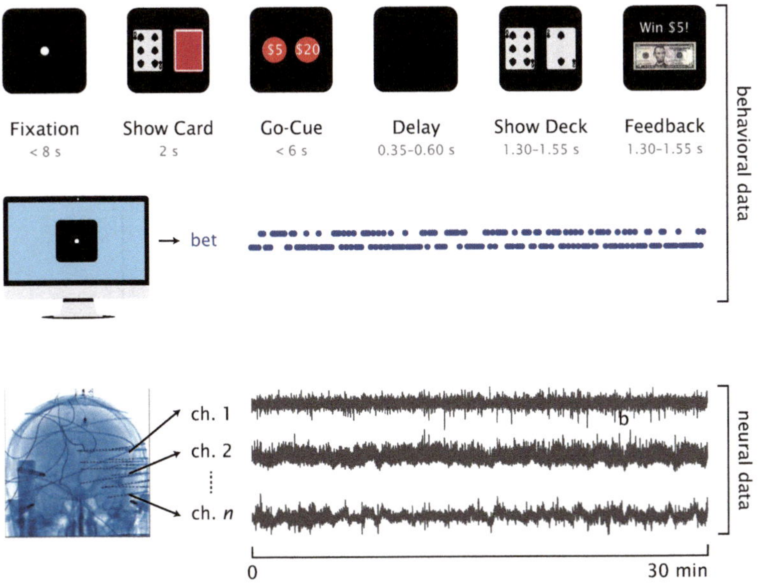

Fig. 7.4 Timeline of the gambling task. After fixation, subjects were shown their card. Once the bets were shown, subjects selected one of the choices and then were shown the computer's card following a delay. Feedback was provided afterwards by displaying the amount won or lost

The computer's card is initially hidden. The screen then shows the two possible choices: a high bet ($20) or a low bet ($5). The participant has 6 s to select one with his cursor. Following selection, the computer's card, which follows the same distribution, is revealed. If the computer's card is larger than the player's card, then the participant loses the amount he bets. If the computer's card is smaller than the player's card, then the participant wins the amount he bet.

For this task, the expected reward and variance of the reward are functions of the player's card and bet. For example, on 10-card trials, the expected reward is higher for a high bet than for a low bet and the variance of reward is small for both decisions. On 6-card trials, the expected reward is zero for both betting decisions; but the variance of reward is higher for a high bet than for a low bet.

In this task, bets and reaction times for each trial, k, were the behavior variables measured. Neural activity was measured with stereotactic EEG depth electrodes. Participants were implanted with 10–14 depth electrodes, each having 10–16 contacts. See Sacré et al. (2016a) for details.

To explore behavioral variability in the data, one can plot behavioral responses to each stimulus type over trials. In our gambling task, we plot the fraction of high bets (smoothed by taking a moving average) on each card-type trials for each patient over his/her session. This is shown in Fig. 7.5a.

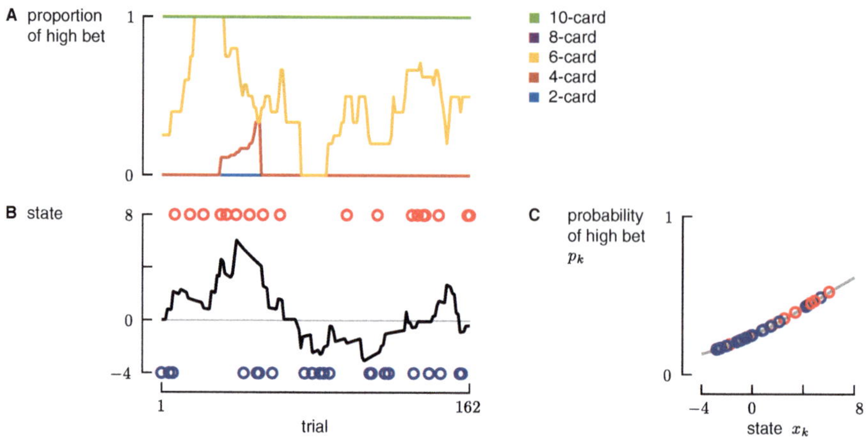

Fig. 7.5 Relationship between betting variability and internal state. **(a)** Moving average of the proportion of high bets over session for different card-type trials (overlapping windows of length $2w+1$, with $w = 10$). **(b)** Estimated state trajectory overlaid with bets on 6-card trials over session for one patient. **(c)** Estimated probability of betting high as a function of estimated state trajectory overlaid with bets (red for a high bet, blue for a low bet) for each 6-card trial

7.3.2.1 Dynamic State-Space Modeling

As seen in Fig. 7.5a, most of the within-participant variability is observed on 6-card trials across all participants. We hypothesized that participants bets on 6-card trials were influenced by past outcomes or a latent state variable that accumulated past outcomes. Specifically, we constructed a fading memory state model of cumulative mismatched expectations that we referred to as "luck" x_k on trial k. The luck variable is the scalar state variable that updates as follows:

$$x_{k+1} = a\,x_k + e_k \qquad x_0 = 0, \tag{7.18}$$

where a is a decay factor ($0 \leq a \leq 1$) and e_k is the mismatched expectations on trial k, that is, the difference between the actual outcome (loss $= -1$, draw $= 0$, or win $= 1$) and expected outcome given the player card pc_k (computed as $\frac{1}{5(pc_k-6)}$). Note that e_k enters the state evolution equation only during trials where expectations are mismatched.

Next, we estimated a in Eq. (7.18) by varying it between 0 and 1 in 0.01 increments and computed the Pearson's correlation coefficient between luck and gamma band power in the orbitofrontal cortex (OFC) at the beginning of each trial *before* the player sees his/her card (see Fig. 7.6). Thus, in this case, the OFC gamma power at the beginning of a trial was first found to be correlated to whether or not the player bets high if he/she receives a 6 card on that trial, and then the state-evolution model was constructed though a grid search.

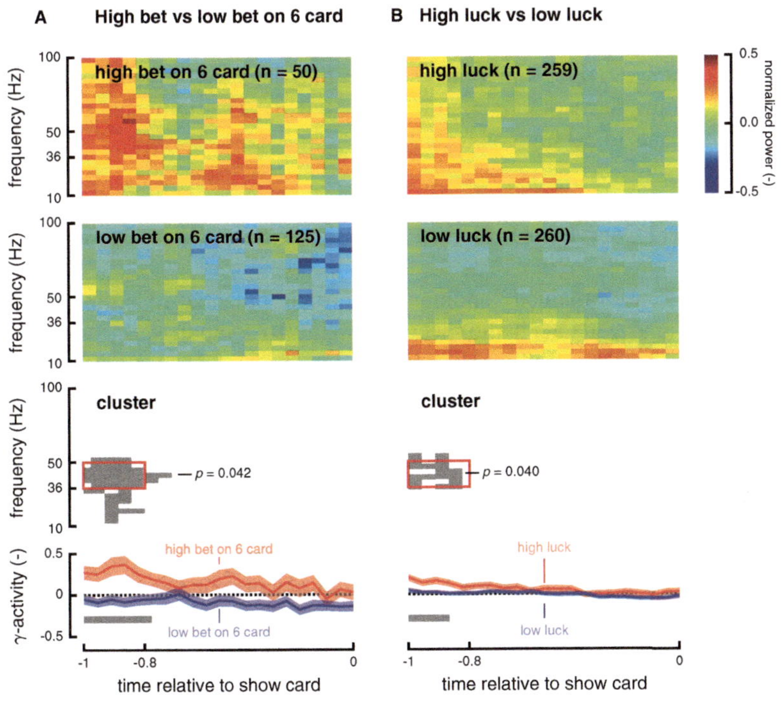

Fig. 7.6 Oscillatory power before the Show Card. **(a)** The average spectrograms show differences between high-bet and low-bet conditions on 6-card trials. One significant cluster ($p = 0.042$) resulted from the cluster-based nonparametric statistical test. The cluster contained frequencies between 36 and 50 Hz at a timing between 1000 and 800 ms before the Show Card. This frequency range matches the traditional lower gamma band. Plots of average oscillatory power (36–50 Hz) over time for 6-card trials resulting in high and low bets show the modulation of the power in the gamma band preceding the Show Card. Time bins with significant differences are marked by the grey bar. Error bars represent one standard error of the mean. The number n denotes the number of trials pooled across patients. **(b)** The average spectrograms show differences between high-luck and low-luck conditions on all trials. One significant cluster ($p = 0.040$) resulted from the cluster-based nonparametric statistical test. The cluster is located in the similar time-frequency region as the cluster emerging from the high-bet and low-bet conditions on 6-card trials. Figure reproduced with permission from Sacré et al. (2016a)

To see whether x_k explains the variability of behavior on 6-card trials, one can overlay the state x_k with bets on all trials across the session as shown in Fig. 7.5b.

To complete the state-space model, the output equation model is then a standard GLM for Bernoulli betting observations:

$$p_k = \frac{1}{1 + e^{-(d_0 + c\,x_k)}}, \qquad \text{for } k \text{ such that } pc_k = 6. \tag{7.19}$$

Equation (7.19) is a standard GLM when x_k is known or estimated a priori. If x_k is not estimated ahead of time, then the EM algorithm can be used. One can also overlay the behavioral data with the output models for $\hat{p}_k$. An example of this is shown in Fig. 7.5c.

In this example, since the output equation is constructed using a standard GLM, the fitted model can be evaluated by checking the significance of the parameter c in front of the state variable. The model showed that the state of "luck" significantly influenced betting decisions ($c = 0.20$, $p = 0.028$). This indicates positive luck biases participants to bet high on 6 cards.

7.3.2.2 Neural Correlates Informed by Dynamic State-Space Model

In this example, we obtained continuous local field potential recordings from the OFC and thus analyzed data in the spectral domain. To compute spectrograms, three orthogonal tapers were used with a 300-ms window sliding at 50-ms steps. Frequencies under 10 Hz were dropped because of the Rayleigh criterion and analyzed upwards to 100 Hz. Afterwards, each frequency bin's power was normalized based on the power across the entire recording session by fitting the log of the power in each frequency bin to a standard normal distribution. The mean and standard deviation used for the normalization were computed from the power between the 5th and 95th percentiles of the data set. This calculation was performed for every electrode's recording with the final normalized power being averaged across all electrodes in the brain region of interest (OFC in our example). In addition, we removed artifacts by identifying time points in the spectrograms for which the median of the absolute value of the power across all frequencies is larger than 2.5. Finally, in order to remove the effect of 60 Hz power-line noise, we ignored the frequency bins between 56.66 and 63.33 Hz in all analyses.

OFC oscillatory power was compared between the set of trials where subjects end up betting high on a 6 card and the set where they end up betting low. The average normalized spectrograms for both high and low bet trials showed that high bet trials have higher 40–50 Hz oscillatory power about 1000 ms preceding the show card epoch (Fig. 7.6a). To determine statistical significance of this effect, we used a cluster-based nonparametric statistical test described above. Clusters here are defined as a set of adjacent time-frequency windows whose activity is different between trials where the subjects end up betting high versus low.

To examine the correlation between OFC activity and the state variable, we separated the trials between high-luck and low-luck conditions (defined as the bottom third and top third of the values taken by luck variable for all patients) and computed the average normalized spectrograms for both conditions. High-luck trials showed higher OFC oscillatory power than low-luck trials (Fig. 7.6b, first and second panels). Interestingly, the cluster-based nonparametric statistical test identified a significant cluster ($p = 0.040$) in the time-frequency vicinity of the cluster identified when separating trials based on high-bet and low-bet conditions on 6-card trials (Fig. 7.6b, third panel).

These findings suggest OFC may play a pivotal role in processing a subject's internal (emotional) state during financial decision-making.

7.4 Discussion

In this discussion, we highlight four important lessons to use a two-step state-space modeling approach described in this chapter to explore links between behavior and neural activity in humans.

Lesson 1: Investigating Variability in Behavioral Data The first lesson is to always start data analyses by exploring the variability in the behavioral data prior to building a model. A good understanding of the variability existing in the behavioral data is the key to a useful model. There are essentially two sources of variability in a participant's behavior. The first source is that the behavior changes as the stimulus changes, which is expected. The second source is that the participant's behavior changes in a "smooth" way over trials during which the same stimulus is applied. This can happen when internal states, such as motivation and attention, vary over trials. If the latter variability is observed in the data, then a state-space modeling framework is appropriate. In the two examples described above, we plot behavior and see both sources of variability and thus move forward with model development. If the second source of variability is not present, then a simple GLM of the behavior may suffice in explaining the first source of behavioral variability, which is how a stimulus impacts behavior.

Lesson 2: Identifying Model Structure The second lesson is to identify a model structure that explains the variability that we observed. The design of the measurement map $h_{\boldsymbol{\theta}}$ is the easiest part at most of the time: it involves a combination of states $\boldsymbol{x}_k$ and inputs $\boldsymbol{u}_k$. The design of the state-transition map $f_{\boldsymbol{\theta}}$ is usually more complex. A useful analysis to guide the design of the state-transition map is to investigate the influence of candidate inputs by quantifying the influence of the value of candidate inputs at the previous trial $k - 1$ on the behavior at trial k.

Lesson 3: Estimating State-Space Model Step by Step The third lesson is to estimate the model parameters and the state for each trial step by step. A good approach is to start by estimating the parameters of the static model, that is, the model where the state is fixed to 0. Then, it is also sometimes useful to estimate the parameters of the dynamic model where we fix the noise in the state evolution to zero. Finally, we can estimate the parameters of the whole dynamic model by using the previous estimates as a first guess for this more complex estimation problem. The decomposition into these different steps helps to interpret the meaning of each parameter and its influence on the state.

Lesson 4: Dealing with Multiple Comparisons in Neural Data Analysis The fourth lesson is to deal with multiple comparisons in neural data analysis. Indeed, we are often interested in looking at the neural activity from multiple brain regions (when available) and at different epochs during the task. A standard approach to tackle this multiple comparisons problem is to correct the significance threshold by controlling the false discovery rate (e.g., $q = 0.05$).

Acknowledgements This MIST work was supported by the National Institutes of Health (NIH R01 MH106700, NIH K12 NS080223, NIH S10 OD018211, NIH R01 NS084142) and the Dana Foundation. This gambling work was supported by the National Science Foundation (NSF EFRI-MC3: #1137237). P.S. was supported by the Kavli Neuroscience Discovery Institute at the Johns Hopkins University.

References

Agarwal, R., Thakor, N. V., Sarma, S. V., & Massaquoi, S. G. (2015). PMv neuronal firing may be driven by a movement command trajectory within multidimensional Gaussian fields. *Journal of Neuroscience, 35*(25), 9508–9525.

Brown, E. N., Barbieri, R., Eden, U. T., & Frank, L. M. (2003). Likelihood methods for neural spike train data analysis. In J. Feng (Ed.), *Computational neuroscience: A comprehensive approach*, Chap. 9 (pp. 253–286). London: Chapman & Hall/CRC.

Carpenter, A. F., Georgopoulos, A. P., & Pellizzer, G. (1999). Motor cortical encoding of serial order in a context-recall task. *Science, 283*(5408), 1752–1757.

Coleman, T., & Sarma, S. (2007). Using convex optimization for nonparametric statistical analysis of point processes. In *Proceedings of IEEE International Symposium on Information Theory, 2007* (pp. 1476–1480).

Coleman, T. P., & Sarma, S. V. (2010). A computationally efficient method for nonparametric modeling of neural spiking activity with point processes. *Neural Computation, 22*(8), 2002–2030.

Cox, D. R., & Isham, V. (1980). *Point processes* (Vol. 12). Boca Raton: CRC Press.

Johnson, M. A., Thompson, S., Gonzalez-Martinez, J., Park, H. J., Bulacio, J., Najm, I., et al. (2014). Performing behavioral tasks in subjects with intracranial electrodes. *Journal of Visualized Experiments* (92), e51947. https://doi.org/10.3791/51947. https://www.jove.com/video/51947/performing-behavioral-tasks-in-subjects-with-intracranial-electrodes.

Johnson, N. L., & Kotz, S. (1970). *Continuous univariate distributions*. New York: Houghton Mifflin.

Louis, L. S. (1991). *Statistical signal processing: Detection, estimation, and time series analysis*. Boston, MA: Addison-Wesley Publishing Company.

Maris, E., & Oostenveld, R. (2007). Nonparametric statistical testing of EEG- and MEG-data. *Journal of Neuroscience Methods, 164*(1), 177–190.

Mauss, I. B., & Robinson, M. D. (2009). Measures of emotion: A review. *Cognition and Emotion, 23*(2), 209–237.

McCullagh, P., & Nelder, J. A. (1989). *Generalized linear models* (2nd ed.). London: Chapman & Hall/CRC.

Moon, T. K. (1996). The expectation-maximization algorithm. *IEEE Signal Processing Magazine, 13*(6), 47–60.

Sacré, P., Kerr, M. S. D., Kahn, K., González-Martínez, J., Bulacio, J., Park, H.-J., et al. (2016a). Lucky rhythms in orbitofrontal cortex bias gambling decisions in humans. *Scientific Reports, 6*, 36206.

Sacré, P., Kerr, M. S. D., Subramanian, S., Kahn, K., González-Martínez, J., Johnson, M. A., et al. (2016b). Winning versus losing during gambling and its neural correlates. In *Proceedings of 50th Annual Conference on Information Science and Systems, Princeton, NJ* (pp. 557–561).

Sacré, P., Kerr, M. S. D., Subramanian, S., Kahn, K., González-Martínez, J., Johnson, M. A., et al. (2016c). The precuneus may encode irrationality in human gambling. In *Proceedings of 38th Annual International Conference on IEEE Engineering in Medicine and Biology Society, Orlando, FL* (pp. 3406–3409).

Santaniello, S., Gale, J. T., Montgomery, E., & Sarma, S. V. (2010). Modeling the motor striatum under deep brain stimulation in normal and MPTP conditions. In *Proceedings of 32nd Annual International Conference on IEEE Engineering in Medicine and Biology Society* (pp. 2065–2068).

Santaniello, S., Montgomery, E. B., Gale, J. T., & Sarma, S. V. (2012). Non-stationary discharge patterns in motor cortex under subthalamic nucleus deep brain stimulation. *Frontiers in Integrative Neuroscience, 6*, 35.

Sarma, S. V., Cheng, M. L., Eden, U. T., Williams, Z., Brown, E. N., & Eskandar, E. N. (2012). The effects of cues on neurons in the basal ganglia in Parkinson's disease. *Frontiers in Integrative Neuroscience, 6*, 40.

Sarma, S. V., Eden, U. T., Cheng, M. L., Williams, Z. M., Hu, R., Eskandar, E. N., et al. (2010). Using point process models to compare neural spiking activity in the subthalamic nucleus of Parkinson's patients and a healthy primate. *IEEE Transactions in Biomedical Engineering, 57*(6), 1297–1305.

Schieber, M. H. (2004). Motor control: Basic units of cortical output? *Current Biology, 14*(9), R353–R354.

Shoham, S., Fellows, M. R., & Normann, R. A. (2003). Robust, automatic spike sorting using mixtures of multivariate *t*-distributions. *Journal of Neuroscience Methods, 127*(2), 111–122.

Sklar, S., Walmer, M., Sacré, P., Schevon, C. A., Srinivasan, S., Banks, G. P., et al. (2017). Neuronal activity in human anterior cingulate cortex modulates with internal cognitive state during multi-source interference task. In *Proceedings of 39th Annual International Conference on IEEE Engineering in Medicine and Biology Society* (pp. 962–965).

Smith, E. H., Banks, G. P., Mikell, C. B., Cash, S. S., Patel, S. R., Eskandar, E. N., et al. (2015). Frequency-dependent representation of reinforcement-related information in the human medial and lateral prefrontal cortex. *Journal of Neuroscience, 35*(48), 15827–15836.

Snyder, D. L., & Miller, M. I. (1991). *Random point processes in time and space*. New York: Springer.

Truccolo, W., Eden, U. T., Fellows, M. R., Donoghue, J. P., & Brown, E. N. (2005). A point process framework for relating neural spiking activity to spiking history, neural ensemble, and extrinsic covariate effects. *Journal of Neurophysiology, 93*(2), 1074–1089.

Van Trees, H. L. (1968). *Detection, Estimation, and Modulation Theory, Part I*. New York, NY: Wiley.

Chapter 8
Brain–Machine Interfaces

Maryam M. Shanechi

8.1 An Overview of Motor BMIs

Motor brain–machine interfaces (BMI) have the potential to restore movement to patients with disabling neurological injury or disease. Motor BMIs allow subjects to control external devices by modulating their neural activity. To do so, BMIs record the neural activity from motor cortical areas, use a mathematical transform called the "decoder" to infer the subject's motor intent and move an external device, and provide visual feedback of the generated movement to the subject (Fig. 8.1), closing the loop. Thus motor BMIs can be viewed as closed-loop control systems in which the brain is the controller and the neuroprosthetic is the plant.

Early studies (Humphrey et al. 1970; Fetz 1969) showed that movement kinematics in non-human primates (NHP) can be estimated from motor cortical neural populations (Humphrey et al. 1970) and that NHPs can volitionally modulate their neural activity based on biofeedback (Fetz 1969). These studies demonstrated the feasibility of motor BMIs. Since then motor BMI studies have shown that rodents, NHPs, and humans can operate external devices using their neural activity (Chapin et al. 1999; Serruya et al. 2002; Taylor et al. 2002; Carmena et al. 2003; Musallam et al. 2004; Santhanam et al. 2006; Hochberg et al. 2006, 2012; Velliste et al. 2008; Kim et al. 2008; Mulliken et al. 2008; Li et al. 2009, 2011; Ganguly and Carmena 2009; Suminski et al. 2010; Mahmoudi and Sanchez 2011; O'Doherty et al. 2011; Gilja et al. 2012, 2015; Orsborn et al. 2012, 2014; Hauschild et al. 2012; Shanechi et al. 2012, 2013a, 2016, 2017; Collinger et al. 2013; Willett et al. 2013; Thakor 2013; McMullen et al. 2014; Aflalo et al. 2015). Moreover, by combining decoding

M.M. Shanechi (✉)
University of Southern California, Los Angeles, CA, USA
e-mail: shanechi@usc.edu

Z. Chen, S.V. Sarma (eds.), *Dynamic Neuroscience*,
https://doi.org/10.1007/978-3-319-71976-4_8

Fig. 8.1 Schematic of BMIs. BMIs are closed-loop control systems in which the brain controls the prosthetic plant by modulating its neural activity. BMIs provide feedback to the subject often in the form of visual feedback

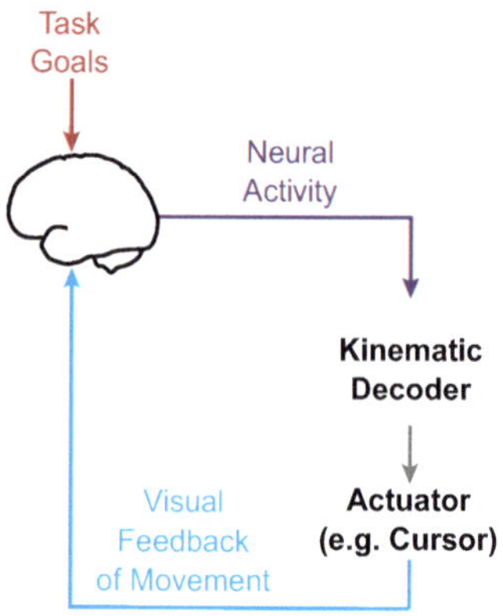

with stimulation, motor BMIs have been able to control the subject's native limb (Moritz et al. 2008; Ethier et al. 2012; Shanechi et al. 2014; Bouton et al. 2016; Capogrosso et al. 2016).

While various invasive and non-invasive neural signal modalities could be used as the control signal in BMIs, the highest levels of performance to date have been achieved by using intracortical spiking activity recorded from penetrating electrodes (Chapin et al. 1999; Serruya et al. 2002; Taylor et al. 2002; Musallam et al. 2004; Santhanam et al. 2006; Carmena et al. 2003; Hochberg et al. 2006, 2012; Velliste et al. 2008; Kim et al. 2008; Mulliken et al. 2008; Li et al. 2009, 2011; Ganguly and Carmena 2009; Suminski et al. 2010; Mahmoudi and Sanchez 2011; O'Doherty et al. 2011; Gilja et al. 2012, 2015; Orsborn et al. 2012, 2014; Hauschild et al. 2012; Shanechi et al. 2012, 2013a, 2014, 2016, 2017; Collinger et al. 2013; Willett et al. 2013; Thakor 2013; Moritz et al. 2008; Ethier et al. 2012; Aflalo et al. 2015; Bouton et al. 2016; Capogrosso et al. 2016). To move these BMI towards clinical viability, a critical component that should be improved is the decoding algorithm.

Most BMI decoders control the continuous kinematics of the neuroprosthetic (Taylor et al. 2002; Serruya et al. 2002; Carmena et al. 2003; Hochberg et al. 2006, 2012; Velliste et al. 2008; Kim et al. 2008; Mulliken et al. 2008; Ganguly and Carmena 2009; Suminski et al. 2010; Li et al. 2009, 2011; Mahmoudi and Sanchez 2011; O'Doherty et al. 2011; Gilja et al. 2012, 2015; Orsborn et al. 2012, 2014; Hauschild et al. 2012; Collinger et al. 2013; Willett et al. 2013; Thakor 2013; Shanechi et al. 2013a, 2016, 2017; McMullen et al. 2014; Aflalo et al. 2015). Some BMI studies have also decoded discrete movement targets (Musallam et al. 2004; Santhanam et al. 2006; Shanechi et al. 2013a; Aflalo et al. 2015) or an entire sequence of targets before movement initiation (Shanechi et al. 2012). Moreover, some work have jointly decoded the target and kinematics of movement (Yu et al. 2007; Srinivasan et al. 2006; Mulliken et al. 2008; Shanechi et al. 2013b,a). In this chapter we focus on BMI decoding algorithms for control of continuous movement given their potential for generalizability to various tasks.

To develop a motor BMI decoder, multiple computational elements should be designed. First, we need to decide on a parametric model structure to characterize the relationship between the spikes and kinematic states. This model structure is referred to as the "encoding model." Second a calibration or training method should be developed to learn the encoding model parameters for each subject. Finally, BMI design could benefit from modeling how movements are generated within the closed-loop system. In this chapter, we discuss these elements and how they can be used to design BMI decoders that use spiking activity. Our emphasis will be on recent designs that have significantly improved performance by using tools from control theory, statistical inference, and adaptive estimation. These designs have the potential to bring BMI systems closer to clinical viability.

8.2 BMI Decoding Structures

To design a decoder, we need to decide on the choice of a parametric model structure to characterize the relationship between the spikes and the kinematic states. This model structure is referred to as the "encoding model." In the vast majority of BMIs, the input has been taken as the binned spike counts (or equivalently the firing rates) computed in bin sizes typically of 50–100 ms length (Serruya et al. 2002; Taylor et al. 2002; Carmena et al. 2003; Velliste et al. 2008; Kim et al. 2008; Mulliken et al. 2008; Li et al. 2009, 2011; Ganguly and Carmena 2009; Suminski et al. 2010; Hochberg et al. 2006, 2012; Gilja et al. 2012, 2015; Orsborn et al. 2012, 2014; Hauschild et al. 2012; Collinger et al. 2013; Willett et al. 2013). Consequently, the encoding model in these systems has been constructed by assuming that movements are represented linearly in the spike counts. These linear encoding models have led to decoders such as the population vector (PV), the optimal linear estimator (OLE), the Wiener filter, and the Kalman filter (KF). However, spikes can be modeled as a time-series of 0's and 1's, representing the absence or presence of a spike at a given time, respectively. This binary time-series can be modeled as a point process (Brown et al. 1998, 2001; Kass and Ventura 2001; Truccolo et al. 2005). Point process models have been studied in offline or numerical simulation studies, for example to model the spiking activity in the hippocampus (Brown et al. 1998) or for offline motor decoding (Eden et al. 2004; Srinivasan et al. 2006; Shanechi et al. 2013b). Recent algorithmic advances have led to closed-loop BMIs that use the spikes directly at their millisecond time-scale using point process modeling (Shanechi et al. 2013a, 2016, 2017), resulting in performance improvements (Shanechi et al. 2017). Here we first review the linear filters that have been vastly used in the BMI field and then discuss the recent BMI architectures that have incorporated the spikes directly using point process modeling.

8.2.1 Wiener and Kalman Filters

Early BMI studies used linear filters that take as input binned spike counts or equivalently the estimated firing rates within a given time bin. These decoders included the Wiener filter (Serruya et al. 2002; Carmena et al. 2003; Mulliken et al. 2008; Ganguly and Carmena 2009; Suminski et al. 2010; Hochberg et al. 2006; Willett et al. 2013) and the closely related PV (Taylor et al. 2002; Velliste et al. 2008) and OLE algorithms (Collinger et al. 2013; Chase et al. 2009; Koyama et al. 2009). These algorithms estimate the kinematics as a linear function of the firing rates over a desired time-window, and have been incorporated in many BMI systems in animal and human studies (Serruya et al. 2002; Carmena et al. 2003; Mulliken et al. 2008; Ganguly and Carmena 2009; Suminski et al. 2010; Hochberg et al. 2006; Willett et al. 2013; Collinger et al. 2013; Chase et al. 2009; Koyama et al. 2009; Taylor et al. 2002; Velliste et al. 2008). However, these decoders do not include a model of the movement in their computations. Since movements have structure, a kinematic model of movement can further help with decoding.

Recursive Bayesian decoders (Kailath et al. 2000) can include such a dynamical model of movement. To date, the most commonly used Bayesian decoder in the BMI field has been the Kalman filter. Kalman filters also use as their input the binned spike counts (Kim et al. 2008; Mulliken et al. 2008; Li et al. 2011, 2009; Hochberg et al. 2012; Gilja et al. 2012; Orsborn et al. 2012, 2014; Hauschild et al. 2012). A recursive Bayesian estimator consists of a prior state model and an observation model. The state model $p(\mathbf{x}_0, \cdots, \mathbf{x}_t)$ characterizes the kinematic states and the observation model $p(\mathbf{y}_0, \cdots, \mathbf{y}_t | \mathbf{x}_0, \cdots, \mathbf{x}_t)$ relates the neural activity to these kinematics. The filter then computes the posterior density, $p(\mathbf{x}_t | \mathbf{y}_0, \cdots, \mathbf{y}_t)$ at time t based on the observations up to that time. Recursive Bayesian filters are derived by solving the Bayes' rule and Chapman-Kolmogorov system of equations (Arulampalam et al. 2002). The recursions consist of a prediction step that predicts the kinematic state at time t from its estimate at time $t-1$, and an update step that corrects this prediction using the neural observation at time t. The prediction step finds the prediction density $p(\mathbf{x}_t | \mathbf{y}_1, \cdots, \mathbf{y}_{t-1})$ as

$$p(\mathbf{x}_t | \mathbf{y}_1, \cdots, \mathbf{y}_{t-1}) = \int p(\mathbf{x}_t | \mathbf{x}_{t-1}) p(\mathbf{x}_{t-1} | \mathbf{y}_0, \cdots, \mathbf{y}_{t-1}) d\mathbf{x}_{t-1} \tag{8.1}$$

where $p(\mathbf{x}_t | \mathbf{x}_{t-1})$ is the known prior state transition model and $p(\mathbf{x}_{t-1} | \mathbf{y}_0, \cdots, \mathbf{y}_{t-1})$ is the previous time-step posterior density. The above is the Chapman-Kolmogorov equation. Using the Bayes' rule, the update step finds the posterior density as

$$p(\mathbf{x}_t | \mathbf{y}_1, \cdots, \mathbf{y}_t) = \frac{p(\mathbf{y}_t | \mathbf{x}_t) p(\mathbf{x}_t | \mathbf{y}_1, \cdots, \mathbf{y}_{t-1})}{\int p(\mathbf{y}_t | \mathbf{x}_t) p(\mathbf{x}_t | \mathbf{y}_1, \cdots, \mathbf{y}_{t-1}) d\mathbf{x}_t} \tag{8.2}$$

where $p(\mathbf{y}_t | \mathbf{x}_t)$ is the observation model. This creates the recursion as the present step posterior is written as a function of the last step posterior. The minimum mean-squared error (MMSE) estimate of the kinematic states $\hat{\mathbf{x}}_t$ is given by the mean of the posterior density at time t, denoted as $\mathbf{x}_{t|t}$ (Kailath et al. 2000).

The Kalman filter is a special case of a recursive Bayesian decoder where the prior state model and the observation model are linear and Gaussian. In this case, the state and observation models are given as

$$\mathbf{x}_t = \mathbf{A}\mathbf{x}_{t-1} + \mathbf{w}_t \tag{8.3}$$

$$\mathbf{y}_t = \mathbf{H}\mathbf{x}_t + \mathbf{q}_t \tag{8.4}$$

This model is also referred to as a state-space model. Here $\mathbf{A}$ is the dynamics matrix, and $\mathbf{w}_t$ and $\mathbf{q}_t$ are zero-mean white Gaussian state and observation noises with covariances $\mathbf{W}$ and $\mathbf{Q}$, respectively. $\mathbf{A}$ can be selected by fitting it to arm movements while obeying physical laws (e.g., position is the integral of velocity), and $\mathbf{H}$ and $\mathbf{Q}$ are the observation model parameters and need to be learned in decoder training (see Sect. 8.3). Denoting the posterior covariance by $\Lambda_{t|t}$ and the prediction mean and covariance by $\mathbf{x}_{t|t-1}$ and $\Lambda_{t|t-1}$, respectively, the Kalman prediction and update steps are

$$\mathbf{x}_{t|t-1} = \mathbf{A}\mathbf{x}_{t-1|t-1} \tag{8.5}$$

$$\Lambda_{t|t-1} = \mathbf{A}\Lambda_{t-1|t-1}\mathbf{A}^{\top} + \mathbf{W} \tag{8.6}$$

$$\mathbf{K}_t = \Lambda_{t|t-1}\mathbf{H}^{\top}(\mathbf{H}\Lambda_{t|t-1}\mathbf{H}^{\top} + \mathbf{Q})^{-1} \tag{8.7}$$

$$\Lambda_{t|t} = (\mathbf{I} - \mathbf{K}_t\mathbf{H})\Lambda_{t|t-1} \tag{8.8}$$

$$\mathbf{x}_{t|t} = \mathbf{x}_{t|t-1} + \mathbf{K}_t(\mathbf{y}_t - \mathbf{H}\mathbf{x}_{t|t-1}) \tag{8.9}$$

where $\mathbf{K}_t$ is the standard Kalman gain. Kalman filters have been extensively used in BMIs (Kim et al. 2008; Mulliken et al. 2008; Li et al. 2011; Hochberg et al. 2012; Gilja et al. 2012; Orsborn et al. 2012, 2014; Hauschild et al. 2012) including clinical trials (Hochberg et al. 2012; Gilja et al. 2015).

The above linear filters all take as input the spike counts computed in bin lengths typically of 50–100 ms. They then model these counts as a Gaussian process tuned to kinematics as in Eq. (8.4). However, spikes happen at a millisecond time-scale. Hence linear filters control the BMI at slower time-scales than the millisecond-by-millisecond time-scale of the spikes. Moreover, from the central limit theorem (Kailath et al. 2000), the Gaussian approximation on the spike counts is reasonable in the limit of a large number of spikes per bin. Thus the accuracy of the assumptions in these linear filters depends on their bin width and may not hold at fast time-scales (i.e, short bin-widths). These filters are thus time-scale dependent. Recent studies have designed BMIs that run at the millisecond time-scale of the spiking activity using point process modeling to improve performance and robustness by allowing for rapid adaptation, control and feedback rates (Shanechi et al. 2013a, 2016, 2017). Before presenting these high-rate BMI designs, we first present the point process model and the corresponding *point process filter* (PPF).

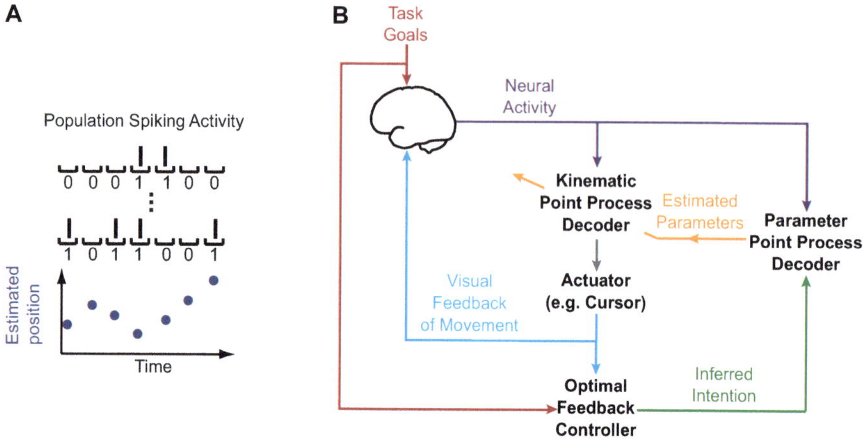

Fig. 8.2 Control-theoretic high-rate BMI using adaptive OFC-PPF. Figure is adapted from Shanechi et al. (2016). (**a**) PPF processes the spikes directly by modeling them as a binary time-series. This allows the PPF to have fast millisecond-by-millisecond adaptation, control, and feedback rates. (**b**) The process of adaptive OFC-PPF BMI architecture is shown. During CLDA, the OFC model is used to infer the motor intent of the subject based on the task goal (e.g., the instructed target in a center-out task) and the visual feedback of the kinematics. A parameter PPF is then used to estimate the parameters of the encoding models based on the OFC-inferred intended kinematics and the simultaneously recorded spiking activity. These estimated parameters are used in the PPF kinematic decoder. After performance converges, adaptation stops and the trained PPF kinematic decoder is used by the subject to control the BMI

8.2.2 *Point Process Filters*

To directly model the spikes, we can bin them in small intervals Δ such that each interval contains at most one spike (typically 1–5 ms). The resulting time-series will be binary and consist of a sequence of 0's and 1's (Fig. 8.2a). This binary time-series can then be modeled as a point process (Brown et al. 1998, 2001; Kass and Ventura 2001; Eden et al. 2004; Truccolo et al. 2005). We denote the spiking activity of an ensemble of C neurons by $\mathbf{N}_1, \cdots, \mathbf{N}_t$ where $\mathbf{N}_t = (N_t^1, \cdots, N_t^C)$ is the binary spike events of the C neurons at time t. We assume that neurons are conditionally independent given the kinematics states. The point process observation model for a neural population is then given by Brown et al. (1998), Brown et al. (2001), Kass and Ventura (2001), Eden et al. (2004), Truccolo et al. (2005)

$$p(\mathbf{N}_t|\mathbf{x}_t) = \prod_c \left(\lambda_c(t|\mathbf{x}_t, \boldsymbol{\phi}_t^c)\Delta\right)^{N_t^c} e^{-\lambda_c(t|\mathbf{x}_t, \boldsymbol{\phi}_t^c)\Delta} \tag{8.10}$$

Here $\lambda_c(t|\mathbf{x}_t, \boldsymbol{\phi}_t^c)$ is the instantaneous firing rate of neuron c at time t, and $\boldsymbol{\phi}_t^c$ is the model parameters for neuron c that need to be estimated in BMI training.

The instantaneous firing rate of motor cortical neurons can be characterized as a modified cosine tuning model (Moran and Schwartz 1999; Truccolo et al. 2005) by

$$\begin{aligned}\lambda_c(t|\mathbf{x}_t, \boldsymbol{\phi}_t^c) &= \lambda_c(t|\mathbf{v}_t, \boldsymbol{\phi}_t^c) = \exp(\beta_t^c + \boldsymbol{\alpha}_t^{c\top}\mathbf{v}_t) \\ &= \exp([1, \mathbf{v}_t^\top]\boldsymbol{\phi}_t^c),\end{aligned} \tag{8.11}$$

where $\mathbf{x}_t = [\mathbf{d}_t, \mathbf{v}_t]^\top$ is the kinematic state where the components represent position and velocity in the two dimensions, respectively. Here $\boldsymbol{\phi}_t^c = [\beta_t^c; \boldsymbol{\alpha}_t^c]$ are the encoding model parameters to be learned (see Sect. 8.3).

The prior kinematic model for the PPF can be written in the same form as (8.3) in the Kalman filter but adjusting for a smaller time-scale. For example, to enforce continuity in the evolution of velocity, this prior model can be written as

$$\mathbf{x}_t = \mathbf{A}\mathbf{x}_{t-1} + \mathbf{w}_t, \mathbf{A} = \begin{bmatrix} 1 & 0 & \Delta & 0 \\ 0 & 1 & 0 & \Delta \\ 0 & 0 & a & 0 \\ 0 & 0 & 0 & a \end{bmatrix}, \mathbf{W} = \text{diag}\begin{bmatrix} 0 \\ 0 \\ w \\ w \end{bmatrix} \tag{8.12}$$

where a and w are fit to the subject's end-point manual kinematics using maximum-likelihood (ML) estimation (Orsborn et al. 2012; Shanechi et al. 2016).

PPF recursions can be derived by solving the general equations in Eqs. (8.1) and (8.2) for the prior model in Eq. (8.12) and the point process observation model in Eq. (8.10) using Laplace-type approximations (Brown et al. 1998; Eden et al. 2004; Truccolo et al. 2005; Shanechi et al. 2013a, 2016). For the log-linear function in Eq. (8.11), PPF can be written as (Shanechi et al. 2013a)

$$\mathbf{x}_{t|t-1} = \mathbf{A}\mathbf{x}_{t-1|t-1} \tag{8.13}$$

$$\Lambda_{\mathbf{x}_{t|t-1}} = \mathbf{A}\Lambda_{\mathbf{x}_{t-1|t-1}}\mathbf{A}^\top + \mathbf{W} \tag{8.14}$$

$$\Lambda_{\mathbf{x}_{t|t}}^{-1} = \Lambda_{\mathbf{x}_{t|t-1}}^{-1} + \sum_{c=1}^{C} \tilde{\boldsymbol{\alpha}}_{t-1|t-1}^c \tilde{\boldsymbol{\alpha}}^{c\top}_{t-1|t-1} \lambda_c(t|\mathbf{v}_{t|t-1}, \boldsymbol{\phi}_{t-1|t-1}^c)\Delta \tag{8.15}$$

$$\mathbf{x}_{t|t} = \mathbf{x}_{t|t-1} + \Lambda_{\mathbf{x}_{t|t}} \sum_{c=1}^{C} \tilde{\boldsymbol{\alpha}}_{t-1|t-1}^c (N_t^c - \lambda_c(t|\mathbf{v}_{t|t-1}, \boldsymbol{\phi}_{t-1|t-1}^c)\Delta) \tag{8.16}$$

where $\boldsymbol{\phi}_{t-1|t-1}^c = [\beta_{t-1|t-1}^c, \boldsymbol{\alpha}_{t-1|t-1}^c]^\top$ are the estimated parameters found during BMI training (Shanechi et al. 2016) (see Sect. 8.3), and $\tilde{\boldsymbol{\alpha}}_{t-1|t-1}^c = [\mathbf{0}, \boldsymbol{\alpha}_{t-1|t-1}^c]^\top$.

PPF was first used on offline decoding of rat locations in a maze based on the hippocampal place cell spiking activity (Brown et al. 1998). Later work used PPF for decoding of movement intention in numerical simulation or offline studies (Brockwell et al. 2004; Srinivasan et al. 2006; Shanechi et al. 2013b; Eden et al. 2004; Truccolo et al. 2005). As we will show in Sect. 8.3, recently BMI work has used adaptive filtering and control-theoretic modeling (Shanechi et al. 2013a,b) to enable the development of closed-loop PPF BMIs (Shanechi et al. 2016, 2017).

8.3 BMI Calibration

Regardless of the decoder structure, its parameters need to be learned for each subject based on data. Thus the decoder calibration or training method is another critical element of a BMI system. The majority of BMIs to date have been trained in open-loop experiments prior to real-time control. In these open-loop experiments, subjects are instructed to move their arms or imagine movements while their neural activity is being recorded. The obtained datasets are then used to fit the model parameters, for example using least-squares or ML methods. However, recent studies have shown that neural representations of movement can be different for movement of a BMI compared to arm movements or to motor imagery (Taylor et al. 2002; Carmena et al. 2003; Ganguly and Carmena 2009). These changes in neural representation have motivated the design of adaptive algorithms that learn the model parameters as subjects control the BMI in closed loop (Taylor et al. 2002; Velliste et al. 2008; Gilja et al. 2012; Orsborn et al. 2012; Collinger et al. 2013; Mahmoudi and Sanchez 2011; Hochberg et al. 2012; Dangi et al. 2014; Shanechi et al. 2016, 2017). These methods are often termed *closed-loop decoder adaptation* (CLDA) algorithms.

CLDA algorithms have been guided by a closed-loop control view of the BMI (Fig. 8.1) and have improved performance compared to open-loop training methods. A CLDA is typically comprised of three elements. First, decoder parameters are initialized, for example, based on arm reaching movements (Gilja et al. 2012) or randomly (Orsborn et al. 2012; Shanechi et al. 2016). The subject then uses the initialized decoder to make BMI movements towards instructed targets. However, these BMI movements are not precise given the suboptimality of the initialized decoder. Hence as the second element, an intention estimation method is used to infer the intended kinematics during adaptation. Finally, as the third element, an algorithm is used to fit the parameters based on the inferred intentions and the recorded neural activity.

A common approach for intention estimation, sometimes termed the CursorGoal method (Gilja et al. 2012; Fan et al. 2014), infers the intended velocity by assuming that the subject aims to go straight towards the target at each time. Hence the direction of the intended velocity is found by rotating the decoded velocity at each time towards the instructed target. CursorGoal does not infer the intended speed and instead sets it equal to the decoded speed. At the target, the intended speed is set to zero.

In addition to the method of intention estimation, a CLDA algorithm should devise a technique to learn the parameters based on the inferred intentions and the recorded neural activity. Most CLDA methods for Kalman filters have been batch-based. These methods collect batches of neural activity on the time-scale of minutes, and refit the decoder parameters within these batches using ML estimation (Gilja et al. 2012; Orsborn et al. 2012; Dangi et al. 2014). Parameter estimates from previous batches are then either replaced with these new estimates (Gilja et al. 2012) or averaged with previous batch estimates either continuously (Dangi et al. 2014), or intermittently using a method termed *SmoothBatch* (Orsborn et al. 2012).

CursorGoal intention estimation combined with batch-based ML methods has been the basis for CLDA training of closed-loop Kalman filters (Gilja et al. 2012, 2015; Orsborn et al. 2012, 2014; Dangi et al. 2014). CLDA methods have improved the performance of Kalman filters compared to open-loop training (Gilja et al. 2012). As we discuss in the next sections, the CursorGoal method of intention estimation does not infer the speed and moreover does not model the closed-loop processes within the BMI. Moreover, batch-based CLDA methods result in a slow rate of adaptation on a time-scale of minutes. To resolve these issues, recent PPF BMIs have explicitly modeled the BMI as a closed-loop controller to infer the intended movement and have enabled a high millisecond-by-millisecond rate of adaptation (Shanechi et al. 2016, 2017). Having provided an overview of the BMI designs to date, we will now discuss this new control-theoretic PPF BMI design.

8.4 Control-Theoretic High-Rate BMIs Using Optimal Feedback-Control Modeling and Adaptive Point Process Filtering

In this section, we present our recent work that has enabled the development of control-theoretic high-rate BMIs resulting in improvement in robustness and performance (Shanechi et al. 2016, 2017). There are two main elements that have led to these designs. First, these designs have explicitly modeled the BMI as an optimal feedback control (OFC) system to better infer the brain's motor intentions. The OFC model has been used as a method of intention estimation during decoder adaptation. Second, these BMIs have enabled adaptation, control, and feedback at the fast millisecond-by-millisecond rate of the spiking activity using point process modeling and PPF parameter and kinematic decoding. We refer to this BMI architecture as the adaptive OFC-PPF.

In the following sections, we first present the computational elements of adaptive OFC-PPF. We then present recent experimental data in Shanechi et al. (2016, 2017) that show how adaptive OFC-PPF improves both the transient and the steady-state operation of BMIs. During the learning phase, adaptive OFC-PPF allows for faster parameter convergence compared to batch-based methods because of the fast adaptation rates. Moreover, it results in a more accurate steady-state decoder because of the OFC method of intention estimation (as compared with CursorGoal). During steady-state operation, the resulting PPF decoder significantly improves performance compared to the state-of-the-art Kalman filter due to the fast control rate, the fast feedback rate, and the point process mathematical encoding model.

8.4.1 *Optimal Feedback-Control Model of Brain Behavior in Closed-Loop BMI Control*

A BMI system can be modeled as an optimal feedback-control system (Shanechi et al. 2013a,b, 2016, 2017). In this model, the brain is the controller and selects the next neural command based on the visual feedback of the prosthetic device and the task goal. This OFC model is inspired by the theories of the natural sensorimotor control (Todorov and Jordan 2002; Todorov 2004; Shadmehr and Krakauer 2008). The OFC model can be constructed by specifying an approximate forward kinematics model, quantifying the task goals as cost functions, and modeling the visual feedback. Adaptive OFC-PPF decoder uses this OFC model of the brain's control behavior to infer the subject's intended kinematics during decoder adaptation.

We model the evolution of the kinematic states $\mathbf{x}_0, \ldots, \mathbf{x}_t$ in the OFC model as

$$\mathbf{x}_t = \mathbf{A}\mathbf{x}_{t-1} + \mathbf{B}\mathbf{u}_{t-1} + \mathbf{w}_{t-1}. \tag{8.17}$$

Here $\mathbf{A}$ and $\mathbf{B}$ are parameters that can be fitted based on the subject's manual movements, $\mathbf{u}_t$ is the brain's control command at time t, and $\mathbf{w}_t$ is a zero-mean white Gaussian state noise with covariance matrix $\mathbf{W}$, which represents the uncertainty in the forward model. We assume that the visual feedback is perfect and instantaneous, i.e., that the subject observes the decoded kinematics $\mathbf{x}_{t|t}$ at each time. This OFC model also assumes that the brain has learned an internal forward model of movement as evidenced in prior studies (Shadmehr and Krakauer 2008; Golub et al. 2012).

To infer the intended kinematics, we construct a cost function that quantifies the goal of the task and then minimize the expected value of this cost function over choices of $\mathbf{u}_t$. In BMI training, model adaptation is typically performed in a supervised session in which subjects are instructed to move the prosthetic (e.g., cursor) towards known targets. Hence the goal in this supervised training task is to reach the target and to stop there while using minimum effort. Again, taking the kinematic state to be $\mathbf{x}_t = [\mathbf{d}_t, \mathbf{v}_t]^\top$, where the components represent the cursor's position and velocity in the two dimensions, and denoting the target position by $\mathbf{d}^*$, we form the cost function as

$$J = \sum_{t=1}^{\infty} \| \mathbf{d}_t - \mathbf{d}^* \|^2 + w_v \| \mathbf{v}_t \|^2 + w_r \| \mathbf{u}_t \|^2, \tag{8.18}$$

where the three terms in the sum enforce positional accuracy, stopping condition, and energetic efficiency, respectively. The scalar weights w_r and w_v in the cost function (8.18) can be approximately chosen such that the OFC model generated movements resemble naturalistic movements. These weights can be then validated and refined experimentally (Shanechi et al. 2016). The optimal control solution $\mathbf{u}_t$ at each time that minimizes the expected cost is given by a linear function of the brain's estimate of the state at that time (Bertsekas 2005). This is the standard linear-quadratic-Gaussian (LQG) solution. Given the assumption of noiseless visual feedback, this estimate is equal to the displayed state on the screen, $\mathbf{x}_{t|t}$ i.e.,

$$\mathbf{u}_t = -\mathbf{L}(\mathbf{x}_{t|t} - \mathbf{x}^*), \tag{8.19}$$

where $\mathbf{x}^* = [\mathbf{d}^*, \mathbf{0}]^\top$ is the target state for position and velocity, and $\mathbf{L}$ is the steady-state solution to the discrete form of the algebraic Riccati equation found recursively and offline (Bertsekas 2005).

To learn the decoder parameters, we use the OFC model to infer the intended velocity. The OFC model just needs knowledge of the task goal and of the visual feedback to the subject, which are both independent of the quality of the kinematic decoder. The visual feedback is just the decoded kinematics $\mathbf{x}_{t|t}$ and the task goal is quantified as in Eq. (8.18). Hence the intended kinematics at each time, denoted by $\tilde{\mathbf{x}}_t = [\tilde{\mathbf{d}}_t; \tilde{\mathbf{v}}_t]^\top$, are found from Eqs. (8.19) and (8.17) as

$$\tilde{\mathbf{x}}_t = (\mathbf{A} - \mathbf{BL})\mathbf{x}_{t|t} + \mathbf{BL}\mathbf{x}^*. \tag{8.20}$$

8.4.2 Spike-Event-Based Adaptation Using OFC-PPF

We can now use the OFC-inferred velocity and the recorded neural activity to estimate the parameters. It is possible to fit the generalized linear model (GLM) parameters using batch-based ML methods (Gilja et al. 2012; Orsborn et al. 2012; Truccolo et al. 2005). For example, we can design a SmoothBatch algorithm that finds the ML estimate of the point process parameters in Eq. (8.11) in batches of 90 s length using GLM methods, and average these batch estimates over time with a half-life of 180 s (Orsborn et al. 2012). However, such a batch-based technique would result in a slow rate of adaptation on the time-scale of minutes. Thus we develop an adaptive algorithm for parameter adaptation with every spike event by developing a parameter PPF decoder that runs in parallel to the kinematics decoder (Fig. 8.2b).

The observation model for the parameter PPF is given by Eq. (8.10). We construct the prior model for the parameters of each neuron as a random-walk to enforce continuity in the evolution of parameters

$$\boldsymbol{\phi}_t^c = \boldsymbol{\phi}_{t-1}^c + \mathbf{q}_{t-1}, \tag{8.21}$$

where $\mathbf{q}_t$ is white Gaussian noise with covariance matrix $\mathbf{Q}$ and accounts for model mismatch. The choice of $\mathbf{Q}$ dictates the learning rate of the adaptive decoder. This learning rate can be selected based on the fundamental tradeoff that it dictates between the parameter error and convergence time as we have shown in Hsieh and Shanechi (2015).

For the parameters, let's denote the one step prediction mean by $\boldsymbol{\phi}^c_{t|t-1} = E(\boldsymbol{\phi}^c_t|\mathbf{N}_{1:t-1})$, the prediction covariance by $\Lambda_{\boldsymbol{\phi}^c_{t|t-1}}$, the minimum mean-squared error (MMSE) estimate by $\boldsymbol{\phi}^c_{t|t}$, and its covariance by $\Lambda_{\boldsymbol{\phi}^c_{t|t}}$. The parameter PPF is again obtained using a Laplace-type (Eden et al. 2004) approximation as

$$\boldsymbol{\phi}^c_{t|t-1} = \boldsymbol{\phi}^c_{t-1|t-1} \tag{8.22}$$

$$\Lambda_{\boldsymbol{\phi}^c_{t|t-1}} = \Lambda_{\boldsymbol{\phi}^c_{t-1|t-1}} + \mathbf{Q} \tag{8.23}$$

$$\Lambda_{\boldsymbol{\phi}^c_{t|t}}{}^{-1} = \Lambda_{\boldsymbol{\phi}^c_{t|t-1}}{}^{-1} + \mathbf{s}_t\mathbf{s}_t^{\top}\lambda_c(t|\tilde{\mathbf{v}}_t, \boldsymbol{\phi}^c_{t|t-1})\Delta \tag{8.24}$$

$$\boldsymbol{\phi}^c_{t|t} = \boldsymbol{\phi}^c_{t|t-1} + \Lambda_{\boldsymbol{\phi}^c_{t|t}}\mathbf{s}_t(N^c_t - \lambda_c(t|\tilde{\mathbf{v}}_t, \boldsymbol{\phi}^c_{t|t-1})\Delta) \tag{8.25}$$

where $\mathbf{s}_t = [1, \tilde{\mathbf{v}}_t^{\top}]^{\top}$ (see Eq. (8.11)), and $\tilde{\mathbf{v}}_t$ (i.e., the intended velocity) is given as in Eq. (8.20). Hence adaptive OFC-PPF estimates each neuron's parameters at each time step using Eqs. (8.22)–(8.25).

It is important to emphasize that adaptive OFC-PPF does not perform joint estimation of parameters and kinematics as is done in the simulation studies in Eden et al. (2004) and Kowalski et al. (2013). Instead, in adaptive OFC-PPF, the parameter decoder is not affected by the kinematics decoder, which could be quite poor initially. Adaptive OFC-PPF uses the OFC model to provide the intended kinematics to the parameter estimator as in Eq. (8.20). This ensures that the poor decoded kinematics do not disrupt the adaptation of the parameters. This disruption can occur as joint estimation requires a prior joint distribution to be placed on the kinematics and parameters. This prior joint distribution, however, cannot be easily defined in BMI experiments since parameters and their uncertainty are initially unknown. Indeed a joint estimator is sensitive to this prior joint distribution and to the relative uncertainty placed on the initial parameters and kinematics (dictated by the relative noise covariances in the prior models of kinematics and parameters in Eqs. (8.17) and (8.21) and by the selected covariances on their initial estimates). For poor initial parameters, if a small amount of noise is used in their prior model, the joint estimator will likely not converge as it assumes that the parameters are closer to the true values than they actually are. The joint estimator will instead mostly update the decoded kinematics while assuming the wrong parameters. In contrast, in adaptive OFC-PPF the decoded kinematics merely provide the visual feedback term in the OFC model since the subject observes these decoded kinematics regardless of their quality. Thus adaptive OFC-PPF convergence is not affected by the initial kinematic decoder quality.

8.5 Adaptive OFC-PPF Improves the Speed and Accuracy of Parameter Estimation

We tested the adaptive OFC-PPF in closed-loop NHP experiments to assess its speed and accuracy for closed-loop parameter adaptation. NHP's performed a self-paced delayed center-out reaching task under BMI control. During BMI control, the arms were confined within a primate chair. Trials involved moving from a center target to one of eight peripheral targets, holding there for 250 ms to receive a reward, and then moving back to the center and holding there for 250 ms to initiate a new trial (Shanechi et al. 2016, 2017). As our main measure of decoder performance we used the success-rate defined as the number of trials reached per minute. We first assessed the benefit of adaptive OFC-PPF for decoder training and adaptation.

8.5.1 OFC Intention Estimation Improves PPF Performance

We assessed the effect of the OFC component of adaptive OFC-PPF. To do so, we first used the adaptive OFC-PPF across 12 days of NHP experiments to train a steady-state PPF decoder. The subjects then used this PPF decoder to perform the center-out movement task (Fig. 8.3a). We also conducted another 12 days of experiments in which we ran the adaptive PPF algorithm, but this time incorporated the CursorGoal method for intention estimation instead of OFC (Fig. 8.3a). We compared the steady-state performance of the resulting PPF decoders. We found that using the OFC intention estimation increased the success rate of the PPF decoder by 26% (Fig. 8.3b), largely due to an improved speed. In particular, reach times were 24% shorter with OFC-PPF compared with CursorGoal-PPF (Shanechi et al. 2016).

8.5.2 Spike-Event-Based Adaptation Improves the Speed of Performance Convergence

We also studied the effect of the time-scale of adaptation. In particular, we compared the speed by which adaptive OFC-PPF converged to proficient control compared to batch-based methods that update the parameters on the slower time-scale of minutes. To do so, we also ran experiments with a SmoothBatch OFC-PPF in which we trained the PPF with slower adaptation time-scales using the batch-based ML method of SmoothBatch (Orsborn et al. 2012), while keeping all the other components of the algorithm the same. SmoothBatch adapts the parameters smoothly once every 90 s. We compared the times it took for the performances of adaptive OFC-PPF and SmoothBatch OFC-PPF to converge (i.e., to reach 90% of the maximum performance). On average across 12 days of experiments with each decoder, while the eventual steady-state success rate was the same, adaptive OFC-PPF converged much faster to this steady-state level compared with SmoothBatch

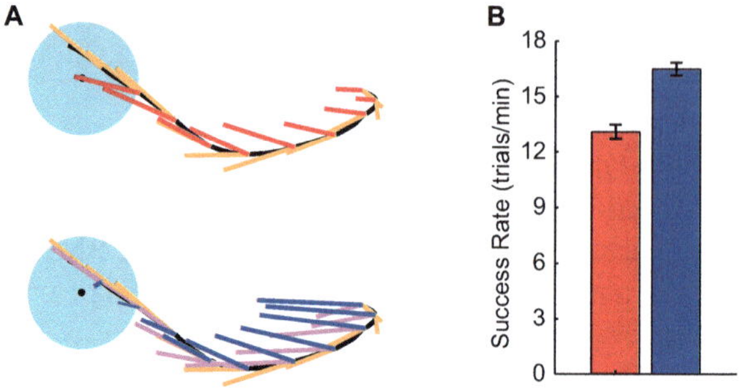

Fig. 8.3 Closed-loop intention estimation when subject is performing a self-paced center-out BMI task. (**a**) Sample decoded trajectories (black), the decoded velocities (orange), and the inferred intended velocities in the center-out task. The intended velocity computed by the CursorGoal method of intention estimation (Gilja et al. 2012) is shown in red. The intended velocity computed by an OFC model in which a in Eq. (8.12) is fit to the subject's arm movements is shown in magenta; the intended velocity computed by an OFC model that assumes the subject's control command directly sets the intended velocity, i.e., that $a = 0$ in Eq. (8.12), is shown in blue. Figure is adapted from Shanechi et al. (2016). (**b**) OFC method of intention estimation (blue bar) improves the performance of the learned PPF decoder compared with CursorGoal (red bar), suggesting that OFC better infers the subject's motor strategy

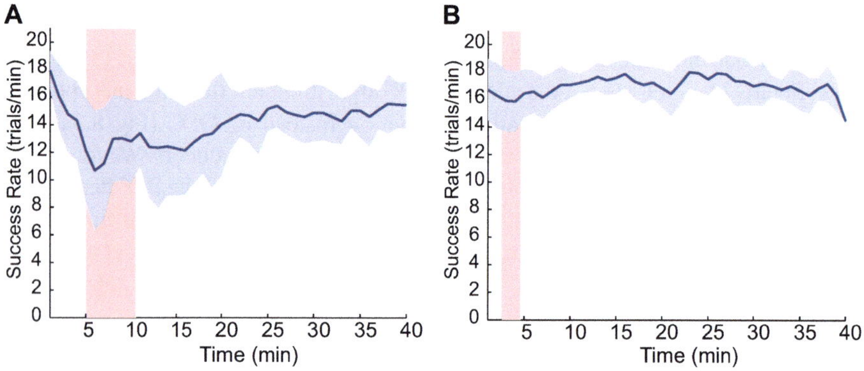

Fig. 8.4 Spike-event-based adaptation enables faster convergence. Average success rate across sessions as a function of time into the adaptive session for SmoothBatch OFC-PPF in (**a**) and adaptive OFC-PPF in (**b**). Blue curves show the mean success rate over 12 days of experiments for each decoder and shading reflects the standard deviation across these days. Success rate is calculated in sliding 2 min windows. Initially, assisted training was provided to the subject to keep them engaged in the task given the poor quality of the initialized decoders. Assistance stopped when the subject's non-assisted success rate exceeded the desired minimum threshold of 5 trials/min (Shanechi et al. 2016). The red bar shows the time range in which the BMI architecture stopped the assisted training across days. Spike-event-based adaptation resulted in faster convergence and less variability compared with SmoothBatch adaptation that had slower adaptation time-scale of 90 s. Figure is adapted from Shanechi et al. (2016)

OFC-PPF. The success rate in SmoothBatch OFC-PPF converged in 18.7 ± 3.2 min (mean $\pm$ s.e.m.) compared to 6.5 ± 0.7 min for adaptive OFC-PPF (Fig. 8.4).

8.5.3 *Adaptive OFC-PPF Is Robust to Parameter Initialization*

We investigated whether adaptive OFC-PPF was robust to parameter initialization. In general, we initialized the decoder with parameters fitted during sessions in which NHPs simply observed the movement of the cursor. We refer to this initialization as the visual feedback seed. These initialized decoders were poor and could not be operated by the subjects (average success rate was zero). Hence across tens of days of experiments, adaptive OFC-PPF could always result in high performance despite the poor initial decoder. As a control, we also conducted 2 days of experiments in which we started the adaptive OFC-PPF once from a visual feedback seed and once from a seed that was obtained by randomly permuting the visual feedback seed across neurons. We found that adaptive OFC-PPF resulted in similar high performance regardless of the seed (Shanechi et al. 2016).

8.6 PPF Outperforms State-of-the-Art Closed-Loop Kalman Filters

Significant progress has been made in the design of BMIs. In the last few years the field has converged to decoding spike counts using Kalman filters (KF) that are trained in closed-loop BMI operation (Gilja et al. 2012; Orsborn et al. 2012, 2014; Hochberg et al. 2012). The PPF BMI is different from the KF BMIs in three major elements, the control rate, the feedback rate, and the mathematical encoding model. The control rate indicates how often neural commands are sent from the brain to the prosthetic and the feedback rate indicates how often feedback of the generated movement is provided to the subject. The state-of-the-art KF is a rate-dependent decoder. KF is only optimal when the spike count within a bin is approximately Gaussian distributed. Typical KF-BMIs run at 10–20 Hz such that the bins are relatively large for the count to satisfy this assumption. In contrast, PPF processes every spike event directly. Thus, PPF enables a fast control and a fast feedback rate by controlling the prosthetic with every spike event and by providing feedback with every spike event (Figs. 8.5 and 8.6). Finally, the mathematical encoding model used in PPF BMIs is a point process unlike KF-BMIs that use a Gaussian encoding model over the spike counts. We thus explored whether the PPF-BMI could outperform the state-of-the-art KF-BMI and the contribution of each of the above three elements to such improvement (Shanechi et al. 2017).

We compared the PPF-BMI to KF-BMI in two NHPs and across tens of days. We found that monkeys could control the PPF-BMI significantly better than the KF-BMI. All performance measures were significantly improved in the PPF-BMI (Shanechi et al. 2017). Success rate on the center-out task in PPF-BMI was 32% and 24% higher than the KF-BMI in the two monkeys, respectively (Fig. 8.5a, b). We also compared the two decoders on other tasks, including a challenging obstacle avoidance task (Fig. 8.5c). Performance was again significantly higher for the PPF-

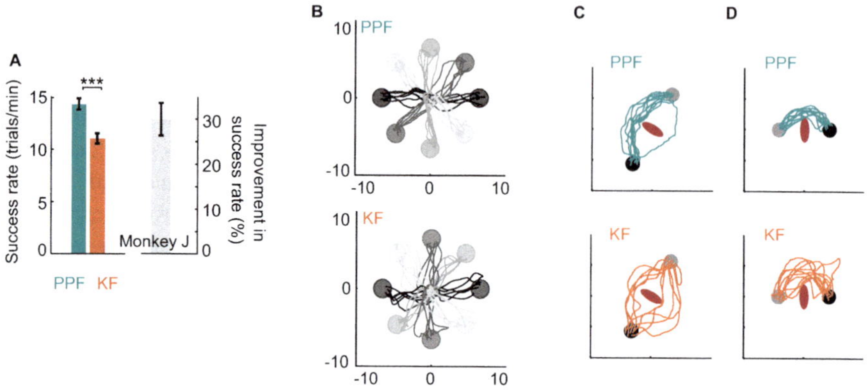

Fig. 8.5 PPF outperforms the state-of-the-art KF. (**a**) Success rates using PPF and KF on the same days and the corresponding percent improvement in success rate when using PPF. Stars indicate $P < 0.001$. (**b**) Random PPF and KF center-out trajectories. (**c, d**) Sample random trajectories on the obstacle avoidance task. Figure is adapted from Shanechi et al. (2017)

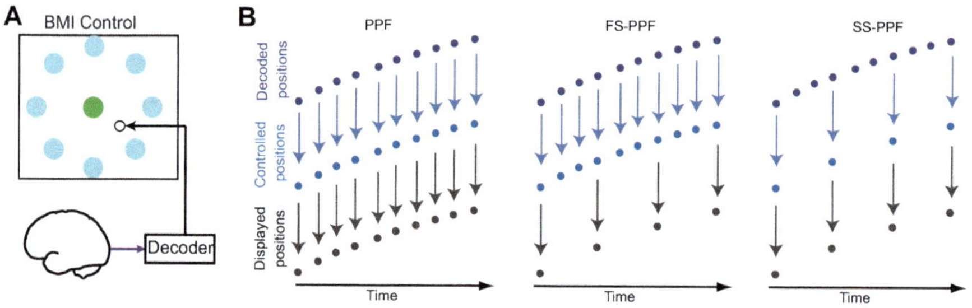

Fig. 8.6 Rapid control and feedback rates enhance neuroprosthetic control. (**a**) Subject performing the self-paced delayed center-out BMI task. (**b**) The process of generating the controlled and feedback positions in PPF, FS-PPF, and SS-PPF is shown for a hypothetical spike train. We changed the control rate by adjusting how often the PPF decoded position was sent to the cursor to control the task. Task success was based on these controlled positions. We changed the feedback rate by adjusting how often the controlled positions were displayed to the subject. PPF consists of both a fast control and a fast feedback rate, FS-PPF consists of a fast control and a slow feedback rate, and SS-PPF consists of a slow control and a slow feedback rate. Figure is adapted from Shanechi et al. (2017)

BMI, for example path-lengths were reduced by 30% using the PPF-BMI. These results show that NHPs can control the PPF-BMI significantly better than the KF-BMIs.

8.7 Rapid Control and Feedback Rates and the Point Process Encoding Model Result in Performance Improvement

We explore the factors that led to the significant improvement of the PPF-BMI over the KF-BMI. In particular, we dissociated the influence of each of the factors that were different between the two BMI decoders: the control rate, the feedback rate, and the point process mathematical encoding model.

To dissociate the effect of control and feedback rates from the mathematical encoding model, we explored these rate effects using the PPF. By manipulating the control and feedback rates, we designed three variants of the PPF-BMI: A BMI with slow control and slow feedback rate, termed SS-PPF; A BMI with fast control rate and slow feedback rate termed FS-PPF, and the default PPF-BMI with fast control and fast feedback rates (Fig. 8.6). Comparing SS-PPF with FS-PPF allowed us to explore the effect of the control rate. Comparing FS-PPF with PPF then elucidated the effect of feedback rate.

We found that increasing the control rate significantly improved BMI performance. Even when feedback rate was slow, allowing the NHP to send control commands faster to the prosthetic improved BMI control. While future studies are required to understand the neural basis of this improvement, we found that this improvement was consistent with the hypothesis that BMI control involved a feedforward control strategy using a high-rate internal forward model (Shanechi et al. 2017). We also found that increasing the feedback rate further improved BMI performance by comparing FS-PPF and PPF. This may suggest that BMI control also relies on a feedback control strategy. Together, these results suggest that BMI control may involve a hybrid of feedforward and feedback control strategies, consistent with theories from motor control (Desmurget and Grafton 2000).

We also assessed the contribution of the point process encoding model to performance improvement. To do so, we compared the KF and PPF at the same control and feedback rates. We used both a fast rate and a slow rate to make the comparisons. We found that regardless of the rate, PPF significantly improved performance over the KF. This result indicates that the point process encoding model can be controlled better than the Gaussian encoding model used in the KF, and more accurately models the spiking activity (Shanechi et al. 2017).

8.8 Conclusion

In this chapter, we first reviewed the computational elements involved in designing a BMI decoder and provided an overview of the most common decoders used in the field. We then focused on our recent work on a control-theoretic high-rate decoder design that explicitly modeled the BMI as an optimal feedback-control system and the spikes as point processes. We showed that this design improved both the transient operation of the BMI during the training and parameter adaptation phase, and the steady-state operation of the BMI. The OFC model resulted in a better estimation of the motor intent during closed-loop parameter adaptation and hence led to more accurate steady-state PPF decoders. The fast rate of closed-loop adaptation in the adaptive PPF improved the speed of performance convergence compared with common batch-based techniques. Finally, at steady state, the PPF BMI significantly improved the performance of the state-of-the-art KF BMIs because of a faster control rate, a faster feedback rate, and the point process encoding model.

Despite significant progress on BMI systems in recent years, these systems still require significant improvement to become clinically viable. For example, while BMI designs can perform 2D computer interface tasks well, they need to be extended to allow for proficient control of dexterous high degree-of-freedom movements, such as those of a robotic limb in 3D space. The tools presented in this chapter can help pave the way for the design of such generalizable BMI architectures in multiple ways.

BMI systems can be viewed as feedback-control systems in which the brain is the controller of the prosthetic plant (Fig. 8.1). This view, for example, has motivated the design of CLDA algorithms (Taylor et al. 2002; Velliste et al. 2008; Collinger et al. 2013; Mahmoudi and Sanchez 2011; Hochberg et al. 2012; Gilja et al. 2012; Orsborn et al. 2012; Dangi et al. 2014; Shanechi et al. 2016, 2017; Hsieh and Shanechi 2015). The closed-loop control view further motivates the explicit use of the tools of control theory to build better BMIs. Using control-theoretic models of BMI, such as the optimal feedback control model presented here (Shanechi et al. 2013a,b, 2016, 2017; Hsieh and Shanechi 2015), will help make BMI designs generalizable to prosthetics with different dynamics and to more complex tasks with various goals (Shanechi et al. 2016). The use of control-theoretic models can also provide algorithmic design guidelines for future neuroprosthetics. As we discussed, the fast control and feedback rates are essential in improving BMI performance. Previous studies have also shown the importance of short delays for neuroprosthetic performance (Willett et al. 2013). A control-theoretic model can characterize these system properties such as delays and rates, and thus predict their influence on closed-loop performance. Control-theoretic models can also be used to understand the effects of BMI learning on neural adaptation and plasticity (Ganguly and Carmena 2009; Jarosiewicz et al. 2008; Chase et al. 2012).

In this chapter we mainly reviewed designs using the spiking activity as the control signal. However, various neural signal modalities can be incorporated in BMIs, including spikes, local field potentials (LFP), and electrocorticogram (ECoG). Recent technological advances have allowed for simultaneous recording of these various scales of neural activity, from small-scale spikes to large-scale LFP and ECoG. Allowing decoders to concurrently extract information from all these modalities could significantly improve BMI performance. Our recent work has developed multiscale state-space models that simultaneously model spikes, LFP, and ECoG using a combination of point process and Gaussian encoding models to improve decoding performance and robustness (Hsieh and Shanechi 2016, 2017; Abbaspourazad and Shanechi 2017).

The brain is highly plastic and non-stationary and can adapt to control a BMI decoder over time (Taylor et al. 2002; Carmena et al. 2003; Ganguly and Carmena 2009; Orsborn et al. 2014), a process referred to as neural adaptation. Here we presented some of the recent closed-loop decoder adaptation techniques that aim to track the changes in neural representation as a result of neural adaptation. Decoder and neural adaptation in BMIs create a "two-learner system" (Orsborn et al. 2014) and can interact. It is critical to model this interaction in future studies, and study the effect that the time-scales and the speed of adaptation have in this interplay. We have shown that even with the fast adaptation time-scale of the adaptive PPF, neural adaptation can still occur and improve the performance of BMI (Shanechi et al. 2016; Orsborn et al. 2014). Another critical design element in an adaptive algorithm is the learning rate, which dictates how fast model parameters are updated based on neural activity. Our recent work has designed principled calibration algorithms to optimally select the learning rate for a desired parameter error and convergence time (Hsieh and Shanechi 2015). This calibration algorithm can also help adjust the speed of adaptation and consequently provide a tool to study the two-learner system. The examination of how the time-scales of neural and decoder adaptation may interfere from a theoretical perspective is also important in future studies (Merel et al. 2015).

Taken together, using the tools of control theory and statistical inference have great potential in improving BMI decoder designs. Such algorithmic advances could improve the performance and robustness of BMI systems. Such advances can help pave the way for BMI clinical viability to improve the quality of life for millions of disabled patients.

References

Abbaspourazad, H., & Shanechi, M. M. (2017). An unsupervised learning algorithm for multiscale neural activity. In *2017 39th Annual International Conference of the IEEE Engineering in Medicine and Biology Society (EMBC)* (pp. 201–204). IEEE.

Aflalo, T., Kellis, S., Klaes, C., Lee, B., Shi, Y., Pejsa, K., et al. (2015). Decoding motor imagery from the posterior parietal cortex of a tetraplegic human. *Science 348*(6237), 906–910.

Arulampalam, M. S., Maskell, S., Gordon, N., & Clapp, T. (2002). A tutorial on particle filters for online nonlinear/non-Gaussian Bayesian tracking. *IEEE Transactions on Signal Processing, 50*(2), 174–188.

Bertsekas, D. (2005). *Dynamic programming and optimal control*. Belmont: Athena Scientific.

Bouton, C. E., Shaikhouni, A., Annetta, N. V., Bockbrader, M. A., Friedenberg, D. A., Nielson, D. M., et al. (2016). Restoring cortical control of functional movement in a human with quadriplegia. *Nature, 533*(7602), 247–250.

Brockwell, A. E., Rojas, A. L., & Kass, R. E. (2004). Recursive Bayesian decoding of motor cortical signals by particle filtering. *Journal of Neurophysiology, 91*, 1899–1907.

Brown, E. N., Barbieri, R., Ventura, V., Kass, R., & Frank, L. (2001). The time-rescaling theorem and its application to neural spike train data analysis. *Neural Computation, 14*, 325–346.

Brown, E. N., Frank, L. M., Tang, D., Quirk, M. C., & Wilson, M. A. (1998). A statistical paradigm for neural spike train decoding applied to position prediction from ensemble firing patterns of rat hippocampal place cells. *Journal of Neuroscience, 18*(18), 7411–7425.

Capogrosso, M., Milekovic, T., Borton, D., Wagner, F., Moraud, E. M., Mignardot, J.-B., et al. (2016). A brain–spine interface alleviating gait deficits after spinal cord injury in primates. *Nature, 539*(7628), 284–288.

Carmena, J. M., Lebedev, M. A., Crist, R. E., O'Doherty, J. E., Santucci, D. M., Dimitrov, D. F., et al. (2003). Learning to control a brain-machine interface for reaching and grasping by primates. *PLoS Biology, 1*(2), e42.

Chapin, J. K., Moxon, K. A., Markowitz, R. S., & Nicolelis, M. A. L. (1999). Real-time control of a robot arm using simultaneously recorded neurons in the motor cortex. *Nature Neuroscience, 2*(7), 664–670.

Chase, S. M., Kass, R. E., & Schwartz, A. B. (2012). Behavioral and neural correlates of visuomotor adaptation obsceneerved through a brain-computer interface in primary motor cortex. *Journal of Neurophysiology, 108*(2), 624–644.

Chase, S. M., Schwartz, A. B., & Kass, R. E. (2009). Bias, optimal linear estimation, and the differences between open-loop simulation and closed-loop performance of spiking-based brain-computer interface algorithms. *Neural Networks, 22*, 1203–1213.

Collinger, J. L., Wodlinger, B., Downey, J. E., Wang, W., Tyler-Kabara, E. C., Weber, D. J., et al. (2013). High-performance neuroprosthetic control by an individual with tetraplegia. *The Lancet, 381*(9866), 557–564.

Dangi, S., Gowda, S., Moorman, H. G., Orsborn, A. L., So, K., Shanechi, M., et al. (2014). Continuous closed-loop decoder adaptation with a recursive maximum likelihood algorithm allows for rapid performance acquisition in brain-machine interfaces. *Neural Computation, 26*(9), 1811–1839.

Desmurget, M., & Grafton, S. (2000). Forward modeling allows feedback control for fast reaching movements. *Trends in Cognitive Sciences, 4*(11), 423–431.

Eden, U. T., Frank, L. M., Barbieri, R., Solo, V., & Brown, E. N. (2004). Dynamic analysis of neural encoding by point process adaptive filtering. *Neural Computation, 16*, 971–998.

Ethier, C., Oby, E. R., Bauman, M. J., & Miller, L. E. (2012). Restoration of grasp following paralysis through brain-controlled stimulation of muscles. *Nature, 485*, 368–371.

Fan, J. M., Nuyujukian, P., Kao, J. C., Chestek, C. A., Ryu, S. I., & Shenoy, K. V. (2014). Intention estimation in brain-machine interfaces. *Journal of Neural Engineering, 11*(1), 016004.

Fetz, E. E. (1969). Operant conditioning of cortical unit activity. *Science, 163*(3870), 955–958.

Ganguly, K., & Carmena, J. M. (2009). Emergence of a stable cortical map for neuroprosthetic control. *PLoS Biology, 7*(7), e1000153.

Gilja, V., Nuyujukian, P., Chestek, C. A., Cunningham, J. P., Yu, B. M., Fan, J. M., et al. (2012). A high-performance neural prosthesis enabled by control algorithm design. *Nature Neuroscience, 15*, 1752–1757.

Gilja, V., Pandarinath, C., Blabe, C. H., Nuyujukian, P., Simeral, J. D., Sarma, A. A., et al. (2015). Clinical translation of a high-performance neural prosthesis. *Nature Medicine, 21*, 1142–1145.

Golub, M. D., Yu, B. M., & Chase, S. M. (2012). Internal models engaged by brain-computer interface control. In *Proceedings of IEEE Conference on EMBS*, San Diego, USA.

Hauschild, M., Mulliken, G. H., Fineman, I., Loeb, G. E., & Andersen, R. A. (2012). Cognitive signals for brain-machine interfaces in posterior parietal cortex include continuous 3D trajectory commands. *Proceedings of the National Academy of Sciences of the United States of America, 109*, 17075–17080.

Hochberg, L. R., Bacher, D., Jarosiewicz, B., Masse, N. Y., Simeral, J. D., Vogel, J., et al. (2012). Reach and grasp by people with tetraplegia using a neurally controlled robotic arm. *Nature, 485*, 372–375.

Hochberg, L. R., Serruya, M. D., Friehs, G. M., Mukand, J. A., Saleh, M., Caplan, A. H., et al. (2006). Neuronal ensemble control of prosthetic devices by a human with tetraplegia. *Nature, 442*, 164–171.

Hsieh, H.-L., & Shanechi, M. (2015). Optimal calibration of the learning rate in closed-loop adaptive brain-machine interfaces. In *Proceedings of IEEE Conference on Engineering in Medicine and Biology Society (EMBC)*, Milan, Italy, pp. 1667–1670.

Hsieh, H. L., & Shanechi, M. M. (2016). Multiscale brain-machine interface decoders. In *Proceedings of IEEE Conference on Engineering in Medicine and Biology Society (EMBC)*, pp. 6361–6364.

Hsieh, H. L., Wong, Y. T., Pesaran, B., & Shanechi, M. M. (2017). Multiscale decoding for reliable brain-machine interface performance over time. In *2017 39th Annual International Conference of the IEEE Engineering in Medicine and Biology Society (EMBC)* (197–200). IEEE.

Humphrey, D. R., Schmidt, E. M., & Thompson, W. D. (1970). Predicting measures of motor performance from multiple cortical spike trains. *Science, 170*(3959), 758–762.

Jarosiewicz, B., Chase, S. M., Fraser, G. W., Velliste, M., Kass, R. E., & Schwartz, A. B. (2008). Functional network reorganization during learning in a brain-computer interface paradigm. *Proceedings of the National Academy of Sciences of the United States of America, 105*(49), 19486–19491.

Kailath, T., Sayed, A. H., & Hassibi, B. (2000). *Linear estimation*. Upper Saddle River: Prentice Hall.

Kass, R. E., & Ventura, V. (2001). A spike-train probability model. *Neural Computation, 13*(8), 1713–1720.

Kim, S.-P., Simeral, J. D., Hochberg, L. R., Donoghue, J. P., & Black, M. J. (2008). Neural control of computer cursor velocity by decoding motor cortical spiking activity in humans with tetraplegia. *Journal of Neural Engineering, 5*, 455–476.

Kowalski, K. C., He, B. D., & Srinivasan, L. (2013). Dynamic analysis of naive adaptive brain-machine interfaces. *Neural Computation, 25*(9), 2373–2420.

Koyama, S., Chase, S. M., Whitford, A. S., Velliste, M., Schwartz, A. B., & Kass, R. E. (2009). Comparison of brain-computer interface decoding algorithms in open-loop and closed-loop control. *Journal of Computational Neuroscience, 29*, 73–87.

Li, Z., O'Doherty, J. E., Hanson, T. L., Lebedev, M. A., Henriquez, C. S., & Nicolelis, M. A. L. (2009). Unscented Kalman filter for brain-machine interfaces. *PLoS ONE, 4*(7), 1–18.

Li, Z., O'Doherty, J. E., Lebedev, M. A., & Nicolelis, M. A. L. (2011). Adaptive decoding for brain-machine interfaces through bayesian parameter updates. *Neural Computation, 23*, 3162–3204.

Mahmoudi, B., & Sanchez, J. C. (2011). A symbiotic brain-machine interface through value-based decision making. *PLOS ONE, 6*(3), e14760.

McMullen, D., Hotson, G., Katyal, K., Wester, B., Fifer, M., McGee, T., et al. (2014). Demonstration of a semi-autonomous hybrid brain-machine interface using human intracranial EEG, eye tracking, and computer vision to control a robotic upper limb prosthetic. *IEEE Transactions on Neural Systems and Rehabilitation Engineering, 22*(4), 784–796.

Merel, J., Pianto, D. M., Cunningham, J. P., & Paninski, L. (2015). Encoder-decoder optimization for brain-computer interfaces. *PLoS Computational Biology, 11*(6), e1004288.

Moran, D. W., & Schwartz, A. B. (1999). Motor cortical representation of speed and direction during reaching. *Journal of Neurophysiology, 82*, 2676–2692.

Moritz, C. T., Perlmutter, S. I., & Fetz, E. E. (2008). Direct control of paralysed muscles by cortical neurons. *Nature, 456*, 639–643.

Mulliken, G. H., Musallam, S., & Andersen, R. A. (2008). Decoding trajectories from posterior parietal cortex ensembles. *The Journal of Neuroscience, 28*(48), 12913–12926.

Musallam, S., Corneil, B. D., Greger, B., Scherberger, H., & Andersen, R. A. (2004). Cognitive control signals for neural prosthetics. *Science, 305*, 258–262.

O'Doherty, J. E., Lebedev, M. A., Ifft, P. J., Zhuang, K. Z., Shokur, S., Bleuler, H., et al. (2011). Active tactile exploration using a brain-machine-brain interface. *Nature, 479*, 228–231.

Orsborn, A. L., Dangi, S., Moorman, H. G., & Carmena, J. M. (2012). Closed-loop decoder adaptation on intermediate time-scales facilitates rapid BMI performance improvements independent of decoder initialization conditions. *IEEE Transactions on Neural Systems and Rehabilitation Engineering, 20*(4), 468–477.

Orsborn, A. L., Moorman, H. G., Overduin, S. A., Shanechi, M. M., Dimitrov, D. F., & Carmena, J. M. (2014). Closed-loop decoder adaptation shapes neural plasticity for skillful neuroprosthetic control. *Neuron, 82*, 1380–1392.

Santhanam, G., Ryu, S. I., Yu, B. M., Afshar, A., & Shenoy, K. V. (2006). A high-performance brain-computer interface. *Nature, 442*, 195–198.

Serruya, M. D., Hatsopoulos, N. G., Paninski, L., Fellows, M. R., & Donoghue, J. P. (2002). Instant neural control of a movement signal. *Nature, 416*, 141–142.

Shadmehr, R., & Krakauer, J. W. (2008). A computational neuroanatomy for motor control. *Experimental Brain Research, 185*(3), 359–381.

Shanechi, M. M., Hu, R. C., Powers, M., Wornell, G. W., Brown, E. N., & Williams, Z. M. (2012). Neural population partitioning and a concurrent brain-machine interface for sequential motor function. *Nature Neuroscience, 15*(12), 1715–1722.

Shanechi, M. M., Hu, R. C., & Williams, Z. M. (2014). A cortical-spinal prosthesis for targeted limb movement in paralysed primate avatars. *Nature Communications, 5*, 3237.

Shanechi, M. M., Orsborn, A. L., & Carmena, J. M. (2016). Robust brain-machine interface design using optimal feedback control modeling and adaptive point process filtering. *PLoS Computational Biology, 12*(4), e1004730.

Shanechi, M. M., Williams, Z. M., Wornell, G. W., Hu, R., Powers, M., & Brown, E. N. (2013a). A real-time brain-machine interface combining motor target and trajectory intent using an optimal feedback control design. *PLoS ONE, 8*(4), e59049.

Shanechi, M. M., Wornell, G. W., Williams, Z. M., & Brown, E. N. (2013b). Feedback-controlled parallel point process filter for estimation of goal-directed movements from neural signals. *IEEE Transactions on Neural Systems and Rehabilitation Engineering, 21*, 129–140.

Shanechi, M., Orsborn, A., Moorman, H., Gowda, S., Dangi, S., & Carmena, J. (2017). Rapid control and feedback rates enhance neuroprosthetic control. *Nature Communications, 8*, 13825.

Srinivasan, L., Eden, U. T., Willsky, A. S., & Brown, E. N. (2006). A state-space analysis for reconstruction of goal-directed movements using neural signals. *Neural Computation, 18*, 2465–2494.

Suminski, A. J., Tkach, D. C., Fagg, A. H., & Hatsopoulos, N. G. (2010). Incorporating feedback from multiple sensory modalities enhances brain-machine interface control. *Journal of Neuroscience, 30*(50), 16777–16787.

Taylor, D. M., Tillery, S. I. H., & Schwartz, A. B. (2002). Direct cortical control of 3D neuroprosthetic devices. *Science, 296*, 1829–1832.

Thakor, N. V. (2013). Translating the brain-machine interface. *Science Translational Medicine, 5*, 210–217.

Todorov, E. (2004). Optimality principles in sensorimotor control. *Nature Neuroscience, 7*, 907–915.

Todorov, E., & Jordan, M. I. (2002). Optimal feedback control as a theory of motor coordination. *Nature Neuroscience, 5*(11), 1226–1235.

Truccolo, W., Eden, U. T., Fellows, M. R., Donoghue, J. P., & Brown, E. N. (2005). A point process framework for relating neural spiking activity to spiking history, neural ensemble, and extrinsic covariate effects. *Journal of Neurophysiology, 93*, 1074–1089.

Velliste, M., Perel, S., Spalding, M. C., Whitford, A. S., & Schwartz, A. B. (2008). Cortical control of a prosthetic arm for self-feeding. *Nature, 453*, 1098–1101.

Willett, F. R., Suminski, A. J., Fagg, A. H., & Hatsopoulos, N. G. (2013). Improving brain–machine interface performance by decoding intended future movements. *Journal of Neural Engineering, 10*(2), 026011.

Yu, B. M., Kemere, C., Santhanam, G., Afshar, A., Ryu, S. I., Meng, T. H., et al. (2007). Mixture of trajectory models for neural decoding of goal-directed movements. *Journal of Neurophysiology, 97*, 3763–3780.

Chapter 9
Control-Theoretic Approaches for Modeling, Analyzing, and Manipulating Neuronal (In)activity

ShiNung Ching

9.1 Introduction

Dynamical systems theory and computational modeling have proven to be powerful approaches in neuroscience, insofar as they enable an increased understanding of the mechanisms that underlie different behavioral regimes. An exemplar of this power can be found in the study of inactivated brain dynamics associated with general anesthesia. In the state of the general anesthesia the brain often manifests oscillatory dynamics with structured spatial organization (Brown et al. 2010). However, the spatial resolution with which these dynamics are observed means that a purely data-driven approach to understanding their origins, especially at a neuronal circuit level, is challenging. In this regard, modeling and dynamical systems analysis can advance our understanding beyond observation alone (Ching and Brown 2014; McCarthy et al. 2012).

An example of how modeling can clarify biophysical mechanisms can be found in the phenomenon of burst suppression, a pattern of the electroencephalogram (EEG) characterized by quasi-periodic alternations of high-voltage activity (burst) and isoelectric silence (suppression) that occurs in deep general anesthesia as well as pathological states of unconsciousness. While the major features of burst suppression (e.g., slow cycling of bursts and suppressions distributed across the scalp) have been relatively well-characterized, a mechanism for how the phenomenology is generated in the cortex remained elusive until a series of modeling results established a hypothesis involving slow homeostatic interactions of neural activity and metabolic substrate (Ching et al. 2012b). These models successfully accounted for the observed features of burst suppression while also generating

S. Ching (✉)
Washington University at St Louis, St Louis, MO, USA
e-mail: shinung@wustl.edu

Z. Chen, S.V. Sarma (eds.), *Dynamic Neuroscience*,
https://doi.org/10.1007/978-3-319-71976-4_9

testable predictions regarding finer aspects of the phenomenon, such as parametric sensitivity of burst suppression ratio to the level of anesthetic drug (Liu and Ching 2017). They also helped to enable the development of new engineering solutions to better manage burst suppression in clinical scenarios.

But perhaps even more powerful that this mechanistic elucidation is the potential for modeling to suggest links between brain dynamics and function. In this regard, dynamical systems analysis can play an instrumental role. This process is exemplified by the characterization of anesthesia-induced synchronized alpha oscillations. More specifically, it has been known that the general anesthetic drug propofol induces 11–14 Hz alpha oscillations that tend to be coherent across the frontal regions of the scalp as measured by EEG (Cimenser et al. 2011; Purdon et al. 2013). Modeling has helped to clarify a potential mechanism for these oscillations, namely the elongation of inhibitory time-scales (due to propofol) leading to synchronization with intrinsic rhythms in the thalamus, thus promoting synchronization (Ching et al. 2010; Vijayan et al. 2013). Such modeling efforts help to create a set of structured hypotheses that can guide experiments. Indeed, recent invasive recordings substantiate the notion of propofol-induced thalamocortical synchronization (Flores et al. 2017). Moreover, the modeling provides a hypothesis not just for the origin oscillations, but the link between the oscillations and the state of unconsciousness. Namely, the notion that thalamocortical synchronization impairs or impedes the normal dynamics and information propagation needed to sustain cognitive function.

This latter hypothesis is, at a conceptual level, related to the formalisms of systems and control theory, which focuses on how the dynamics of a system affects the ability to manipulate it via exogenous inputs (Sontag 2013). In the above anesthesia context, the thalamocortical system, by virtue of being in a synchronous dynamical regime, is harder to "control," (for example via ascending excitation or afferent sensory activity). This chapter will discuss a line of study related to the formalization of this idea with specific focus on mathematically linking neuronal circuit dynamics to control and information processing properties. While focus will be directed at understanding inactivated neuronal regimes, such as during general anesthesia, the overall framework is quite generalizable and other examples of theoretical neuroscience studies will be presented, including efforts to understand the effects of neural plasticity on circuit sensitivity. Such efforts have the potential to lead not just to new hypothesis, but advances in practical approaches for data-driven inference of brain states as well as optimization of extrinsic control strategies to manipulate brain activity, and examples of these applications will also be discussed.

9.2 Neural Trajectories and Control-Theoretic Analysis of Brain Dynamics

Quantitative electrophysiology has traditionally been centered on so-called spectral analysis, in which a signal is decomposed into constituent frequency components (e.g., EEG bands) by expansion in the Fourier basis. As a companion to such

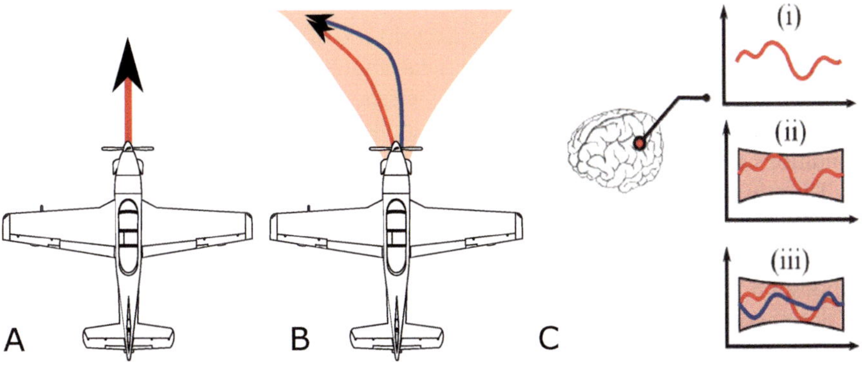

Fig. 9.1 Control-theoretic analysis. In systems theory, dynamics are analyzed as a property of the underlying physical system and not activity per se. (**a**) A plane is observed flying in a straight line. That observation alone does not describe the dynamics. (**b**) In reality, the aircraft could, at any time, execute a set of possible trajectories, *constrained* and *subject to* its dynamics. Within this "reachable set," certain trajectories might be harder to realize than others, e.g., the red versus blue maneuvers. (**c**) A substantial focus in brain medicine has centered on observing and statistically characterizing brain activity, e.g., via electrophysiological recordings. A particular observation (e.g., (i)), is only one possible realization within the reachable set of "neural trajectories" (e.g., (ii)). Within the reachable set, certain trajectories (patterns) may be harder to realize than others (e.g., (iii)). We may hypothesize that states of brain inactivity, such as general anesthesia, are associated with neuronal dynamics that ultimately contracts the set of reachable neural trajectories

descriptive analyses, EEG activity can be viewed through the lens of systems theory, which does not determine the patterns embedded in the signal per se, but rather attempts to directly reveal the characteristics of the neural substrate (i.e., the neural circuit dynamics) that underlie those patterns. In particular, *control theory* can help to reveal the input to output transformations mediated by the dynamics in question, thus allowing inferences on signal propagation and, potentially, information processing.

Dynamical systems and control theory, a branch of mathematics and engineering, centers on the characterization of complex systems subject to exogenous stimuli. The central question in control theory is: how easy is it to control a given system, for example an airplane (see Fig. 9.1)? Answering these questions is possible because, for such engineered systems, we have excellent models for how these systems behave subject to the laws of physics that is, we understand their dynamics.

Although our understanding of neuronal dynamics is grossly incomplete compared to engineered systems, we can nevertheless formulate tractable, biophysical, dynamical-systems models that capture key spatiotemporal features of neuronal activity at microscopic and macroscopic scales, for example those discussed above as well as neural field models that describe the EEG activity induced by certain classes of anesthetic drugs (Ching and Brown 2014). Such models can be coupled with systems-theoretic analysis in order to describe electrical activity in the inactivated brain. By virtue of using a generative model, this approach has the advantage of directly yielding mechanistic characterizations: for instance, are the circuit dynamics more, or less, labile (see Fig. 9.1d)? In addition to mechanistic interpretations, practical solutions for neuromonitoring can also be realized by opening a wider space of model-based metrics upon which to build biomarkers; and the overall paradigm can be readily applied to a wide range of neurological conditions.

Thus, in the sense of systems theory, states of altered arousal including general anesthesia can be fundamentally characterized by a contraction in lability, i.e., the dynamic range of realizable neuronal trajectories. Said more mathematically: suppose $\mathscr{A}$ is the set of all neural activation patterns that are realized in the course of normal cognition and function. Now, suppose that $\mathscr{B}$ is the set of all realizable patterns during general anesthesia or perhaps, injury. Then, $\mathscr{B} \subseteq \mathscr{A}$. In other words, the inactivated brain, in addition to being less active, is also less labile—it simply cannot do as much. Such a hypothesis is in line with anecdotal observations that EEG waveforms of brain injured patients are less dynamic in a manner that is not restricted to rhythmicity (though, oscillations may be a particularly important manifestation of said dynamics).

We will proceed to discuss various ways in which these systems-theoretic concepts can be examined in the context of brain dynamics.

9.3 Reachability and Controllability as Surrogates for Information Processing

9.3.1 Anesthesia as a Contraction in Reachability

To illustrate the major premise of this chapter we can return to the examples discussed in the Introduction section, namely dynamical systems-based models for neuronal networks in the context of general anesthetic drugs. Specifically, in Ching et al. (2010), a biophysical model was developed to explain the genesis of 11–13 Hz alpha oscillations that occur concomitant with loss of consciousness, due to the anesthetic drug propofol (see time-frequency spectrogram in Fig. 9.2a). The model attributes this electrophysiological phenomenon to the effect of propofol on GABAergic inhibition in thalamocortical networks (see Fig. 9.2b). Specifically, amplifying the weight of this connection, w_{IE}, caused the synchronization of neurons into 11–13 Hz oscillations (see Fig. 9.2b, c). In the sense of dynamical systems, this effect can be understood as the emergence of a stable limit cycle in the network via a bifurcation with respect to w_{IE}. While such analysis is itself interesting, even more intriguing is considering the extent to which these emergent dynamics might be causal to the associated *behavior*, i.e., loss of consciousness. In considering this, we go beyond intrinsic dynamics and consider the inputs that impinge on the network.

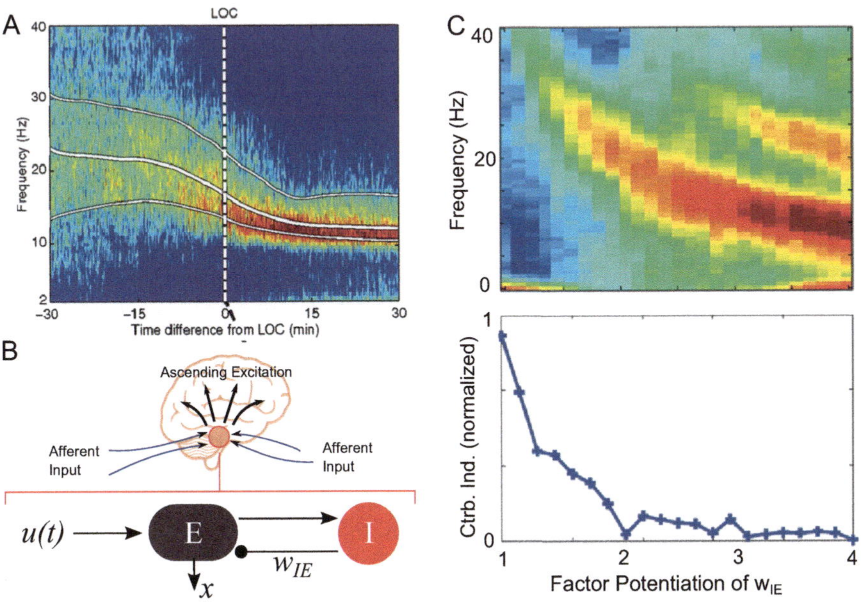

Fig. 9.2 (**a**) The general anesthetic propofol produces narrowband 11–13 Hz EEG activity concomitant with loss of consciousness (LOC) (spectrogram adapted from Purdon et al. (2013). (**b**) We have developed dynamical systems models for thalamic networks, a central relay center in the brain, wherein the actions of propofol amount to potentiation of inhibitory synaptic connections, here schematized through w_{IE}. (**c**) Model output spectrogram from Ching et al. (2010) for parametric modulation of w_{IE}, noting similarity with part (**a**). The nonlinear controllability index, obtained from a low dimensional version of the model, indicates a precipitous loss of controllability at parameterizations concomitant with LOC

Specifically, we note that the neuronal region in question, the thalamus, does not exist in isolation—in fact, it is a primary relay center in the brain (see Fig. 9.2b), receiving feedforward excitation from a host of afferent sensory modalities, as well as feedback excitation from higher cortical areas. Thus, we can pose the following **systems-theoretic hypothesis**: that loss of consciousness is associated with changes in the controllability of the network with respect to these different inputs. Or, said more strongly, that a contraction in the reachable space of thalamic networks leads to a loss of the information processing required in order to support cognition.

To discuss an analysis framework compatible with the above hypothesis, we will consider models of the general form:

$$\dot{x}_i = f_{\alpha_i}(x_i) + g(\mathbf{x}) + \sum_{j=1}^{\Omega} \rho_{ij} r_j, \quad \dot{r}_j = \tau_i^u \left(r_{\text{base}_j} - r_j \right) + \sum_{k=1}^{\kappa} \gamma_{jk} h(F_k). \tag{9.1}$$

Here, the variables $x_i \in \mathbb{R}^n$ describe the state of the ith neuron in the modeled network, $\mathbf{x} = [x_1, \ldots, x_M]$ describes all neurons in the network and the vector

$\mathbf{F}(t) \equiv (F_1(t), F_2(t), \ldots, F_\kappa(t))$ denotes a time-varying input (stimulus) in a κ-dimensional space of features. The variables $r_i \in \mathbb{R}^n$ describe receptor neuronal activity, such that each receptor neuron is tuned via $h(\cdot)$ to specific features via the weights γ_{jk}. The function $f_{\alpha_i}(\cdot)$ describes the intrinsic neuronal dynamics, parametrized by $\alpha_i \in \mathbb{R}^q$. The function $g(\cdot)$ governs network structure and dynamics.

The main question, enabled by this modeling formalism, is how the dynamics of such a neuronal network mediate the transformation between the afferent stimulus contained in $\mathbf{F}(t)$ and the consequent network trajectory $\mathbf{x}(t)$. In control theory, these properties are captured through the notion of ***controllability***, and its generalization, reachability. ***Controllability*** is a general systems-theoretic property that describes the ability of an input to "steer" a system along arbitrary state trajectories (Kalman 1959; Khalil and Grizzle 2002). In the context of neural activity, controllability asks: could a particular spiking pattern or brain activation trajectory, chosen at random, be induced, or "reached," via an input? If any pattern can be reached, then the network is controllable. In this sense, controllability describes a system's expressiveness. The larger the space of ***reachable trajectories***, the more diverse the range of inputs that can, for instance, be encoded. Importantly, there exists a fundamental tradeoff between controllability and the ability of a system to reject disturbances or noise (Freudenberg et al. 2003; Khalil and Grizzle 2002). Systems-theoretic analysis of ***sensitivity***—the resiliency of trajectories to external perturbations—can characterize this tradeoff.

However, attempting to perform formal reachability analysis on high dimensional networks of the form (9.1) is analytically challenging. Indeed, exact analysis is usually only possible by making potentially strong assumptions about linearity of dynamics or focussing on structural aspects of the underlying dynamical system (Liu et al. 2011). Further, controllability is not a monolithic concept—systems may be technically "controllable," even though the inputs required to effect the control may be, for all intents and purposes, infeasible (Cowan et al. 2012).

Nevertheless, for certain small network configurations, a controllability analysis can be carried out. As a demonstration of the concept, we considered small thalamic network motifs. Despite their limited scale, the dynamics of such motifs are informative because they tend to be overrepresented in larger network topologies (Milo et al. 2002).

We proceeded to perform a (local) nonlinear controllability analysis on a 2-neuron motif (Fig. 9.2b) from the model in Ching et al. (2010). Note that the controllability of a nonlinear dynamical system is characterized through the Lie algebra generated by its vector field (Khalil and Grizzle 2002; Hermann and Krener 1977). In the case of a linear system, this amounts to a rank condition on the well-known controllability matrix (Chen 1995). Here, an analogous matrix is computed via evaluation of Lie brackets from network outputs. Specifically, the network Lie brackets can be computed analytically (using symbolic manipulation software) and evaluated to construct the nonlinear controllability matrix (Khalil and Grizzle 2002). The singular values of this matrix are used to compute the **controllability index** (Haynes and Hermes 1970) through which a relativistic notion of controllability for different network parameterizations can be studied. Indeed, such an approach has previously been used to show the effect of symmetries in small networks in Whalen et al. (2015). Figure 9.2b shows this controllability index as a function of w_{IE}, i.e., the modeled action of propofol. The result affirms the conceptual hypothesis that controllability deteriorates rapidly with increasing drug dose. In other words, according to the model, the *dynamics of the network under propofol are not as labile as they are without the drug* , which is indeed consistent with the motivating premise, i.e., contracted reachability as a hypothetical systems-theoretic mechanism for unconsciousness.

9.3.2 Controllability and Plasticity

The systems-theoretic analytical approach can be generalized to many questions in theoretical neuroscience. For example, the notion of controllability is directly related to the sensitivity of networks to afferent inputs, which is a key issue in sensory neuroscience (see, Fig. 9.3 and recall Eq. (9.1)). Consider, for instance, a behavioral experiment involving repetitive presentations of a sensory stimulus (Fig. 9.3b). We posit the following question: with each successive presentation, how "different" should a competing stimulus be in order for discernment to occur. Here, difference can be understood in two ways: (i) the intensity, or energy, of the competing stimulus; and (ii) its orientation, or novelty, with respect to the background.

These questions can be accessed by using analyses adapted from the systems-theoretic notions of controllability and reachability, or, again, the ability of a dynamical system to be "steered" with respect to exogenous inputs. Further, it is straightforward to examine how such control-theoretic properties change as a function of time, due to the dynamics of neural plasticity (Kumar and Ching 2016). Such an analysis proceeds by using contemporary controllability metrics (Pasqualetti et al. 2014) as well as assays of sensitivity to orientation and stimulus novelty (Menolascino and Ching 2017). This latter measure describes the difference in angular orientation of the current input $u(t)$ with respect to a past input $v(t)$ in terms of the inner product

$$\int_0^T u(t) \cdot v(t-T)\mathrm{d}t \tag{9.2}$$

Equipped with this measure, we obtain the minimally novelty input (relative to prior inputs) that induces a given state transfer. In this way, the metric can characterize the smallest change in orientation required for a discernible trajectory change. By combining this analysis with conventional minimum energy-based analysis (i.e, the smallest $\int_0^T \|u(t)\|_2^2 \mathrm{d}t$ required for the same state change), we can comprehensively assess controllability in a time- and stimulus-dependent fashion.

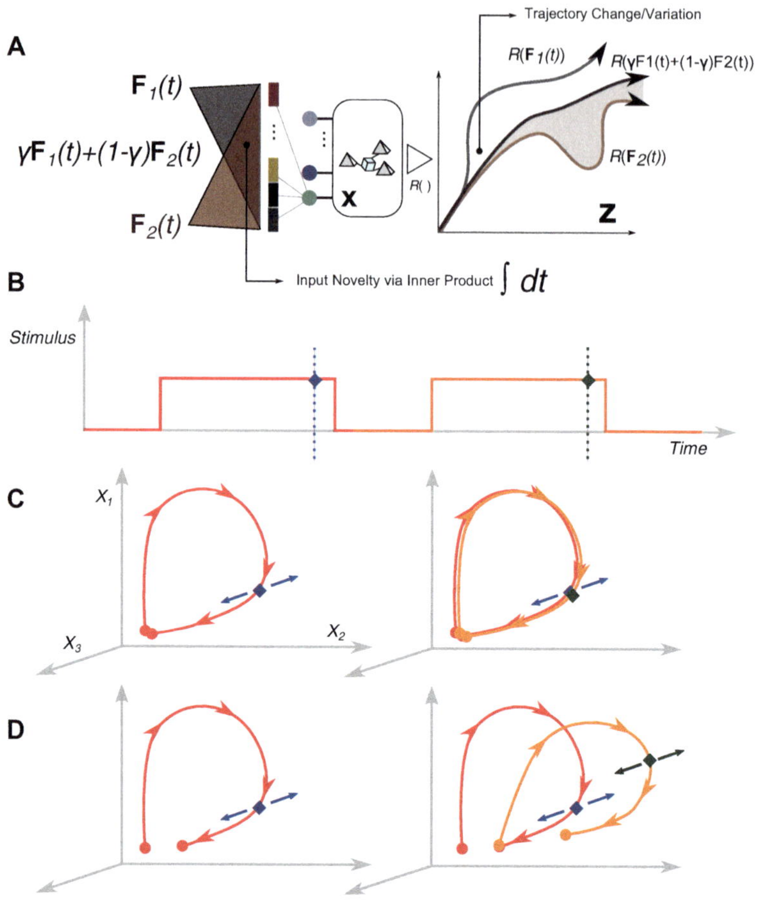

Fig. 9.3 (**a**) We consider control-theoretic properties of prototypical sensory networks whose inputs **F**(t) exist in a high-dimensional feature-space. (**b**) A repetitive stimulus induces trajectories (**c, d**) without, and in the presence of, plasticity. We ask how "different" a competing stimulus should be, at moments in time (blue, green points) in order to induce an altered (discernible) trajectory

To illustrate this analysis, we proceed to deploy this paradigm to study recurrent, *E–I* rate networks wherein the dynamics of the ith neuron are described by:

$$\tau_i \frac{dr_i}{dt} = -r_i + \frac{1}{1 + \exp\left(-\left(\sum_{j \neq i=1}^{N} (-1)^{\alpha} w_{i,j} r_j + I_i(t)\right)\right)} \tag{9.3}$$

Here, r_i is the time-varying firing rate of the neuron, τ_i is the time constant in milliseconds, $w_{i,j}$ is the synaptic weight (connectivity strength) of the synapse from neuron j to neuron i, $I_i(t)$ is the external input to the neuron i, and N is the total number of neurons in the network. $\alpha = 1$ if the neuron j is inhibitory, and $\alpha = 2$ if the neuron j is excitatory. Activity-dependent dynamical evolution of the synaptic weight $w_{i,j}$ is described by the Oja rule

$$\tau_{i,j}\frac{dw_{i,j}}{dt} = r_i r_j - \beta r_i^2 w_{i,j}. \tag{9.4}$$

Here, $\tau_{i,j}$ is the time constant in milliseconds and $\beta > 0$ is a constant. For our study, we consider 20% of the neurons in the network as inhibitory. Further, we exclude E–E synaptic connections. With this setup of the recurrent network, we first obtain the baseline firing rate of neurons as well as the synaptic weights by stimulating these neurons with constant input currents (i.e., $I_i(t) = I_i$) over a sufficiently long period of time such that the steady state of the network is reached. The time constants τ_i and $\tau_{i,j}$ are chosen from uniform distributions $\mathscr{U}(40, 60)$ and $\mathscr{U}(1000, 1500)$, respectively. The initial synaptic weights are chosen from a uniform distribution $\frac{1}{N}\mathscr{U}(0, 1)$, which are then normalized to ensure that $\sum_{j=1}^{N} w_{i,j}^2 = 1$ for all $i = 1, 2, \ldots, N$. The parameter β is set to 1.

Next, we study the effect of the long-term plasticity on the network dynamics by repeating a designed, constant, stimulus, presented on top of a nonzero baseline input. In each successive trial, the stimulus is presented for a specified time duration (T_1), then turned off (for a duration T_2). Figure 9.4a shows the firing rate trajectories of neurons, projected into a three-dimensional space using principal component analysis (PCA), in a recurrent network of 10 neurons. Here, trajectories become overlapped after about 8 presentations.

We proceed to characterize the network controllability at the end of each stimulus trial. To do so, we locally linearize Eq. (9.3) with respect to r_i (state dynamic matrix) and $I_i(t)$ (input matrix), $i = 1, 2, \ldots, N$. We then compute the minimum (average) novelty and energy required to drive the linearized network with fixed synaptic weights a unit distance in the direction of minimum-energy eigenvector of the Controllability Gramian. Figure 9.4b, c shows novelty-based and minimum energy-based characterizations, respectively. As shown here, in this particular case, the long-term plasticity reconfigures the network (in terms of synaptic weights) such

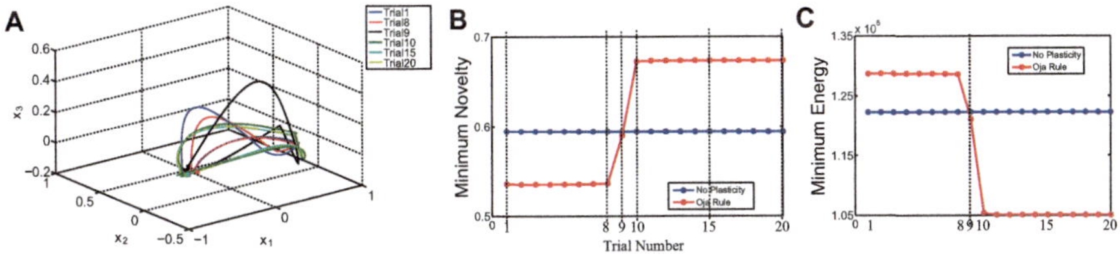

Fig. 9.4 Effects of long-term plasticity on the controllability of a rate-based sensory network. (**a**) Firing rate trajectories over many stimulus trials in a network of 10 neurons, projected in a three-dimensional space. (**b**) Minimum novelty as a function of trial number of the repetitive presentations of a stimulus in the presence and the absence of plasticity. (**c**) Minimum energy as a function of trial number

that the network becomes "easier" to control (less energy and novelty are required). Moreover, a sudden change is observed in both novelty and the minimum energy after 8 stimulations. We confirm such drastic changes in the controllability in our simulations by studying various networks, both by changing network parameters and the size of the network (figures not shown).

These simulation results reveal the significance of plasticity in modifying the control-theoretic properties of large neuronal networks. The utility of this theory will ultimately be in the prediction and explanation of behavioral responses in complex, overlapping stimulus environments. Many additional details regarding these results can be found in Kumar and Ching (2016).

9.3.3 Controllability Analysis in Neuroscience

For the reasons outlined in this section, controllability analysis is emerging as a powerful tool in the analysis of brain networks. Particular applications include the analysis of large scale brain networks parameterized from diffusion tensor imaging and functional magnetic resonance imaging (fMRI) (Gu et al. 2015). Perhaps the major challenge in this line of research will be reconciling the theoretical formalism with the analytical challenges that often necessitate making simplifying assumptions on the dynamics of the networks in question. One path to obviating these challenges may be to focus analysis on smaller circuit configurations at either region or neuronal scale (Ching and Ritt 2013; Li et al. 2013; Whalen et al. 2015).

In the context of small spiking networks, the use of statistical models, such as the popular point-process family (Truccolo et al. 2005), may prove fruitful in understanding controllability. This class of models allows for more relaxed, probabilistic notions of controllability such as the notion of a viable pattern set. Specifically, consider an arbitrary M-dimensional point process-generalized linear model (PP-GLM) defined over I intervals. Given a probability threshold ρ, the ρ-Viable Pattern Set, $\Sigma(\rho; M, I)$, is the set of patterns for which there exists $u \in \mathcal{U}$, i.e.,

$$\Sigma(\rho; M, I) = \{N \mid P(N \mid u) > \rho,\ u \in \mathcal{U}\}, \tag{9.5}$$

where $\mathcal{U}$ denotes the set of admissible inputs. That is, $\Sigma(\rho; M, I)$ amounts to a reachable set of spiking patterns for a specific probability/likelihood level, subject to input constraints. By performing optimization over ρ, such an analysis can be used to assay the controllability of small network configurations. For example, Fig. 9.5 illustrates the approximate reachable set size as a function of connectivity strength for symmetric and asymmetric network configurations. Two observations are of note in this figure. First, a small amount of connectivity (via the connectivity strength) is advantageous for controllability, beyond which controllability decreases monotonically. Second, an asymmetric topology is, in general, more controllable than a symmetric topology, consistent to dynamical studies in 3-neuron motifs (Whalen et al. 2015).

Fig. 9.5 (a) Symmetric and assymetric morifs, (b) Generalized linear model (GLM)-based control analysis can disassociate symmetric and asymmetric connectivity

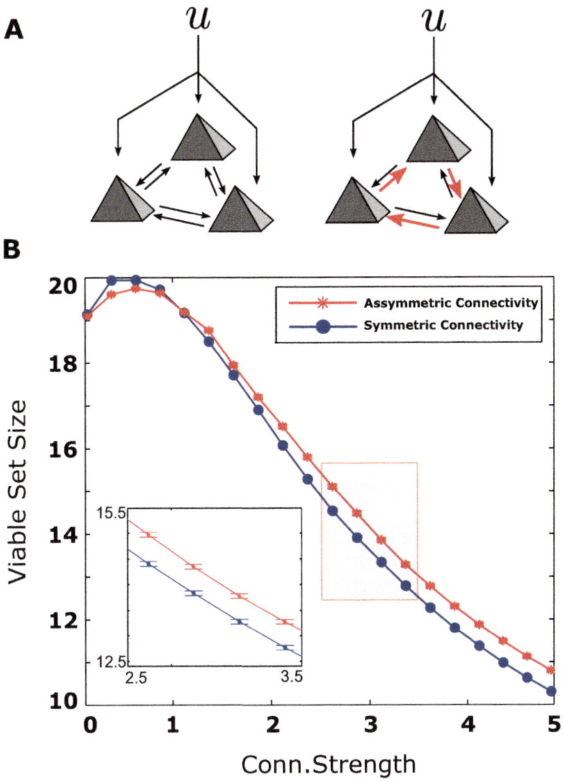

Many additional details regarding these results on statistical characterizations of controllability at the spatial scale of spiking networks can be found in Nandi et al. (2017a).

9.4 Model-Based Control of Dynamics and Brain Network Activity

The above analysis frameworks serve as an important precursor to actual neurocontrol, i.e., the direct manipulation of brain activity through exogenous stimulation. The applications of such manipulation are broad, including: neurostimulation to abate pathological dynamical phenomena such as seizure-like activity (Ching et al. 2012a; Ehrens et al. 2015) or subcortical synchronization in the context of motor disorders (Santaniello et al. 2015); stimulation to induce structured patterns of neural activity in studies of neural coding (Ching and Ritt 2013; Nandi et al. 2017a; Ritt and Ching 2015) using electrical or optical stimulation; or identification of structural connections within the brain through active probing (Lepage et al. 2013a,b).

9.4.1 Spike Control in Small Network Motifs

One example of a neurocontrol problem pertinent to optogenetic neurocontrol (Ching and Ritt 2013) involves inducing structured spike patterns in small networks of neurons. Consider the well-studied leaky integrate-and-fire (IF) neuron model in which the dynamics of the membrane potential $v_i(t)$ of the ith neuron (in a population) is given by:

$$C_i \frac{\mathrm{d}v_i(t)}{\mathrm{d}t} = \frac{(V_{\text{rest}} - v_i(t))}{R_i} + \beta_i u(t) \tag{9.6}$$

where V_{rest} is the resting potential, R_i is the total membrane resistance, C_i is the total membrane capacitance, $u(t)$ is the external (control) input. If at time t_s, any neuron reaches the threshold voltage V_T, i.e. $v(t_s) = V_T$, a spike is said to be generated. After each spike, the membrane potential is reset via

$$v(t_s^+) = V_{\text{rest}} \tag{9.7}$$

Thus, the dynamics of Eq. (9.6) are linear, complicated largely by the presence of the discontinuous state reset Eq. (9.7). Note that the input is not indexed by i which highlights the notion of underactuation, wherein a single (stimulating) input impinges on the entire population, which is a likely scenario in currently available neurostimulation technologies.

Mathematically, we can formulate the control design of $u(t)$ to create precise spiking in populations of IF neurons. Consider M uncoupled IF neurons of the form (9.6), with re-arranged parameters described by:

$$\begin{aligned} \begin{pmatrix} \dot{v}_1 \\ \vdots \\ \dot{v}_M \end{pmatrix} &= \begin{pmatrix} -a_1 & \dots & 0 \\ & \ddots & \\ 0 & \dots & -a_M \end{pmatrix} \begin{pmatrix} v_1 \\ \vdots \\ v_M \end{pmatrix} + \begin{pmatrix} b_1 \\ \vdots \\ b_M \end{pmatrix} u \\ &= f(V, u) = AV + bu \end{aligned} \tag{9.8}$$

where, from Eq. (9.6), $a_i = \frac{1}{R_i C_i}$, $b_i = \frac{\beta_i}{C_i}$, $a_i, b_i > 0$ for $i = 1, \dots, M$ and the origin translated to eliminate V_{rest}. Suppose our goal is to induce a spike in a given neuron, while other neurons stay silent. Without loss of generality for a target spike in Neuron 1 in a population of M neurons we can set up the following regularized time optimal problem:

$$\begin{aligned} &\underset{u \in \mathcal{U}}{\text{minimize}} \quad \mathbb{J}(u) = \int_0^\tau \mathrm{d}t + \frac{1}{2}\eta \, (\mathbf{w}V(\tau))^\top \mathbf{w} \, V(\tau) \\ &\text{s.t.} \qquad\qquad v_1(\tau) = V_T \end{aligned} \tag{9.9}$$

where $\mathbf{w} = [0 \; k_2 \dots \; k_M]$, $k_i \geq 0$, $\forall \; i = 2 \dots M$, the admissible set $\mathcal{U} = [U_1, U_2]$ and η is the regularization constant.

Along with the selectivity if we want to minimize the energy of the control $u(t)$ as well, we can add one more term in the integral of the objective. Once again without loss of generality for a target spike in Neuron 1 in a population of M neurons, we formulate the following regularized minimum time-energy optimal control problem:

$$\begin{aligned} \underset{u \in \mathscr{U}}{\text{minimize}} \quad & \mathbb{J}(u) = \int_0^{\tau} \left(1 + \frac{1}{2}\rho u^2\right) \mathrm{d}t + \frac{1}{2}\eta \left(\mathbf{w}V(\tau)\right)^{\top} \mathbf{w}V(\tau) \\ \text{s.t.} \quad & v_1(\tau) = V_T \end{aligned} \tag{9.10}$$

where ρ is the second regularization constant for the trade-off between the time and energy in the objective and the admissible set $\mathscr{U} = \mathbb{R}$.

Details regarding the solutions to these and related problems can be found in Ching and Ritt (2013), Nandi et al. (2017a,b). We illustrate the numerical results for a population of $M = 7$ neurons with $L = 2$ inputs and random parametrization for the resistance and capacitance of each cell, so that

$$\begin{aligned} E[R] &= 0.5\,\mathrm{G}\Omega,\ \sigma[R] = 0.05\,\mathrm{G}\Omega \\ E[C] &= 300\,\mathrm{pF},\ \sigma[C] = 2\,\mathrm{pF} \\ V_T &= 30\,\mathrm{mV}, \beta = 2 \\ U_1 &= -2.5\,\mathrm{nA},\ U_2 = 2.5\,\mathrm{nA}\ (\text{for } P2) \\ \mathbf{w} &= [0\ 1 \ldots 1]. \end{aligned} \tag{9.11}$$

In Fig. 9.6c, we plot the solution of Eq. (9.10) demonstrate the effect of regularization for the selective spiking problem. In the left panel ($\eta = 0$) and along with the intended spikes in Neuron 1, we observe collateral spiking in the population. In the right panel ($\eta \neq 0$) and we see that selectivity is improved by adding regularization. These solutions are obtained numerically by solving Eq. (9.10) as two point boundary value problem. By embedding this method in a greedy algorithm, we are able to effect control of longer spiking patterns, as illustrated in Fig. 9.6d.

9.4.2 Control via Pharmacology

Finally, we can return to anesthesia and brain inactivation. As we noted in the introduction, an increased understanding of dynamics can aid in the development of new engineering solutions for dosing neural pharmacology. Such designs can be similarly formulated as an exogenous neurocontrol problem (Kumar et al. 2016), for applications including closed-loop control of medical coma (Chemali et al. 2013; Ching et al. 2013; Liberman et al. 2013).

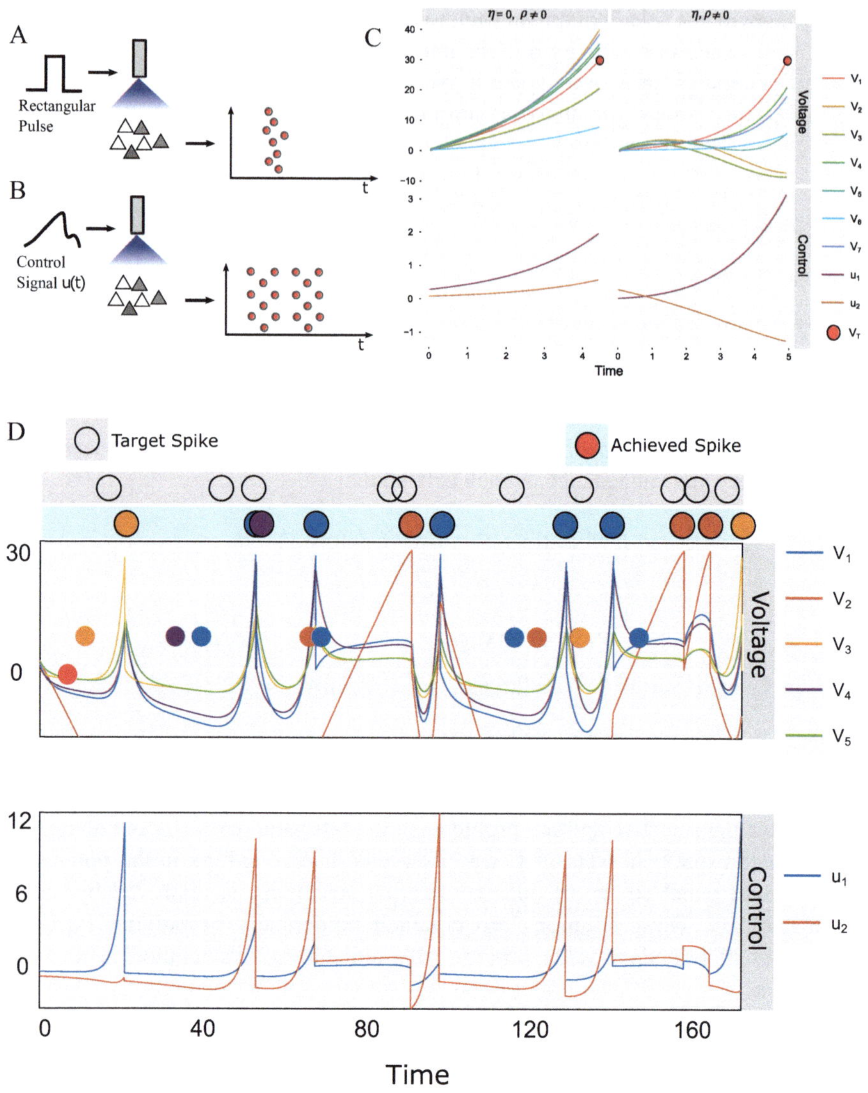

Fig. 9.6 (**a**) Typical neurostimulation paradigms in neuroscience involve applying perturbative pulses to cell populations. (**b**) Principled control design can enable the creation of more structured patterns/trajectories of spiking activity. (**c**) *Left:* The voltage trajectories and controls are shown as a function of time with no regularization on the terminal states ($\eta = 0, \rho = 0.1$) in *P3*. In this case four collateral spikes are induced along with Neuron 1. *Right:* Voltage trajectories and controls are shown for the regularized problem with ($\eta = 5V_T, \rho = 0.1$). In this case too the selective spiking in Neuron 1 is ensured. Note that for both the cases $\eta = 0, \eta = 5V_T$, the optimal controls take the form of exponential kernels. (**d**) Circles in the top two rows: Target and achieved spikes placed at corresponding times, color coded to represent neuron indices. *Top Panel:* Neuron voltages excited by the one step greedy design. *Bottom Panel:* Optimized control generated stepwise for each spike

Almost generically, neurally active drugs such as general anesthetics act by binding to targets at the molecular receptor level, leading to altered synaptic transmission (Ching and Brown 2014). Thus, control design for pharmacology can be conceived of in terms of drug–receptor interaction wherein the final goal is to achieve a certain state in the *receptor space*, i.e., the fraction of receptors bound for each of N receptors. The premise is that brain dynamics and corresponding behavioral outcomes can be more accurately and more generally mapped in the receptor space, rather than to particular site concentration of a specific drug. Consequently, dosing strategies involving potentially complex combinations of drugs can be studied and optimized.

The problem at hand is nontrivial since individual drugs can target multiple receptors and the same type of receptor can be targeted by multiple drugs. For example, propofol, an anesthetic drug that produces a spectrum of behavioral effects from paradoxical excitation to unconsciousness, targets at least $GABA_A$ and HCN1 receptors (Ching et al. 2010; Vijayan et al. 2013). As in the case of neurostimulation, attempting to modulate binding to both receptors independently is an underactuated control problem, where the number of inputs (drugs, here just one) is less than the dimensionality of the system to be controlled (receptors, here two). On the other hand, certain classes of drugs overlap in their primary receptor targets. In such a case, the system at hand has fewer degrees of freedom than it has inputs. Multivariate control-theoretic methods can be used to handle both of these types of design scenarios (Kumar et al. 2016) and we highlight a few salient details of this approach herein.

Specifically, we demonstrate the use of our design methodology to induce a well-defined dynamical regime in a biophysical neuronal network. Specifically, we consider the phenomenon of paradoxical excitation due to the anesthetic drug propofol (McCarthy et al. 2008, 2012), wherein, at low drug concentration, an excitable behavioral state is manifest and, further, is associated with relatively electrophysiological activity in the "β" (13–20 Hz) band.

We considered the biophysical neuronal network model for paradoxical excitation formulated in McCarthy et al. (2008). In this model, the concentration of propofol maps directly to the dynamics of the $GABA_A$ inhibitory synaptic current. In order to implement our design, we use existing in vitro descriptions of propofol affinity (Eghbali et al. 2003; Jin et al. 2009) to model the intermediate dynamics of the drug binding to the $GABA_A$ receptors themselves. Because of the inherently diffusive nature though which drugs permeate in the body, we can in general assume a linear n compartment pharmacokinetics model of the form:

$$\frac{\mathrm{d}X(t)}{\mathrm{d}t} = AX(t) + Bu(t). \tag{9.12}$$

Here, the ith component of $X(t) \in \mathbb{R}_+^{n\times 1}$ represents the concentration of the drug in the ith compartment, $A \in \mathbb{R}^{n\times n}$ models the rates of diffusion (pharmacokinetics parameters) between the compartments, $u(t) \in \mathbb{R}_+^{1\times 1}$ is the rate of infusion of the drug to the central compartment (infusion site), and $B \in \mathbb{R}_+^{n\times 1}$ is a scaling constant.

The concentration $x_{n+1}(t)$ of the drug at the effect site is given by

$$\frac{dx_{n+1}(t)}{dt} = k_{eo}(x_1(t) - x_{n+1}(t)). \tag{9.13}$$

Here, k_{eo} is a rate constant and $x_1(t)$ is the effect site concentration.

Interaction of a drug with its molecular targets is typically described by the following kinetic reaction:

$$[Ce] + [R] \Leftrightarrow [k^{(1)}][k^{(2)}][CeR] \tag{9.14}$$

Here $[Ce] = x_{n+1}$ is the concentration of the drug at the effect site, $[R]$ is the number of free receptors, and $[CeR]$ is the number of bound receptors. The constants $k^{(1)}$ and $k^{(2)}$ are the rates of binding and unbinding, respectively. Thus, the rate of change of $[CeR]$ is expressed as

$$\frac{d[CeR]}{dt} = k^{(1)}x_{n+1}[R] - k^{(2)}[CeR]. \tag{9.15}$$

By letting $y = [CeR]/([CeR] + [R])$ denote the fraction of receptors that are bound at time t (and since the total number of receptors (i.e., $[CeR] + [R]$) is fixed) we can rewrite Eq. (9.15) as

$$\frac{dy(t)}{dt} = k^{(1)}x_{n+1}(t)(1 - y(t)) - k^{(2)}y(t). \tag{9.16}$$

In the presence of multiple drugs binding to a single receptor, the dynamics of the fraction of bound receptors of type j with respect to drug i (i.e., $y_{i,j}$) at time t is then given by

$$\frac{dy_{i,j}(t)}{dt} = k_{i,j}^{(1)}x_{n+1,i}(t)\left(1 - \sum_{l=1}^{m} y_{l,j}(t)\right) - k_{i,j}^{(2)}y_{i,j}(t). \tag{9.17}$$

Here $x_{n+1,i}$ is the concentration of drug i at the effect site at time t, $k_{i,j}^{(1)}$ and $k_{i,j}^{(2)}$ are the rate constants of binding and unbinding, respectively, of receptors of type j with drug i, and m is the total number of drugs. At time t, the total fraction of bound receptors of type j is given by

$$y_j(t) = \sum_{i=1}^{m} y_{i,j}(t). \tag{9.18}$$

Equations (9.12), (9.13), (9.17), and (9.18) represent a nonnegative dynamical system which completely describes the dynamics from drug infusion to receptor binding.

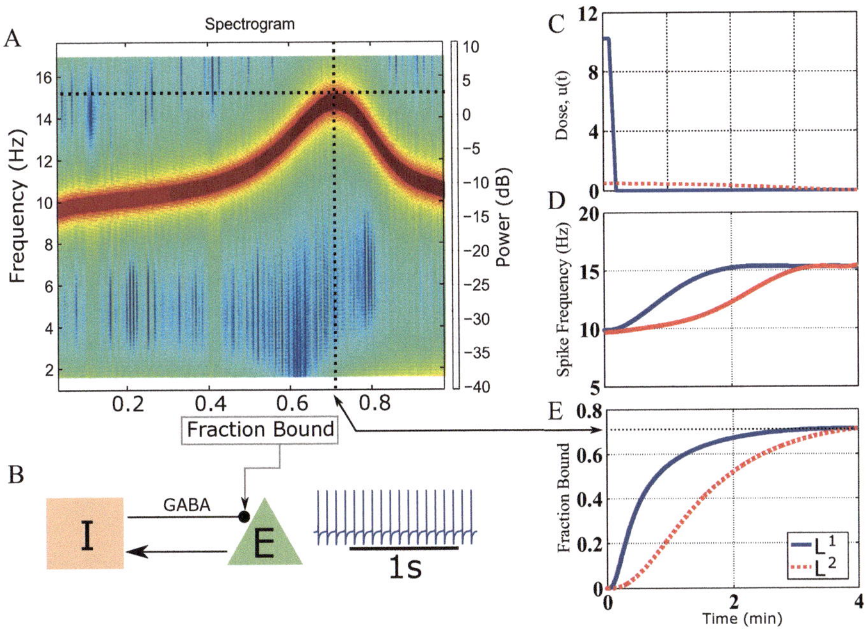

Fig. 9.7 Design of optimal dosing to target paradoxical excitation with propofol. (**a**) Spectrogram of the model output as a function of the fraction of $GABA_A$ receptors bound by propofol. (**b**) The model consists of reciprocally coupled Excitatory and Inhibitory neurons, each modeled using voltage gated conductance equations. The fraction of receptors bound modulates the GABA-ergic synaptic conductance and decay time. The design objective here is to induce the paradoxically elevated firing rate, corresponding to approximately 70% of receptors bound. (**c**) The optimal dose trajectories. (**d**) Model output frequency and (**e**) fraction bound trajectories are shown for ℓ_1 and ℓ_2-based cost functions

What remains to be specified is a mapping from $y(t)$ to some physiological measurable, such as the aforementioned "β" power. If a one-to-one mapping can be found, then the control design can proceed through a variety of contemporary control design methods. Figure 9.7a shows the neuronal firing rate of this model as a function of the fraction of $GABA_A$ receptors bound by propofol. As shown, the model exhibits an increase in neuronal firing rate over a narrow window of fraction bound. This increase is the putative "paradoxical" excitation, since ostensibly, inhibition is increasing monotonically relative to binding.

Here, for illustrative purposes, we formulate a design objective to induce this paradoxical state noting from Fig. 9.7a that it corresponds to approximately a 0.715 fraction bound.

We proceed to design several different control strategies by optimizing different objective functions. Figure 9.7c shows the sparse dosing of propofol obtained by minimizing the ℓ_1 norm based cost function (i.e., $p_1 \sum_{l=0}^{N_p(k)-1} |u_1(k+l \mid k)|$ with $p_1 = 1$). As shown in this figure, the optimal strategy is temporally sparse and, indeed, assumes the form of a bolus—a quick "push" of concentration at the beginning of the design window that ensures that the receptor trajectory rises to the target (i.e., 0.715) according to the intrinsic dynamics (Fig. 9.7d, e). Figure 9.7c also depicts the dosing solution obtained by minimizing the ℓ_2 norm based quadratic cost function (i.e., $\sum_{l=0}^{N_p(k)-1} p_2^2 u_2(k+l \mid k)^2$). As shown in this figure, propofol is administered at a relatively low rate, but continuously, leading to a more gradual binding that nevertheless achieves the desired state at the specified time (Fig. 9.7d, e).

The two solutions are, clearly, qualitatively different, however, the absolute quantity of drug used is commensurate in both cases. Hence, the final decision on strategy may be determined by secondary constraints, such as wanting to rise through intermediate states as quickly as possible (e.g., in the bolus-type strategy).

9.5 Conclusion

With the advent of new recording modalities, observing the activity in neural circuits is possible with ever-increasing spatial and temporal precision. Such capacity offers tantalizing possibilities for understanding the processes that govern cognition, as well as clinically important brain states such as general anesthesia and disorders of consciousness. However, to fully realize this potential will require not just the technology itself, but also accompanying theoretical innovations within mathematics and engineering. Perhaps most obvious is the need for advances in data science and statistics, in order to enable direct interpretation and extraction of salient features from neural recordings. However, in parallel, theoretical approaches in dynamical systems modeling and control-theoretic analysis can be powerful tools for generating mechanistic insight and hypotheses. Further, such methods can directly interface with technologies for *manipulating* brain activity that can work in a experiment-theory-validation loop towards both scientific and clinical endpoints. This chapter has provided several examples that demonstrate the conceptual advances that can be realized through such an interdisciplinary framework. In this regard, the further integration of neuroscience with engineering and mathematics—especially in the early training of students—promises to unlock even greater successes, as evidenced by the breadth and significance of research in this monograph.

Acknowledgements My career has been immeasurably enriched by the training I received from Professor Emery N. Brown, to whom this monograph is dedicated. Thank you Emery, for the opportunity to pursue research at the interface of engineering and neuroscience.

This chapter brings together several ideas and results from the past several years and I would like to acknowledge the important contributions of students and collaborators who were involved in published work. I would also like to acknowledge the Burroughs-Wellcome Fund Career Award at the Scientific Interface; Awards ECCS 1509342, CMMI 1537015 and CMMI 1653589 from the National Science Foundation; Awards R21NS096590, R21EY027590 from the National Institutes of Health; and Award 15RT0189 from the Air Force Office of Scientific Research, for support related to the research discussed in this chapter.

References

Brown, E. N., Lydic, R., & Schiff, N. D. (2010). General anesthesia, sleep, and coma. *New England Journal of Medicine, 363*, 2638–2650.

Chemali, J., Ching, S., Purdon, P. L., Solt, K., & Brown, E. N. (2013). Burst suppression probability algorithms: state-space methods for tracking EEG burst suppression. *Journal of Neural Engineering, 10*, 056017.

Chen, C.-T. (1995). *Linear system theory and design*. Oxford: Oxford University Press.

Ching, S., & Brown, E. N. (2014). Modeling the dynamical effects of anesthesia on brain circuits. *Current Opinion in Neurobiology, 25*, 116–122.

Ching, S., Brown, E. N., & Kramer, M. A. (2012a). Distributed control in a mean-field cortical network model: implications for seizure suppression. *Physical Review E, 86*, 021920.

Ching, S., Cimenser, A., Purdon, P. L., Brown, E. N., & Kopell, N. J. (2010). Thalamocortical model for a propofol-induced alpha-rhythm associated with loss of consciousness. *Proceedings of National Academy of Sciences USA, 107*, 22665–22670.

Ching, S., Liberman, M. Y., Chemali, J. J., Westover, M. B., Kenny, J. D., Solt, K., et al. (2013). Real-time closed-loop control in a rodent model of medically induced coma using burst suppression. *Anesthesiology, 119*, 848–860.

Ching, S., Purdon, P. L., Vijayan, S., Kopell, N. J., & Brown, E. N. (2012b). A neurophysiological-metabolic model for burst suppression. *Proceedings of National Academy of Sciences USA, 109*, 3095–3100.

Ching, S., & Ritt, J. T. (2013). Control strategies for underactuated neural ensembles driven by optogenetic stimulation. *Frontiers in Neural Circuits, 7*, 54.

Cimenser, A., Purdon, P. L., Pierce, E. T., Walsh, J. L., Salazar-Gomez, A. F., Harrell, P. G., et al. (2011). Tracking brain states under general anesthesia by using global coherence analysis. *Proceedings of National Academy of Sciences USA, 108*, 8832–8837.

Cowan, N. J., Chastain, E. J., Vilhena, D. A., Freudenberg, J. S., & Bergstrom, C. T. (2012). Nodal dynamics, not degree distributions, determine the structural controllability of complex networks. *PLoS ONE, 7*(6), e38398.

Eghbali, M., Gage, P. W., & Birnir, B. (2003). Effects of propofol on $GABA_A$ channel conductance in rat-cultured hippocampal neurons. *European Journal of Pharmacology, 468*(2), 75–82.

Ehrens, D., Sritharan, D., & Sarma, S. V. (2015). Closed-loop control of a fragile network: Application to seizure-like dynamics of an epilepsy model. *Frontiers in Neuroscience, 9*, 58.

Flores, F. J., Hartnack, K. E., Fath, A. B., Kim, S.-E., Wilson, M. A., Brown, E. N., et al. (2017). Thalamocortical synchronization during induction and emergence from propofol-induced unconsciousness. *Proceedings of National Academy of Sciences USA, 114*, E6660–E6668.

Freudenberg, J. S., Hollot, C. V., Middleton, R. H., & Toochinda, V. (2003). Fundamental design limitations of the general control configuration. *IEEE Transactions on Automatic Control, 48*(8), 1355–1370.

Gu, S., Pasqualetti, F., Cieslak, M., Telesford, Q. K., Yu, A. B., Kahn, A. E., et al. (2015). Controllability of structural brain networks. *Nature Communications, 6*, 8414.

Haynes, G., & Hermes, H. (1970). Nonlinear controllability via lie theory. *SIAM Journal on Control, 8*(4), 450–460.

Hermann, R., & Krener, A. J. (1977). Nonlinear controllability and observability. *IEEE Transactions on Automatic Control, 22*(5), 728–740.

Jin, Y.-H., Zhang, Z., Mendelowitz, D., & Andresen, M. C. (2009). Presynaptic actions of propofol enhance inhibitory synaptic transmission in isolated solitary tract nucleus neurons. *Brain Research, 1286*, 75–83.

Kalman, R. (1959). On the general theory of control systems. *IRE Transactions on Automatic Control, 4*(3), 110–110.

Khalil, H. K., & Grizzle, J. (2002). *Nonlinear systems* (Vol. 3). Upper Saddle River: Prentice Hall

Kumar, G., & Ching, S. (2016). The geometry of plasticity-induced sensitization in isoinhibitory rate motifs. *Neural Computation, 28*, 1889–1926.

Kumar, G., Kim, S. A., & Ching, S. (2016). A control-theoretic approach to neural pharmacology: Optimizing drug selection and dosing. *Journal of Dynamic Systems, Measurement, and Control, 138*(8), 084501.

Lepage, K. Q., Ching, S., & Kramer, M. A. (2013). Inferring evoked brain connectivity through adaptive perturbation. *Journal of Computational Neuroscience, 34*, 303–318.

Lepage, K. Q., Kramer, M. A., & Ching, S. (2013). An active method for tracking connectivity in temporally changing brain networks. In *Proceedings of IEEE Engineering in Medicine and Biology Conference* (pp. 4374–4377).

Li, J.-S., Dasanayake, I., & Ruths, J. (2013). Control and synchronization of neuron ensembles. *IEEE Transactions on Automatic Control, 58*(8), 1919–1930.

Liberman, M. Y., Ching, S., Chemali, J., & Brown, E. N. (2013). A closed-loop anesthetic delivery system for real-time control of burst suppression. *Journal of Neural Engineering, 10*, 046004.

Liu, S., & Ching, S. (2017). Homeostatic dynamics, hysteresis and synchronization in a low-dimensional model of burst suppression. *Journal of Mathematical Biology, 74*, 1011–1035.

Liu, Y.-Y., Slotine, J.-J., & Barabasi, A.-L. (2011). Controllability of complex networks. *Nature, 473*, 167–173.

McCarthy, M. M., Brown, E. N., & Kopell, N. (2008). Potential network mechanisms mediating electroencephalographic beta rhythm changes during propofol-induced paradoxical excitation. *Journal of Neuroscience, 28*(50), 13488–13504.

McCarthy, M. M., Ching, S., Whittington, M. A., & Kopell, N. (2012). Dynamical changes in neurological diseases and anesthesia. *Current Opinion in Neurobiology, 22*, 693–703.

Menolascino, D., & Ching, S. (2017). Bispectral analysis for measuring energy-orientation tradeoffs in the control of linear systems. *Systems & Control Letters, 102*, 68–73.

Milo, R., Shen-Orr, S., Itzkovitz, S., Kashtan, N., Chklovskii, D., & Alon, U. (2002). Network motifs: simple building blocks of complex networks. *Science, 298*(5594), 824–827.

Nandi, A., Kafashan, M., & Ching, S. (2017a). Control analysis and design for statistical models of spiking networks. *IEEE Transactions on Control of Network Systems*. doi: 10.1109/TCNS.2017.2687824

Nandi, A., Schättler, H., & Ching, S. (2017b). Selective spiking in neuronal populations. In *Proceedings of American Control Conference* (pp. 2811–2816). New York: IEEE.

Pasqualetti, F., Zampieri, S., & Bullo, F. (2014). Controllability metrics, limitations and algorithms for complex networks. *IEEE Transactions on Control of Network Systems, 1*(1), 40–52.

Purdon, P. L., Pierce, E. T., Mukamel, E. A., Prerau, M. J., Walsh, J. L., Wong, K. F. K., et al. (2013). Electroencephalogram signatures of loss and recovery of consciousness from propofol. *Proceedings of National Academy of Sciences USA, 110*, E1142–E1151.

Ritt, J. T., & Ching, S. (2015). Neurocontrol: Methods, models and technologies for manipulating dynamics in the brain. In *Proceedings of American Control Conference* (pp. 3765–3780). New York: IEEE

Santaniello, S., McCarthy, M. M., Montgomery, E. B., Gale, J. T., Kopell, N., & Sarma, S. V. (2015). Therapeutic mechanisms of high-frequency stimulation in Parkinson's disease and neural restoration via loop-based reinforcement. *Proceedings of National Academy of Sciences USA, 112*, E586–E595.

Sontag, E. D. (2013). *Mathematical control theory: Deterministic finite dimensional systems*. New York: Springer.

Truccolo, W., Eden, U. T., Fellows, M. R., Donoghue, J. P., & Brown, E. N. (2005). A point process framework for relating neural spiking activity to spiking history, neural ensemble, and extrinsic covariate effects. *Journal of Neurophysiology, 93*, 1074–1089.

Vijayan, S., Ching, S., Purdon, P. L., Brown, E. N., & Kopell, N. J. (2013). Thalamocortical mechanisms for the anteriorization of α rhythms during propofol-induced unconsciousness. *Journal of Neuroscience, 33*, 11070–11075.

Whalen, A. J., Brennan, S. N., Sauer, T. D., & Schiff, S. J. (2015). Observability and controllability of nonlinear networks: The role of symmetry. *Physical Review X, 5*(1), 011005.

Chapter 10
From Physiological Signals to Pulsatile Dynamics: A Sparse System Identification Approach

Rose T. Faghih

10.1 Introduction

Observed pulsatile physiological experimental data such as neuroendocrine data and electrodermal activity (EDA) data are stimulated by a well-known sequence of impulsive brain signals. An important question in neural signal analysis involves determining the timing and amplitude of brain impulsive signals from single or concurrent time series of pulsatile physiological experimental data. Solution of this problem has important implications for understanding normal and pathological neuroendocrine and affective states. Transmission of information via pulsatile/intermittent signaling is very different from continuous signaling and some disorders are associated with dysregulation of the physiological pulsatile activity. Hence, understanding the underlying nature of pulsatile release of physiological data via mathematical formalization can be beneficial to understanding the pathological states and could lay the basis for a physiologically based approach for administering medications.

As pointed out in Faghih et al. (2014), current data analysis methods for observed pulsatile experimental data either assume that the timing of the impulses belongs to a certain class of stochastic processes (Johnson 2003) or are based on pulse detection techniques (Vidal et al. 2012). The problem of recovering the number, timing, and amplitude of brain impulsive profiles as well as the model parameters of the underlying physiological interactions from a limited number of observations is ill-posed and there could be multiple solutions. One method used for analyzing observed pulsatile experimental data is to assume a point process model for the pulses and embedding a birth-death process in a Markov chain Monte

R.T. Faghih (✉)
University of Houston, Houston, TX, USA
e-mail: rtfaghih@uh.edu

Z. Chen, S.V. Sarma (eds.), *Dynamic Neuroscience*,
https://doi.org/10.1007/978-3-319-71976-4_10

Carlo (MCMC) algorithm (Johnson 2003). Another method for analyzing observed pulsatile experimental data is using a Bayesian approach to solve for the pulses, and the underlying model parameters (Keenan et al. 2005). Moreover, observed pulsatile experimental data has been analyzed via pulse detection algorithms and removing peaks with heights smaller than some threshold (Vidal et al. 2012). Pulse detection methods might work well for the type of pulsatile data in which one pulse decays significantly before the next pulse occurs; however, for pulsatile experimental data in which one pulse can occur before the previous one has significantly decayed to near zero, the extraction of the pattern of pulses may be less clear. By taking advantage of the sparse nature of the pulsatile physiological data and adding more constraints for recovering the underlying impulsive profile, we illustrate that we can deconvolve the observed pulsatile physiological data (Faghih et al. 2014). We formulate this question as a non-convex optimization problem, and solve it using a coordinate descent algorithm that has a principled combination of (i) compressed sensing for recovering the amplitude and timing of the impulses, and (ii) generalized cross validation for finding the number of impulses (Faghih et al. 2014). Here, the key is using the characteristic of the sparsity of underlying brain impulsive profile (i.e., there are a small number of impulses that are important) to recover the timing and amplitude of individual brain pulses using compressed sensing techniques. Compressed sensing is an approach for perfect reconstruction of sparse signals using fewer measurements than required by the Shannon/Nyquist sampling theorem (Boufounos et al. 2007). When only a small number of coefficients in an impulsive profile are large (i.e., most coefficients are small or zero), small coefficients can be discarded, and a sparse representation of the impulsive profile can be recovered using optimization or greedy algorithms (Boufounos et al. 2007). In finding the number of impulses, there is a trade-off between capturing the residual error and the sparsity. We use generalized cross-validation (GCV) (Golub et al. 1979) to find the number of pulses such that there is a balance between the residual error and the sparsity (Faghih et al. 2014). As a result of the deconvolution, we can recover the physiologically plausible brain impulsive profiles that upon interactions with biological or measurement processes lead to observed pulsatile experimental data (Faghih et al. 2014).

Analyzing two sets of simultaneously recorded pulsatile experimental data (i.e., stimulator and final output of the physiological process that eventually affects other tissues) that are released due to the same impulsive brain regulator as well as interactions of the two signals controlled via feedback loops is more challenging. Determining the number, timing, and amplitudes of pulsatile events from simultaneously recorded data is challenging because of several factors (Faghih et al. 2015a): (i) stimulator pulse activity, (ii) kinematics of stimulator and final output of the physiological process, (iii) the sampling interval, and (iv) the measurement error. To analyze simultaneously recorded pulsatile experimental data, Van Cauter proposed a method for recovering episodic pulsatile data fluctuations (Van Cauter 1981) for each of the two pulsatile profiles (Refetoff et al. 1985; Linkowski et al. 1985). Then, by analyzing the timing of the detected pulse peaks from each of the two pulsatile profiles and the respective durations those pulses overlapped (Refetoff et al. 1985;

Linkowski et al. 1985), they detected concomitant pulses. It has been illustrated that (Faghih et al. 2015a), by combining a minimal physiological model with sparse recovery techniques to recover the stimulator secretory events from concurrent data, we can recover physiologically plausible timing and amplitudes for the underlying pulses by concurrent analysis of two hormone profiles.

Due to the distinct information that gets relayed to target cells via pulsatile signaling (Faghih et al. 2015b), it is important to understand the physiology underlying pulsatile release of observed pulsatile experimental data. The intermittent control that is observed in pulsatile control of observed pulsatile experimental data is not one of the traditional control-theoretic methods normally used in control engineering. This type of control is a special case of bang-bang control, in which an action leads to instantaneous changes in the states of the system (Sethi and Thompson 2006). Impulse control occurs when there is not an upper bound on the control variable and an infinite control is exerted on a state variable in order to cause a finite jump (Sethi and Thompson 2006). To characterize the pulsatile control underlying cortisol secretion, we utilize a mathematical formulation for a controller that gives rise to the desired impulses as opposed to continuous control and achieves impulse control (Faghih et al. 2015b). We postulate that this controller is minimizing the number of secretory events that result in cortisol secretion, which is a way of minimizing the energy required for cortisol secretion; this controller maintains the pulsatile experimental data within a specific range while following the first-order dynamics underlying the observed pulsatile physiological experimental data (Faghih et al. 2015b). This novel approach results in pulse control where the pulses and the obtained pulsatile signal have rhythms that are in agreement with the known desired physiological variations. The proposed formulation is a first step in developing intermittent bio-inspired controllers for controlling pathological states related to pulsatile signals such as cortisol.

10.1.1 Chapter Structure

In this chapter, two important questions that deal with pulsatile physiological signals are addressed: (i) analyzing signals with pulsatile dynamics using a minimal multi-rate physiological model (Sect. 10.2), (ii) designing intermittent inputs for achieving desired pulsatile dynamics (Sect. 10.3). To further motivate the potential applications, in Sect. 10.1.2, we discuss the pulsatile dynamics in neuroendocrine system (particularly, cortisol secretion), followed by a brief discussion of pulsatile dynamics in electrodermal activity in Sect. 10.1.3, and end the introduction with examples of potential applications in Sect. 10.1.4. The multi-rate formulation for analyzing signals with pulsatile dynamics in Sect. 10.2 can be applied to deconvolution of single time series (Sect. 10.2.3) as well as concurrent time-series (Sect. 10.2.4). We illustrate that this approach can be applied to deconvolution of single physiological time-series with pulsatile dynamics by deconvolving cortisol and skin conductance data (Sect. 10.2.3). Moreover, we illustrate that this approach can be applied to concurrent physiological time-series with pulsatile dynamics by deconvolving simultaneously recorded cortisol and adrenocorticotropic hormone (ACTH) time-series (Sect. 10.2.4). Moreover, we illustrate that intermittent input design for achieving desired pulsatile dynamics can be accomplished for both constant (Sect. 10.3.2) and time-varying (Sect. 10.3.3) demand and holding costs. In particular, by considering circadian demand and holding costs, we illustrate that we can obtain pulsatile dynamics that are in agreement with experimental pulsatile cortisol data (Sect. 10.3.3). We conclude this chapter with discussion and some concluding remarks (Sect. 10.4). This chapter summarizes and generalizes the concepts that appeared in Rose T. Faghih's PhD thesis (Faghih 2014). Detailed discussion of the specific applications that deal with analyzing cortisol, EDA, and concurrent ACTH and cortisol data has appeared in Faghih et al. (2014, 2015a,c), respectively. A detailed discussion of characterizing pulsatile dynamics of cortisol release via intermittent input design is discussed in detail in Faghih et al. (2015b).

10.1.2 Pulsatile Dynamics in Neuroendocrine Systems

Hormones are chemical messengers that are released from the endocrine glands into the circulation, and relay information to cells and control a wide range of physiologic functions (Kettyle and Arky 1998). The neuroendocrine system consists of several glands that produce hormones in a hierarchical manner, and some hormones are secreted in pulsatile episodes as opposed to a continuous manner (Vis et al. 2010). For some hormones, neural interactions in the hypothalamus result in release of hormone-releasing hormones from the hypothalamus, which have an impulsive profile. Then, these impulsive hormone-releasing hormones induce secretion of pituitary hormones from the pituitary in a pulsatile manner, and the pituitary hormones lead to release of hormones from target glands. These hormones, which are absorbed from the blood, implement regulatory functions in different tissues and have a feedback effect on release of hormone-releasing hormones and pituitary hormones. A key factor in the pulsatile neuroendocrine systems is the pulsatile feedback control of hormone release (Kettyle and Arky 1998). In order to understand the physiology and effects of drugs, quantification of pulsatile episodes of hormone release is crucial.

Since secretion of most endocrine hormones is driven by a similar control mechanism, understanding how one pulsatile hormone is secreted adds insight to how the rest of the pulsatile hormones are secreted. As a prototype, we focus on the hypothalamic-pituitary-adrenal (HPA) axis and cortisol secretion. A similar control feedback system underlies the release of growth hormone, thyroid hormone, estrogen, and testosterone. In a healthy person, these hormones have pulsatile dynamics with regular periodic patterns. For example, the 24-h cortisol profile consists of episodic release of 15 to 22 pulses with varying amplitudes in a regular circadian pattern; the lowest amplitude occurs between 8 PM and 2 AM, followed an increase throughout the late night, reaching the highest amplitude between 8 AM and 10 AM, and then, declining throughout the day (Brown et al. 2001). Moreover, these pulsatile hormones have an ultradian rhythm that allows for amplitude and frequency encoding of information (Walker et al. 2010; Lightman and Conway-Campbell 2010). Pulsatile signaling is an efficient way of transmitting information that leads to rapid changes in hormone concentration and allows for target receptor recovery (Walker et al. 2010). The transcriptional program prompted by hormone pulses is considerably different from constant hormone treatment (Stavreva et al. 2009), and characterization of pulsatile dynamics underlying hormone release has great potential for optimal treatment of hormonal disorders. In this chapter, we characterize the pulsatile dynamics underlying cortisol secretion.

10.1.2.1 Hypothalamic-Pituitary-Adrenal Axis

Cortisol is a steroid hormone that regulates the body's metabolism and response to stress and inflammation (Brown et al. 2001). Stress includes physical stress (e.g., infection and thermal exposure) and psychological stress (e.g., fear and anticipation) (Gupta et al. 2007). Cortisol relays rhythmic impulsive brain signals to synchronize bodily systems with environmental variations (Savić and Jelić 2005). To release cortisol, first in the hypothalamus, corticotropin releasing hormone (CRH) is released in pulses. Then, stimulated by CRH, ACTH is synthesized and released from the anterior pituitary (Dallmant and Yates 1969). Furthermore, via stimulation of adrenal glands by ACTH, cortisol is produced and secreted (Kyrylov et al. 2005; Brown et al. 2001). The secreted cortisol diffuses into the circulation and is absorbed by different tissues to implement regulatory functions as a steroid hormone. Then, cortisol is cleared from the circulation by the liver (Brown et al. 2001). Moreover, cortisol has a feedback effect on the hypothalamus and anterior pituitary (Gupta et al. 2007; Kyrylov et al. 2005; Brown et al. 2001). Since dysregulation of cortisol pulsatility is linked to some psychiatric and metabolic disorders (Walker et al. 2010), and cortisol pulsatile activity is essential for target cell gene expression (McMaster et al. 2011; Walker et al. 2012), in this chapter, we mainly focus on characterizing the pulsatile dynamics underlying release of cortisol.

10.1.3 Pulsatile Dynamics in Electrodermal Activity

Another example of a physiological signal with pulsatile dynamics is EDA. While this chapter mainly focusses on the HPA axis, to illustrate that the applications of the methods presented in this chapter are beyond neuroendocrine hormones, we briefly discuss EDA and in Sect. 10.2.3 analyze skin conductance data. EDA is a measure of neurophysiologic arousal and is composed of separate, discrete, and temporally short bursts triggered by sympathetic nervous system activity (Faghih et al. 2015c; Wallin 1981). Temporal and spatial summation of spikes triggered by sudomotor nerve lead to a skin conductance response (SCR), and increased

SCR frequency or amplitude is associated with increased sympathetic nervous system activity (Lidberg and Wallin 1981); hence, skin conductance data is collected during psychophysical tasks relevant for anxiety disorders (Faghih et al. 2015c). Deconvolution of skin conductance data helps better understand brain activity in complex behaviors and reduce the dimensionality problems in presence of large scale data in stimulus-response affective experiments where the goal is to recover one's emotional response. An example of such experiments is fear conditioning and extinction experiments discussed in Faghih et al. (2015c), in which the underlying stimulus (visual cue or mild electrical shock) was recovered.

10.1.4 Potential Applications of Characterization of Pulsatile Physiological Signals

Identification of the amplitude and timing of pulsatile physiological experimental data allows for quantifying normal and abnormal pulsatile activity to better understand pathological states, and can potentially be used in designing optimal approaches for treating disorders linked to dysregulation of pulsatile activity of such signals. To better motivate the potential applications of characterization of pulsatile physiological signals, we briefly discuss disorders linked to pulsatile physiological signals analyzed in this chapter.

10.1.4.1 Neuroendocrine Disorders

Neuroendocrine disorders may exist in the form of endocrine gland hyposecretion (hormone deficiency), endocrine gland hypersecretion (hormone excess), or tumors of endocrine glands, and may be treated using surgery, tablets, or injections. The dosage (amount and timing) of medications used for treatment is not based on a systemic perspective and is not done optimally, and can have side effects. A desired treatment should use an optimal dosage by employing a model that predicts the dose-response. Many endocrine disorders affect the patient's performance in various ways, and it is important to have a general model for hormone secretion to potentially be able to use an optimal approach in treating hormonal disorders to minimize the side-effects of the medication. Normal endocrine secretion is necessary for cardiovascular health, and the cardiovascular system benefits from correcting endocrine disorders (Rhee and Pearce 2011).

Cortisol Disorders and Beyond

Cortisol is crucial in neurogenesis, metabolism, stress response, cognition, and response to inflammation (Sarabdjitsingh et al. 2012), and diseases that are linked to abnormalities in the HPA axis include diabetes, visceral obesity and osteoporosis, life-threatening adrenal crises, and disturbed memory formation (Vinther et al. 2011; Conrad et al. 2009). The disorders linked to cortisol might be due to changes in the pulsatile episodes or changes in sensitivity of the adrenal glands to ACTH (Young et al. 2001). A sparse system identification approach allows us to investigate the role of the amplitude and frequency of the pulses as well as the sensitivity of the hypothalamus, the pituitary, or the gland to the stimulus. Moreover, a model that predicts the dose-response can eventually lead to an optimal dosage (amount and timing) treatment protocol.

One instance of a cortisol disorder is adrenal deficiencies that might be due to impairment of the adrenal glands, impairment of the pituitary gland or the hypothalamus. An example of a disorder caused by adrenal deficiency is Addison's disease. Persistent vomiting, anorexia, hypoglycemia, unexplained weight loss, fatigue, and muscular weakness can be caused by adrenal deficiency (Ten et al. 2001). A patient who suffers from Addison's disease takes cortisone once or twice a day for their cortisol deficiency which does not seem optimal as in a healthy subject there are 15 to 22 secretory events that lead to the observed cortisol levels over 24 h. It is possible to personalize the medication and use intermittent control to mimic the physiology of a healthy subject so that patients maintain hormonal levels (e.g., cortisol levels) that are similar to healthy subjects. Similarly, such bio-inspired controllers can be used for controlling other hormones (e.g., growth hormone in children with growth failure, or gonadal hormones in women with infertility). Furthermore, inspired by the pulse controller proposed in this research, in brain-machine interface (BMI) design, it is possible to design pulse controllers instead of continuous controllers to improve the battery life of the brain implant.

Mental Health Disorders

As pointed out, another example of a physiological signal with pulsatile dynamics is EDA and skin conductance data normally collected during psychophysical tasks relevant for anxiety disorders (Faghih et al. 2015c). Since an SCR can be considered as a potential indicator of an arousal event, and in different types of stress-related disorders such as post-traumatic stress disorder (PTSD), sparse recovery of arousal events from skin conductance data could potentially be used as a predictor for changes in brain functions to distinguish between healthy subjects and different types of mental disorders. Moreover, it could potentially be used to investigate whether treatment is working and if the clinical symptoms are improving.

10.2 A Multi-Rate Formulation for Analyzing Signals with Pulsatile Dynamics

Motivated by the applications of analyzing pulsatile physiological signals, we start with proposing a general model and sparse recovery approach for analyzing such signals, and apply our framework to a few examples of pulsatile experimental data. The first-order kinetics underlying the physiological processes that lead to the observed pulsatile profile can be represented in the form of a continuous-time state-space model from which a multi-rate discrete analog of the system can be obtained. The state-space system takes the form:

$$\dot{\boldsymbol{x}}(t) = \boldsymbol{A}\boldsymbol{x}(t) + \boldsymbol{B}u(t) \tag{10.1}$$

$$\boldsymbol{y}(t) = \boldsymbol{C}\boldsymbol{x}(t) + \boldsymbol{\nu}(t) \tag{10.2}$$

where $\boldsymbol{A}$ is the state or system matrix (physiological rates and gains such as infusion and decay rates or negative feedback are defined in matrix $\boldsymbol{A}$ as unknown θ_j's for $i = 1, 2, \ldots, n$, where there are n such rates and gains in $\boldsymbol{A}$ and the rest of entries of $\boldsymbol{A}$ are nonzero), $\boldsymbol{B}$ is the input matrix (it indicates how the input affects the state), $\boldsymbol{C}$ is the output matrix (it indicates which states are observed), $\boldsymbol{x}$ represents the state vector (e.g. different hormone concentrations that have a pulsatile profile), $\boldsymbol{y}$ is the output vector that represents the observed pulsatile time-series, $\boldsymbol{\nu}(t)$ represents the measurement noise, $u(t)$ is the input or the control which in our formulation represents an abstraction of the discrete impulsive secretory events that lead to the observed pulsatile experimental data. $u(t)$ takes the form:

$$u(t) = \sum_{i=1}^{m} q_i \delta(t - \tau_i) \tag{10.3}$$

where q_i denotes the amplitude of a secretory event initiated at time τ_i, and m denotes the number of the secretory events. Our goal is to estimate the model parameters (θ_j's), the number of the secretory events (m), and the amplitudes (q_i for $i = 1, 2, \ldots, m$) and timing (τ_i for $i = 1, 2, \ldots, m$) of the secretory events using the observed noisy measurements of the pulsatile time-series collected in w-unit time intervals (e.g., w might take a value of 10-min intervals for cortisol data or might take a value of 5-ms intervals for skin conductance data).

Assuming that the input and the states are constant over one-unit time intervals ($T = 1$), by letting $\boldsymbol{\phi} = e^{AT}$, and $\boldsymbol{\Gamma} = \int_0^T e^{A(T-\tau)} d\tau$, we can represent the system in discrete form:

$$\boldsymbol{x}[k+1] = \boldsymbol{\phi}\boldsymbol{x}[k] + \boldsymbol{\Gamma} u[k] \tag{10.4}$$

$$\boldsymbol{y}[k] = \boldsymbol{C}\boldsymbol{x}[k] + \boldsymbol{\nu}[k] \tag{10.5}$$

Considering that pulsatile time-series are observed every w unit times (e.g., every 10 min for cortisol data or every 5 ms for skin conductance data), assuming that we would like to consider a one-unit time resolution for the input, by letting $\boldsymbol{A_d} = \boldsymbol{\phi}^w$, $\boldsymbol{B_d} = \left[\boldsymbol{\phi}^{(w-1)}\boldsymbol{\Gamma}\ \boldsymbol{\phi}^{(w-2)}\boldsymbol{\Gamma}\ \ldots\ \boldsymbol{\Gamma}\right]$, $\boldsymbol{v_d}[k] = \boldsymbol{v}[wk]$, $z[k] = \boldsymbol{x}[wk]$, $\boldsymbol{u_d}[k] = \left[u[wk]\ u[wk+1] \cdots u[wk+w-1]\right]^\top$, we can represent the multi-rate system as:

$$z[k+1] = \boldsymbol{A_d}z[k] + \boldsymbol{B_d}\boldsymbol{u_d}[k] \tag{10.6}$$

$$\boldsymbol{y}[k] = \boldsymbol{C}z[k] + \boldsymbol{v_d}[k] \tag{10.7}$$

where $\boldsymbol{A_d}$ and $\boldsymbol{B_d}$ are functions of $\boldsymbol{\theta} = \left[\theta_1 \cdots \theta_n\right]^\top$. Then, using the state transition matrix, and considering that the system is causal and N data points are available, we can represent the system as:

$$\boldsymbol{y}[k] = \boldsymbol{F}[k]z_0 + \boldsymbol{D}[k]\boldsymbol{u} + \boldsymbol{v_d}[k] \tag{10.8}$$

where $\boldsymbol{F}[k] = \boldsymbol{CA_d}^k$, $z_0 = z[0]$, $\boldsymbol{D}[k] = \boldsymbol{C}\left[\boldsymbol{A_d}^{k-1}\boldsymbol{B_d}\ \boldsymbol{Ad}^{k-2}\boldsymbol{B_d} \cdots \boldsymbol{B_d}\ \underbrace{0 \cdots 0}_{N-k}\right]$, and $\boldsymbol{u} = \left[\boldsymbol{u_d}[0]\ \boldsymbol{u_d}[1] \cdots \boldsymbol{u_d}[k-1] \cdots \boldsymbol{u_d}[N-1]\right]^\top$. $\boldsymbol{u}$ represents the entire input over the duration of the study, sampled every unit time (e.g., every minute for cortisol, every ms for skin conductance data). We consider de novo synthesis for unobserved states in this formulation (zero initial condition). Moreover, we assume that the initial observed levels of observed states are the initial conditions, and hence let $z_0 = z[0]$. Then, let $\mathbf{y} = \left[y[1]\ y[2] \cdots y[N]\right]^\top$, where y represents all the data points. Moreover, let $\boldsymbol{F_\theta} = \left[\boldsymbol{F}[0]\ \boldsymbol{F}[1] \cdots \boldsymbol{F}[N-1]\right]^\top$, $\boldsymbol{D_\theta} = \left[\boldsymbol{D}[0]\ \boldsymbol{D}[1] \cdots \boldsymbol{D}[N-1]\right]^\top$, and $\boldsymbol{v_y} = \left[\boldsymbol{v_d}[1]\ \boldsymbol{v_d}[2] \cdots \boldsymbol{v_d}[N]\right]^\top$. Hence, we can represent this system as:

$$\mathbf{y} = \boldsymbol{F_\theta}z_0 + \boldsymbol{D_\theta}\boldsymbol{u} + \boldsymbol{v_y} \tag{10.9}$$

where $\boldsymbol{F_\theta}$ and $\boldsymbol{D_\theta}$ are functions of $\boldsymbol{\theta}$ and the sparse vector $\boldsymbol{u}$. $\mathbf{y}_l$, $\boldsymbol{F}_{\theta_l}$, $\boldsymbol{D}_{\theta_l}$, and $\boldsymbol{v}_l$ correspond to the rows of $\mathbf{y}$, $\boldsymbol{F_\theta}$, $\boldsymbol{D_\theta}$, and $\boldsymbol{v}_\mathrm{y}$, respectively that correspond to the lth observed state for $l = 1, 2, \ldots, L$ where L is the number of observed states; then, for $l = 1, 2, \ldots, L$, the system can equivalently be represented as:

$$\mathbf{y}_l = \boldsymbol{F}_{\theta_l}z_0 + \boldsymbol{D}_{\theta_l}\boldsymbol{u} + \boldsymbol{v}_l \tag{10.10}$$

where $\mathbf{y}_l$ represents the observed pulsatile time-series for $l = 1, 2, \ldots, L$, collected at w-unit time intervals, and z_0 is a vector of the initial conditions of the states. $\boldsymbol{u}$ represents the entire input over the entire experiment. Elements of $\boldsymbol{u}$ take nonzero values q_i at times τ_i for $i = 1, 2, \ldots, m$ when there is a secretory event, and are zero otherwise. $\boldsymbol{F}_{\theta_l}$ and $\boldsymbol{D}_{\theta_l}$ are functions of θ_j for $j = 1, 2, \ldots, n$.

10.2.1 Optimization Formulation for Deconvolution of Pulsatile Signals

We can formulate this problem as an optimization problem:

$$\min \sum_{l=1}^{L} \frac{1}{\sigma_l^2} \left\| \mathbf{y}_l - \boldsymbol{F}_{\theta_l} \mathbf{z}_0 - \boldsymbol{D}_{\theta_l} \boldsymbol{u} \right\|_2^2 \tag{10.11}$$

s.t.

$$\boldsymbol{u}_{\min} \le \|\boldsymbol{u}\|_0 \le \boldsymbol{u}_{\max}$$
$$\boldsymbol{u} \ge 0$$
$$\boldsymbol{S\theta} \le \boldsymbol{q}$$

where $\boldsymbol{S}$ and $\boldsymbol{q}$ depend on the physiological constraints. σ_l represent the standard deviation of the measurement errors for each $\mathbf{y}_l$ pulsatile time series. In this optimization problem, solving for $\boldsymbol{u}$ is a combinatorial problem, which is generally NP-hard, and is solved using greedy algorithms and ℓ_p-optimization algorithms. The greedy algorithms include *Matching Pursuit (MP), Orthogonal MP, Iterative Hard Thresholding, Hard Thresholding Pursuit, Gradient Descent with Sparsification*, and *Compressive Sampling Matching Pursuit* (He et al. 2012). In the ℓ_p-optimization algorithms, the ℓ_0-norm is approximated by an ℓ_p-optimization problem where $0 < p < 2$ (He et al. 2012). The ℓ_p-optimization algorithms are more accurate than the greedy algorithms, but computationally more expensive (He et al. 2012). It is possible to cast the above optimization problem as:

$$\min_{\substack{\boldsymbol{u} \ge 0 \\ \boldsymbol{S\theta} \le \boldsymbol{q}}} J_\lambda(\theta, \boldsymbol{u}) = \sum_{l=1}^{L} \frac{1}{\sigma_l^2} \left\| \mathbf{y}_l - \boldsymbol{F}_{\theta_l} \mathbf{z}_0 - \boldsymbol{D}_{\theta_l} \boldsymbol{u} \right\|_2^2 + \lambda \|\boldsymbol{u}\|_p^p \tag{10.12}$$

where the ℓ_p-norm ($0 < p \le 2$) is an approximation to the ℓ_0-norm and λ is chosen such that the sparsity of $\boldsymbol{u}$ is between $\boldsymbol{u}_{\min}$ to $\boldsymbol{u}_{\max}$. Then, using a coordinate descent approach, this optimization problem can be solved iteratively using the following steps until convergence is achieved:

1.

$$\boldsymbol{u}^{(l+1)} = \underset{\boldsymbol{u} \ge 0}{\operatorname{argmin}} \; J_\lambda(\boldsymbol{\theta}^{(l)}, \boldsymbol{u}) \tag{10.13}$$

2.

$$\boldsymbol{\theta}^{(l+1)} = \underset{\boldsymbol{S\theta} \le \boldsymbol{q}}{\operatorname{argmin}} \; J_\lambda(\boldsymbol{\theta}, \boldsymbol{u}^{(l+1)}) \tag{10.14}$$

In order to solve Eq. (10.13), an iteratively reweighted least squares (RWLS) algorithm called FOCUSS which enforces a certain degree of sparsity can be employed (Zdunek and Cichocki 2008). In order to solve Eq. (10.14), different optimization methods such as Levenberg-Marquardt or interior point method can be used. One should select the appropriate optimization algorithm based on their data such that none of the estimated θ_j values stagnates at the boundary conditions. Since this optimization problem is non-convex, there are multiple local minima; a reasonable procedure is to utilize different initializations, and choose the local minimum that minimizes the problem in Eq. (10.12) and provides the best goodness of fit. Our approach uses a coordinate descent approach; the convergence properties of coordinate descent algorithms are well studied and discussed in Attouch et al. (2010).

10.2.2 Sparse Input Recovery from Pulsatile Signals

The sparse input can be recovered using an extension of the FOCUSS algorithm. The FOCUSS algorithm uses a reweighted norm minimization approach and minimizes the ℓ_2-norm at each iteration to find the solution; the iteration refines the initial estimate to the final localized energy solution (Gorodnitsky and Rao 1997). In the FOCUSS algorithm, assuming that a gradient factorization exists, the stationary points of Eq. (10.13) satisfy $\boldsymbol{u} = \boldsymbol{P_u D_\theta^\top} (\boldsymbol{D_\theta P_u D_\theta^\top} + \lambda \boldsymbol{I})^{-1} \mathbf{y}_\theta$ (Gorodnitsky and Rao 1997), where $\boldsymbol{P_u}$ =diag($|\mathsf{u}_\mathsf{i}|^{2-p}$), and $\mathbf{y}_\theta = \mathbf{y} - \boldsymbol{F}_\theta \mathbf{z_0}$. By iteratively updating λ and $\boldsymbol{u}$ until convergence, we can solve for the sparse vector $\boldsymbol{u}$. In the optimization problem in Eq. (10.12), λ balances between the sparsity of $\boldsymbol{u}$ and the residual error $\|\mathbf{y}_\theta - \boldsymbol{D}_\theta \boldsymbol{u}\|_2$. The sparsity of $\boldsymbol{u}$ increases with λ.

A version of the FOCUSS algorithm called FOCUSS+ (Murray 2005) allows for solving for $\boldsymbol{u}$ such that the maximum sparsity of $\boldsymbol{u}$ is n (for example, $n = 22$ for 24-h cortisol data) and $\boldsymbol{u}$ is nonnegative. This algorithm uses a heuristic approach for updating λ, which tunes the trade-off between the sparsity and the residual error by increasing λ to a maximum regularization $\lambda_{\max}$ as the residual error decreases.

For $r = 0, 1, 2, \ldots$, FOCUSS+ works as follows:

1. $\boldsymbol{P}_{\boldsymbol{u}}^{(r)}$ =diag($|\mathsf{u}_\mathsf{i}^{(r)}|^{2-p}$)
2. $\lambda^{(r)} = \left(1 - \frac{\|\mathbf{y}_\theta - \boldsymbol{D}_\theta \boldsymbol{u}^{(r)}\|_2}{\|\mathbf{y}_\theta\|_2}\right)\lambda_{\max},\ \lambda > 0$
3. $\boldsymbol{u}^{(r+1)} = \boldsymbol{P}_{\boldsymbol{u}}^{(r)} \boldsymbol{D}_\theta^\top (\boldsymbol{D}_\theta \boldsymbol{P}_{\boldsymbol{u}}^{(r)} \boldsymbol{D}_\theta^\top + \lambda^{(r)} \boldsymbol{I})^{-1} \mathbf{y}_\theta$
4. $\mathsf{u}_\mathsf{i}^{(r+1)} \leq 0 \rightarrow \mathsf{u}_\mathsf{i}^{(r+1)} = 0$
5. After more than half of the selected number of iterations, if $\|\boldsymbol{u}^{(r+1)}\|_0 > n$, select the largest n elements of $\boldsymbol{u}^{(r+1)}$ and set the rest to zero.
6. Iterate

FOCUSS+ usually converges within 10 to 50 iterations (Murray 2005). In this algorithm, the sparsity is determined by λ (the sparsity of $\boldsymbol{u}$ increases with λ), and λ balances between sparsity and the residual error. Since FOCUSS+ (Murray 2005) uses a heuristic approach for finding λ, it can overfit the data and find a less sparse solution.

Considering the ill-posedness of deconvolution problems, small variations in the data can result in large changes in the solution, and a balanced choice of regularization is required to filter out the effect of noise. Tikhonov regularization, truncated singular value decomposition, and the method of L-curve are well-known methods used when dealing with such problems (Hansen 1999); among these methods, the L-curve method appears to be the most commonly used; however, the L-curve method is computationally expensive, requiring computation of the solution for several samples of the parameter (Zdunek and Cichocki 2008). Zdunek et al. point out that GCV is usually more accurate in estimating the regularization parameter than the L-curve method (Zdunek and Cichocki 2008). In the GCV technique, the optimal choice of regularization minimizes the predictive mean-squared error. Hence, to balance between sparsity and the residual error, the GCV-FOCUSS+ (Faghih et al. 2014) algorithm can be used. The GCV-FOCUSS+ algorithm is based on FOCUSS+ (Murray 2005) that solves for nonnegative $\boldsymbol{u}$ such that $\boldsymbol{u}$ has a certain maximum sparsity n (i.e., $n = 22$ for the HPA axis), and uses the GCV (Golub et al. 1979) technique for estimating the regularization parameter. In particular, GCV-FOCUSS+ is closely related to a special version of the FOCUSS algorithm (Zdunek and Cichocki 2008), which uses the GCV technique for updating the regularization parameter λ. Choosing an optimal λ value that balances between the noise and sparsity is important in detecting the sparsity level. If λ is too small, overfitting can occur and noise can be detected as signal; on the other hand, if λ is too large, it leads to data underfitting, and as a result, the signal will not be constructed completely.

The GCV function is defined as:

$$G(\lambda) = \frac{N\|(\boldsymbol{I} - \boldsymbol{H}_\lambda)\mathbf{y}_{,,}\|^2}{(\text{trace}(\boldsymbol{I} - \boldsymbol{H}_\lambda))^2}, \tag{10.15}$$

where N is the number of data points, and $\boldsymbol{H}_\lambda$ is the influence matrix. For the FOCUSS algorithm, $\boldsymbol{H}_\lambda = \boldsymbol{D}_\theta \boldsymbol{P}_{\text{u}} \boldsymbol{D}_\theta^\top (\boldsymbol{D}_\theta \boldsymbol{P}_{\text{u}} \boldsymbol{D}_\theta^\top + \lambda \boldsymbol{I})^{-1}$. The GCV technique was employed for estimating the regularization parameter for the FOCUSS algorithm through singular value decomposition (Zdunek and Cichocki 2008):

$$G(\lambda) = \frac{N \sum_{i=1}^{N} \zeta_i^2 \left(\frac{\lambda}{\sigma_i^2 + \lambda}\right)^2}{\left(\sum_{i=1}^{N} \frac{\lambda}{\sigma_i^2 + \lambda}\right)^2}, \tag{10.16}$$

where $\boldsymbol{\zeta} = \boldsymbol{R}^\top \mathbf{y}_\theta = \left[\zeta_1\ \zeta_2 \cdots \zeta_L\right]^\top$ and $\boldsymbol{D}_\theta \boldsymbol{P}_{\boldsymbol{u}}^{1/2} = \boldsymbol{R}\boldsymbol{\Sigma}\boldsymbol{Q}^\top$ with $\boldsymbol{\Sigma} = diag\{\sigma_i\}$; $\boldsymbol{R}$ and $\boldsymbol{Q}$ are unitary matrices and σ_i's are the singular values of $\boldsymbol{D}_\theta \boldsymbol{P}_{\boldsymbol{u}}^{1/2}$. Furthermore, $G(\lambda)$ is minimized such that λ is bounded between some minimum and maximum values ($\lambda_{\min}$ and $\lambda_{\max}$) using an implementation of the golden section (GS) search (Zdunek and Cichocki 2008). Although the GS search only finds a local extremum, considering that $G(\lambda)$ is unimodal, the GS search always finds the desired solution given a large range for λ (Zdunek and Cichocki 2008). We recommend using a range of zero to 10 for λ.

For $r = 0, 1, 2, \ldots,$ GCV-FOCUSS+ works as follows:

1. $\boldsymbol{P}_{\boldsymbol{u}}^{(r)} = \mathrm{diag}(|\mathsf{u}_{\mathsf{i}}^{(r)}|^{2-p})$
2. $\boldsymbol{u}^{(r+1)} = \boldsymbol{P}_{\boldsymbol{u}}^{(r)} \boldsymbol{D}_{\theta}^{\top} (\boldsymbol{D}_{\theta} \boldsymbol{P}_{\boldsymbol{u}}^{(r)} \boldsymbol{D}_{\theta}^{\top} + \lambda^{(r)} \boldsymbol{I})^{-1} \mathbf{y},$
3. $\mathsf{u}_{\mathsf{i}}^{(r+1)} \leq 0 \rightarrow \mathsf{u}_{\mathsf{i}}^{(r+1)} = 0$
4. $\lambda^{(r+1)} = \underset{0 \leq \lambda \leq 10}{\mathrm{argmin}} \quad G(\lambda) = \frac{N \|(\boldsymbol{I} - \boldsymbol{D}_{\theta} \boldsymbol{P}_{\boldsymbol{u}}^{(r)} \boldsymbol{D}_{\theta}^{\top} (\boldsymbol{D}_{\theta} \boldsymbol{P}_{\boldsymbol{u}}^{(r)} \boldsymbol{D}_{\theta}^{\top} + \lambda \boldsymbol{I})^{-1}) \mathbf{y}_{\theta} \|^2}{(\mathrm{trace}(\boldsymbol{I} - \boldsymbol{D}_{\theta} \boldsymbol{P}_{\boldsymbol{u}}^{(r)} \boldsymbol{D}_{\theta}^{\top} (\boldsymbol{D}_{\theta} \boldsymbol{P}_{\boldsymbol{u}}^{(r)} \boldsymbol{D}_{\theta}^{\top} + \lambda \boldsymbol{I})^{-1}))^2}.$
5. Iterate until convergence

10.2.3 Application 1: Deconvolution of Single Time-Series Pulsatile Data

As pointed out, cortisol data is an example of pulsatile physiological experimental data. In order to model the first-order kinetics underlying cortisol synthesis in the adrenal glands, cortisol infusion to the blood, and cortisol clearance by the liver, Eq. (10.1) can be described as:

$$\frac{dx_1(t)}{dt} = -\theta_1 x_1(t) + u(t) \tag{10.17}$$

$$\frac{dx_2(t)}{dt} = \theta_1 x_1(t) - \theta_2 x_2(t) \tag{10.18}$$

where x_1 is the cortisol concentration in the adrenal glands and x_2 is the serum cortisol concentration. θ_1 and θ_2, respectively, represent the infusion rate of cortisol from the adrenal glands into the blood and the clearance rate of cortisol by the liver (Faghih et al. 2014). $u(t)$ is an abstraction of the hormone pulses that result in cortisol secretion as defined in Eq. (10.3). In this formulation, Eq. (10.17) represents the first-order kinetics underlying cortisol synthesis in the adrenal glands while Eq. (10.18) represents the first-order kinetics underlying cortisol infusion to the blood, and cortisol clearance by the liver. Figure 10.1 shows an example of experimental cortisol data from a healthy female participant (Klerman et al. 2001), and model-predicted cortisol estimates, and the estimated amplitude and timing of hormone pulses (Faghih et al. 2014). Details about the data can be found in Klerman et al. (2001) and details about the analysis and more examples of cortisol deconvolution can be found in Faghih et al. (2014). The circadian amplitudes of the recovered pulses demonstrate the known circadian variation of cortisol; the recovered pulses are small at the beginning of the scheduled sleep, and there is a large pulse towards the end of the sleep period. There are multiple small and medium sized pulses during the wake period. The number of detected pulses are within the physiologically plausible range (Brown et al. 2001; Veldhuis et al. 1989; Faghih et al. 2011; Faghih 2010).

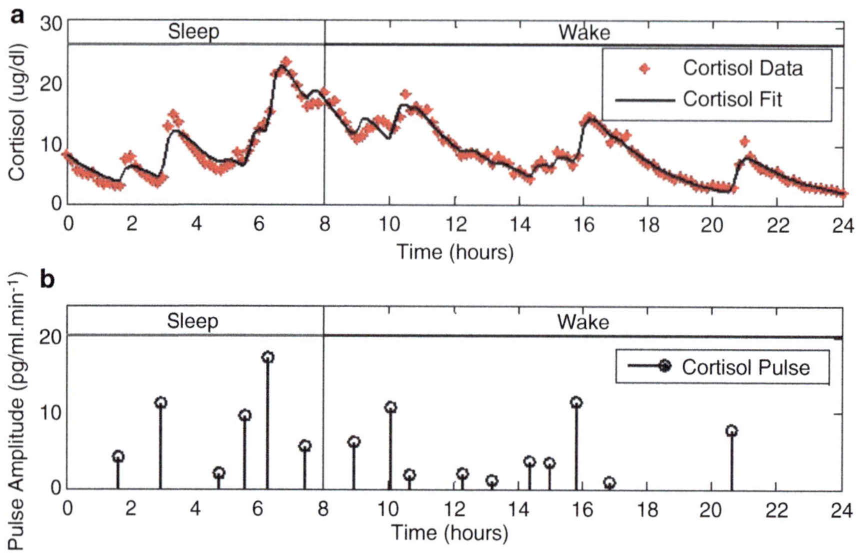

Fig. 10.1 Estimated deconvolution of the experimental 24-h cortisol levels in a healthy female participant. (**a**) Top panel shows the measured 24-h cortisol time series (red stars), and the estimated cortisol levels (black curve). (**b**) Bottom panel shows the estimated pulse timing and amplitudes (black vertical lines with dots) for one of the participants. The shaded gray area corresponds to sleep period and the white area corresponds to wake period. The same data collected in Klerman et al. (2001) and analyzed in Faghih (2014)

One should note that alternatively, we could consider the same model for analyzing skin conductance data, in which case, x_1 is a hidden state variable that stimulates the skin conductance levels and x_2 represents the skin conductance levels; θ_1 and θ_2 are time constants in the model. $u(t)$ is an abstraction of the discrete and temporally short bursts triggered by sympathetic nervous system activity (Faghih et al. 2015c). Figure 10.2 shows an example of experimental skin conductance data from a healthy female participant during the anger phase of an emotion study (Vyzas and Picard 1999), model-predicted skin conductance estimates, and the estimated amplitude and timing of potential arousal impulses of the brain. Details about the data can be found in Vyzas and Picard (1999). Another example of deconvolution of skin conductance data experiments is fear conditioning and extinction experiments discussed in Faghih et al. (2015c).

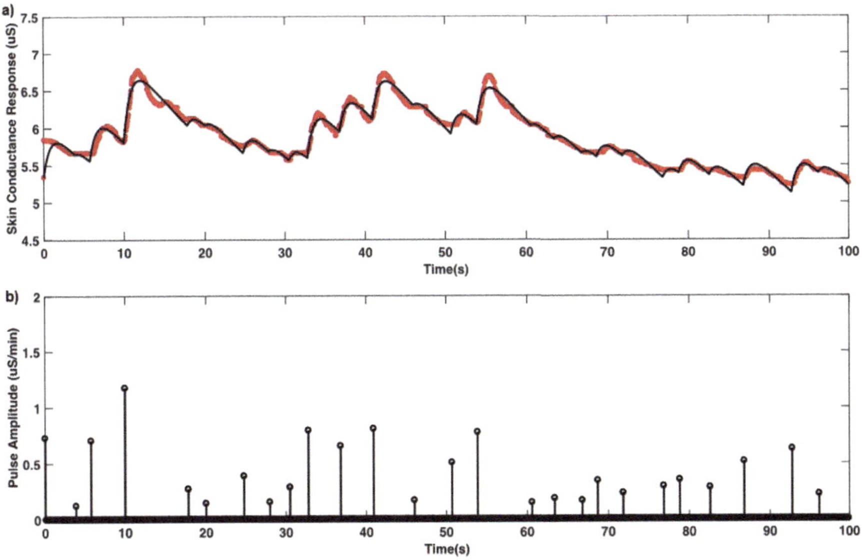

Fig. 10.2 Estimated deconvolution of the experimental electrodermal activity in a healthy female participant. (**a**) Top panel shows the measured 24-h skin conductance time series (red stars), and the estimated skin conductance levels (black curve) during the anger phase of an emotion study. (**b**) Bottom panel shows the estimated pulse timing and amplitudes (black vertical lines with dots) for one of the participants during the anger phase of an emotion study. The same data collected and used in Vyzas and Picard (1999)

10.2.4 Application 2: Deconvolution of Concurrent Pulsatile Data

Various models of the HPA axis and cortisol secretion have been proposed where cortisol synthesis in the adrenal glands is modeled based on the first-order kinetics of cortisol secretion (Brown et al. 2001; Faghih 2010; Faghih et al. 2011; Gupta et al. 2007; Vinther et al. 2011; Conrad et al. 2009). Mathematical models for concurrent ACTH and cortisol measurements include Faghih et al. (2015a), Peters et al. (2007), Lönnebo et al. (2007), Van Cauter (1981), Refetoff et al. (1985), and Linkowski et al. (1985). Equations (10.19)–(10.21) model the HPA axis and cortisol and ACTH release and are based on the model in Faghih et al. (2015a):

$$\frac{dx_1(t)}{dt} = -\theta_1 x_1(t) - \theta_2 x_3(t) + u(t) \text{ (Anterior Pituitary)} \tag{10.19}$$

$$\frac{dx_2(t)}{dt} = \theta_3 x_1(t) - \theta_4 x_2(t) \text{ (Adrenal Glands)} \tag{10.20}$$

$$\frac{dx_3(t)}{dt} = \theta_4 x_2(t) - \theta_5 x_3(t) \text{ (Serum)} \tag{10.21}$$

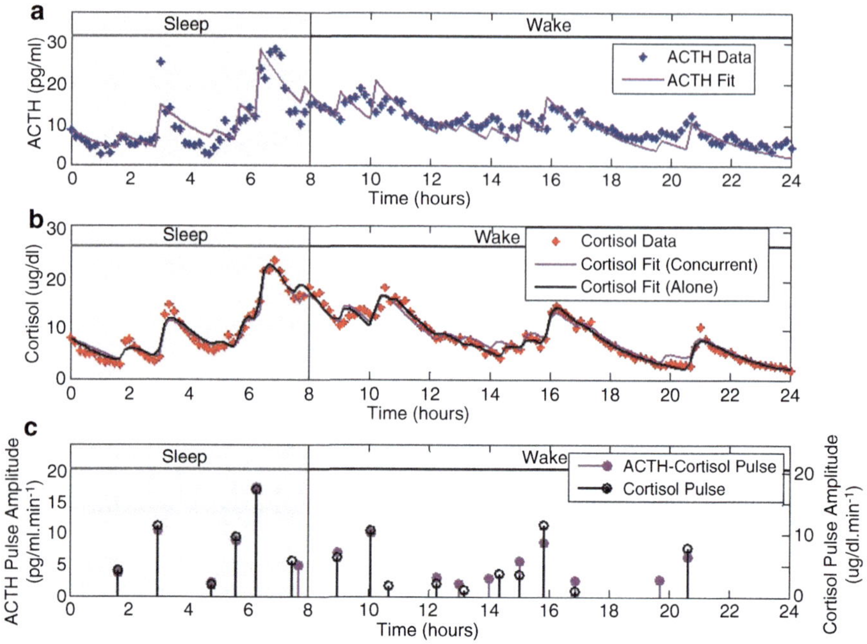

Fig. 10.3 Estimated deconvolution of experimental 24-h concurrent ACTH and cortisol levels in a healthy female participant. (**a**) Top panel shows the measured 24-h ACTH time series (blue stars), and the estimated ACTH levels (purple curve). (**b**) Middle panel shows the measured 24-h cortisol time series (red stars), the estimated cortisol levels (purple curve) using concurrent measurements of ACTH and cortisol, and the estimated cortisol levels (black curve) from deconvolution of only cortisol measurements. (**c**) Bottom panel shows the estimated pulse timing and amplitudes (purple vertical lines with dots) using concurrent measurements of ACTH and cortisol, and the estimated pulse timing and amplitudes (black vertical lines with dots) using only cortisol measurements for one of the participants. The shaded gray area corresponds to sleep period and the white area corresponds to wake period. The same data were collected in Klerman et al. (2001) and analyzed in Faghih (2014)

where x_1 is the serum ACTH concentration, x_2 is the cortisol concentration in the adrenal glands, x_3 is the serum cortisol concentration, and θ_3 is the ACTH gain. θ_1 and θ_2 represent the infusion rate of ACTH from the anterior pituitary to the blood and the cortisol negative feedback gain, respectively. θ_4 and θ_5 represent the coefficients corresponding to infusion of cortisol into the circulation from the adrenal glands and clearance of cortisol by the liver, respectively. $u(t)$ is an abstraction of the secretory events in the anterior pituitary that result in ACTH release and consequent cortisol release. Figure 10.3 shows an example of experimental ACTH and cortisol data from a healthy female participant (Klerman et al. 2001), model-predicted ACTH and cortisol estimates, and the estimated amplitude and timing of hormone pulses. The model-predicted ACTH and cortisol estimates (purple curves), and the estimated amplitude and timing of hormone pulses (purple vertical lines with dot) were obtained by deconvolving concurrent ACTH and cortisol data (Faghih et al. 2015a). The model-predicted cortisol estimates (black curves), and the estimated amplitude and timing of hormone pulses (black vertical lines with dots) were obtained by deconvolving cortisol data. As illustrated in Fig. 10.3, the timing of most of the significant pulses recovered from concurrent measurements of ACTH and cortisol is in agreement with the timing of most significant pulses recovered only from cortisol measurements for most participants. Details about the data can be found in Klerman et al. (2001) and details about the analysis and more examples of concurrent deconvolution of ACTH and cortisol can be found in Faghih et al. (2015a). These examples suggest that using only cortisol data one can find the timing of most of the significant secretory events in the HPA axis, namely the timing of pulses that are crucial in both cortisol and ACTH pulsatile profiles.

10.3 Impulsive Input Design for Achieving Desired Pulsatile Dynamics

Motivated by the potential applications of bio-inspired intermittent controllers, we assume that the first-order kinetics underlying a physiological process, the demand for the observed pulsatile profile, and the upper bound on the desired pulsatile profile are known, and the goal is to find an impulsive brain signal that achieves a pulsatile profile that satisfies the physiological constraints. In this formulation, we assume that the first-order kinetics that lead to the pulsatile profile take the form of Eqs. (10.1) and (10.2), where the system is known (i.e., $\boldsymbol{A}$, $\boldsymbol{B}$, and $\boldsymbol{C}$ are known). Moreover, the demand for the pulsatile profile is defined by a known time-varying function $\boldsymbol{h}(t)$ and should be satisfied. Furthermore, the upper bound on the pulsatile profile that the body can produce or a holding cost so that the pulsatile profile would not be much above the demand is a known time-varying function $\boldsymbol{q}(t)$ and should be satisfied. Both $\boldsymbol{h}(t)$ and $\boldsymbol{q}(t)$ are slowly varying compared to the pulsatile dynamics. The impulsive profile (control) that results in the pulsatile profile $\boldsymbol{u}(t)$ is nonnegative. Assuming that the body is minimizing the number of resources (control), our goal is to construct an optimization formulation that can lead to an impulsive profile that achieves the desired pulsatile dynamics given the underlying physiological process as well as demand and holding cost constraints. Hence, one possible optimization formulation for intermittent control of the pulsatile profile is as follows (Faghih et al. 2015b):

$$\min_{\boldsymbol{u}} \|\boldsymbol{u}\|_0 \tag{10.22}$$

s.t.

$$\begin{aligned}
&\boldsymbol{u}(t) \geq 0\\
&\dot{\boldsymbol{x}} = \boldsymbol{A}\boldsymbol{x}(t) + \boldsymbol{B}\boldsymbol{u}(t)\\
&\boldsymbol{y}(t) = \boldsymbol{C}\boldsymbol{x}(t)\\
&\boldsymbol{h}(t) \leq \boldsymbol{y}(t) \leq \boldsymbol{q}(t)
\end{aligned}$$

The state-space system in Eqs. (10.1) and (10.2) can alternatively be represented as Eq. (10.9) where $\boldsymbol{F}_\theta$ and $\boldsymbol{D}_\theta$ are known. Then by letting $\mathbf{h} = \left[\boldsymbol{h}_1\ \boldsymbol{h}_2 \cdots \boldsymbol{h}_N\right]^\top$ where $\boldsymbol{h}_k$ is the pulsatile profile demand at k and $\mathbf{q} = \left[\boldsymbol{q}_1\ \boldsymbol{q}_2 \cdots \boldsymbol{q}_N\right]^\top$ where $\boldsymbol{q}_k$ is the upper bound at k for $k = 1$ to N. Hence, we can alternatively solve the discrete analog of the formulation in Eq. (10.22) as (Faghih et al. 2015b):

$$\min_{\mathbf{u}} \|\mathbf{u}\|_0 \tag{10.23}$$

$$\begin{aligned} \text{s.t.} & \\ & \mathbf{u} \geq 0 \\ & \mathbf{y} = \boldsymbol{F}_\theta \mathbf{z_0} + \boldsymbol{D}_\theta \mathbf{u} \\ & \mathbf{h} \leq \mathbf{y} \leq \mathbf{q} \end{aligned}$$

10.3.1 Algorithm for Impulsive Input Design

Given that ℓ_0 problems are generally NP-hard, instead an ℓ_1-norm relaxation of such problems can be solved. In solving ℓ_1-norm problems, there is a dependence on the amplitude of the coefficients over which the ℓ_1-norm is minimized, and there is more penalty on larger coefficients than on smaller ones. However, it is possible to strategically construct a reweighted ℓ_1-norm such that nonzero coefficients are penalized in a way that the cost further resembles the ℓ_0-norm (Faghih et al. 2015b). If large weights are put on small entries, the solution concentrates on entries with small weights, and a cost function that is more similar to an ℓ_0-norm cost function can be solved such that nonzero entries are discouraged in the recovered signal (Candes et al. 2008). To find such weights for ℓ_1-norm cost function, Candes et al. have proposed an iterative algorithm for enhancing the sparsity using reweighted ℓ_1 minimization, which solves $\min\limits_{\mathbf{u}}\|\mathbf{u}\|_0$ (Candes et al. 2008). This algorithm is based on Fazel's *log-det heuristic* algorithm for minimizing the number of nonzero entries of a vector (Fazel 2002). The convergence of this log-det heuristic algorithm has been studied in Lobo et al. (2007). The algorithm for designing an impulsive profile that achieves desired pulsatile dynamics is as follows (Faghih et al. 2015b):

1. For the future time period τ where τ is an integer multiple of unit-time, initialize the diagonal matrix $\boldsymbol{O}^{(0)}$ with entries $o_i^{(0)} = 1, i = 1, \ldots, \tau + 1$ on the diagonal and zeros elsewhere.
2. Solve

$$\mathbf{u}^{(\ell)} = \arg\min_{\mathbf{u}} \|\boldsymbol{O}^{(\ell)}\mathbf{u}\|_1$$

$$\begin{aligned} \text{s.t.} & \\ & \mathbf{u} \geq 0 \\ & \mathbf{y} = \boldsymbol{F}_\theta \mathbf{z_0} + \boldsymbol{D}_\theta \mathbf{u} \\ & \mathbf{h} \leq \mathbf{y} \leq \mathbf{q} \end{aligned}$$

3. Update the weights $o_i^{(\ell+1)} = \frac{1}{|u_i^{(\ell)}|+\epsilon}$, $i = 1, \ldots, \tau + 1$.
4. Go to step 5 on convergence or when ℓ reaches a certain number of iterations. Otherwise, increment ℓ and go to step 2.
5. After a time period $\frac{\tau}{2}$, go to step 1 to update $\mathbf{u}$ for the next time period τ.
6. Repeat this process for the desired time period for which obtaining pulsatile dynamics is desired.

The idea is that by solving $\mathbf{u}^{(\ell+1)} = \arg\min_{\mathbf{u}} \sum_{i=1}^{\tau+1} \frac{|u_i|}{|u_i^{(\ell)}|+\epsilon}$ iteratively, the algorithm attempts to solve for a local minimum of a concave penalty function that is more similar to the ℓ_0-norm (Candes et al. 2008). The parameter ϵ is used to ensure that weights on the recovered zero entries will not be set to ∞ at the next step, which would prevent us from obtaining estimates at the next step (Faghih et al. 2015b). ϵ should be slightly larger than the expected nonzero amplitudes of the signal that is to be recovered, and a value of at least 0.001 is recommended (Candes et al. 2008). This algorithm does not always find the global minimum and as $\epsilon \to 0$, the likelihood of stagnating at an undesirable local minimum increases. For ϵ values closer to zero, the iterative reweighted ℓ_1-norm algorithm stagnates at an undesirable local minimum (Candes et al. 2008). Based on our empirical observations for convergence of the algorithm, we use $\ell = 10$ when running the algorithm for this formulation (Faghih et al. 2015b).

10.3.2 Special Case 1: Impulsive Input Design for Constant Demand and Holding Cost

Assuming that the demand and holding cost are constant, the optimal solution is achieved when the initial condition starts at the holding cost; then, the state decays to the lower bound that satisfies the demand, followed by an impulse that causes a jump in the state which brings it back to the holding cost, and then again the state decays to the lower bound that satisfies the demand and the same jump to the holding cost occurs again, and the same process keeps repeating (Faghih et al. 2015b). Figure 10.4 shows that solving the optimization problem (10.22) results in impulse control for a constant demand of 6 and a constant holding cost of 14 by considering the model in Eqs. (10.17) and (10.18) where $\theta_1 = 0.0585$ $\theta_2 = 0.0122$, $\epsilon = 0.01$, and $\tau = 360$. There are 12 constant impulses obtained over a 24-h period, which occur periodically. This example is just a simple toy problem illustrating that the optimization formulation in Eq. (10.22) can achieve intermittent control using a low energy input.

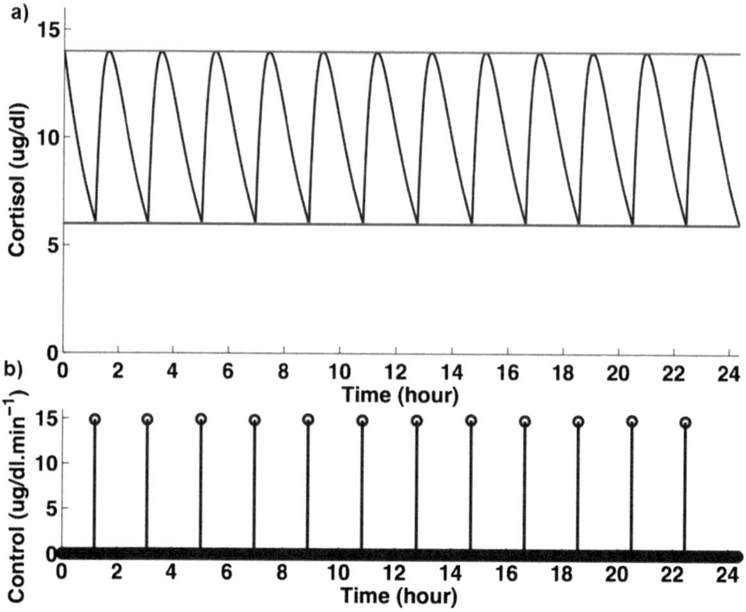

Fig. 10.4 Pulsatile profile and impulsive control obtained for constant demand and holding cost. (**a**) Top panel displays the optimal pulsatile profile (black curve), constant holding cost/upper bound (red curve), and constant demand/lower bound (blue curve). (**b**) Bottom panel displays the intermittent control. The optimization problem obtained 12 impulses over 24 h by assuming one-minute intervals (the obtained control takes 12 non-zero values out of 1440 possibilities, i.e., impulses, while it is zero everywhere else). This example is presented as a toy problem and does not have any physiological implications for cortisol secretion as it does not include the circadian rhythm observed in cortisol secretion. This figure is from the open-access paper (Faghih et al. 2015b) distributed under Creative Commons Attribution License

10.3.3 Special Case 2: Impulsive Input Design for Circadian Demand and Holding Cost

Since inducing constant CRH levels results in pulsatile cortisol release (Walker et al. 2010) while constant ACTH levels do not result in pulsatile release of cortisol (Spiga et al. 2011), Walker et al. suggest a sub-hypothalamic pituitary-adrenal system in intermittent control of cortisol secretion (Walker et al. 2012). Hence, the dynamics in the anterior pituitary control pulsatile secretion of cortisol (Faghih et al. 2015b). In healthy humans, cortisol levels have a regular circadian pattern and we can assume that the body is satisfying a circadian demand for cortisol as well as a circadian holding cost. Figure 10.5 shows that solving the optimization problem (10.22) by considering the model in Eqs. (10.17) and (10.18)

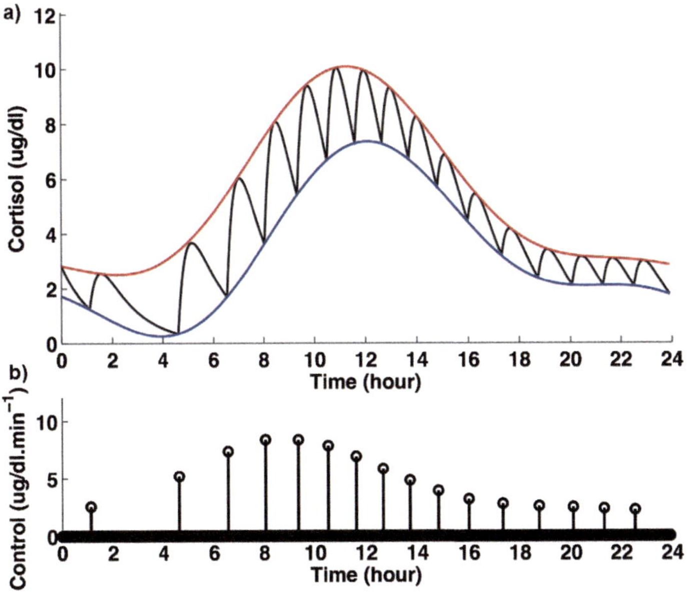

Fig. 10.5 Pulsatile profile and impulsive control obtained for circadian demand and holding cost. (**a**) Top panel displays the optimal pulsatile profile (black curve), circadian holding cost/upper bound (red curve), and circadian demand/lower bound (blue curve). (**b**) Bottom panel displays the intermittent control. The optimization problem obtained 16 impulses over 24 h by assuming one-minute intervals (the obtained control takes 16 non-zero values out of 1440 possibilities, i.e., impulses, while it is zero everywhere else). This figure has been modified from the open-access paper (Faghih et al. 2015b) distributed under Creative Commons Attribution License

where $\theta_1 = 0.0585$ and $\theta_2 = 0.0122$ and parameters $\epsilon = 0.0055$ and $\tau = 360$ for two-harmonic bounds with a circadian rhythm, the obtained control is intermittent control. A lower bound of

$$h(t) = 3.2478 - 0.7813\sin\left(\frac{2\pi t}{1440}\right) - 2.8144\cos\left(\frac{2\pi t}{1440}\right) - 0.2927\sin\left(\frac{2\pi t}{720}\right) + 1.3063\cos\left(\frac{2\pi t}{720}\right)$$

and an upper bound of

$$q(t) = 5.3782 + 0.3939\sin\left(\frac{2\pi t}{1440}\right) - 3.5550\cos\left(\frac{2\pi t}{1440}\right) - 0.5492\sin\left(\frac{2\pi t}{720}\right) + 1.0148\cos\left(\frac{2\pi t}{720}\right)$$

were used for this simulation (Faghih et al. 2015b). There are 16 impulses over a 24-h period with time-varying circadian amplitudes and ultradian timings (Faghih et al. 2015b); the obtained control is within the physiologically plausible range of 15 to 22 pulses (Brown et al. 2001; Veldhuis et al. 1989). There are more impulses with higher amplitudes during the day than at night. Around 6 AM, cortisol levels increase and are at higher values between 10 AM to 12 PM, followed by a gradual decrease to low values at night. The state starts at the circadian holding cost and decays to the lower bound that satisfies the circadian demand at which point an impulse causes a jump to reach the circadian holding cost. Then, the state decays again to the lower bound that satisfies the circadian demand and this process repeats. This example illustrates that the optimization formulation in Eq. (10.22) can achieve intermittent control of cortisol release, and result in a physiologically plausible pulsatile cortisol profile similar to those observed in healthy human data (Faghih et al. 2015b).

10.4 Discussion and Concluding Remarks

Understanding and modeling the physiological processes underlying release of pulsatile physiological signals is a challenging problem for various factors: (i) simultaneous release and clearance of pulsatile signals, (ii) the unknown timing and amount of the underlying impulsive brain profile, (iii) potential consecutive impulses, (iv) variations in impulsive profiles and model parameters of the physiological processes depending on the participant's state (e.g., mental state or sleep-wake state), (v) inter-individual variation, even among healthy individuals, and (vi) unknown process and measurement noise.

Modeling concurrent measurements of pulsatile physiological signals is even a more challenging problem for several reasons: (i) significant periods of the two pulsatile physiological signals might be different from each other due to some nonlinearities in the input-output relation of the concurrent pulsatile physiological signals or due to noise, (ii) one pulsatile physiological signal might decay faster than the other one, and when the data has a low resolution, in the high frequency pulsatile signal, the response to a smaller impulse might have already decayed out while the response might be observed in the low frequency pulsatile signal, (iii) there might be pulsatile activity in data from one pulsatile signal without a response in the other pulsatile signal which makes it challenging to model the interactions using a simple linear model.

Data analysis methods for modeling pulsatile experimental physiological data either assume the timing of the impulses belongs to a certain class of stochastic processes (Johnson 2003) or use pulse detection algorithms (Vidal et al. 2012). While these procedures work well for the cases that pulses are readily identifiable, analyzing pulsatile physiological data is more challenging when the timing of the pulses is not as clearly defined. In this chapter, we modeled brain impulsive profiles that result in cortisol and skin conductance time series as well as concurrent

ACTH and cortisol data. This was achieved using a coordinate descent approach to estimate the parameters underlying the physiological processes and recover the sparse impulsive brain signals. To recover the accurate number of underlying pulses, it is important to select a regularization parameter that balances between capturing the sparsity and the residual error. In the algorithm presented in this chapter, generalized cross-validation was used to choose the number of pulses. This algorithm works well even when the brain impulses are not easily identifiable without making assumptions about the inter-arrival times of the impulses, and timings of pulses can belong to various classes of distributions of inter-arrival times. The algorithm provided is a general framework that can be implemented on single as well as multiple time series and is not limited to the examples presented; it can be applied to other pulsatile physiological signals such as other hormones (e.g., growth hormone, thyroid hormone, and gonadal hormones). This algorithm makes it possible to capture the variation in the timing and amplitude of the underlying brain impulsive profile as well as the parameters underlying the physiological processes underlying the observed experimental pulsatile physiological signal. Using the multi-rate approach presented here, it is possible to detect the underlying brain impulsive profile with a higher resolution than the pulsatile experimental data. While the algorithm presented here uses a deterministic approach for deconvolution of signals with pulsatile dynamics, it is possible to consider a Gaussian distribution for the ℓ_2-norm and a Laplace distribution for the ℓ_1-norm in the cost function in Eq. (10.12) to cast the parameter estimation problem as a Bayesian estimation problem and obtain confidence intervals for the model parameters underlying the physiological process and the brain impulsive profile.

For the case of endocrine hormones, currently stimulation tests are used for diagnosis of hormonal disorders due to problems in the pituitary or the hormone producing glands. Using data from multiple healthy human subjects, and a minimal physiological model and the deconvolution algorithm presented here, we could find a range for model parameters for the healthy population to recognize the cause of the disorder for specific patients. For example, if a patient has elevated cortisol levels, the model can distinguish whether it is due to ACTH synthesis, or cortisol clearance by the liver as the algorithm provided here is a novel way of recovering the model parameters underlying the physiological processes and the brain secretory events simultaneously.

Some physiological signals are released in pulses and intermittent signaling might be an optimal approach for relaying information as opposed to continuous signaling. Here, we presented an optimization formulation for a physiologically plausible controller that achieves intermittent control. In the proposed formulation, we assumed that the body satisfies demand and holding cost constraints as well as the first-order dynamics underlying the release of the pulsatile physiological signal. We have illustrated an example in which the proposed optimization formulation yields impulse control for cortisol release with physiologically plausible number, timing, and amplitude of secretory events and cortisol profile. One should note that the iterative algorithm for enhancing the sparsity by reweighted ℓ_1 minimization (Candes et al. 2008) used to solve the optimization formulation does not always find the global minimum; while here this method was used to solve examples of optimization problems formulated in Eq. (10.22), for arbitrary choices of ϵ and τ, the algorithm for enhancing the sparsity by reweighted ℓ_1 minimization (Candes et al. 2008) might stagnate at local minima and not achieve the optimal solution. However, problem (10.22) can be solved using other methods as well. One should note that abrupt changes in the physiological process could also result in impulse control. For example, if the infusion rate of cortisol starts from a constant level and decreases abruptly to a new constant level, a very large level of cortisol should be produced in a short time to maintain the desired cortisol level (Faghih et al. 2015b). Another example could be when both the infusion and the clearance rates could change abruptly to different levels periodically with the overall effect of clearing cortisol faster or infusing cortisol to the circulation more slowly to require a very large cortisol secretory event in a short time to maintain the desired cortisol levels (Faghih et al. 2015b). Then, impulse control can be achieved as long as there is not an upper bound on the control variable; a mathematical example of a model with a time-varying rate that achieves impulse control is given in Sethi and Thompson (2006). Another possibility is that the timing of the impulses are functions of the states and are activated when a resetting condition is satisfied (Faghih et al. 2015b). A mathematical example of such a model is given in Wang and Balakrishnan (2008) where the cost function minimizes the energy in the input and the state. Also, another possibility is that different costs are associated with the control at different times of the day (Faghih et al. 2015b).

While in this chapter we presented examples that dealt with intermittent input design for constant and circadian demand and holding cost, the proposed optimization formulation can be tailored to include the constraints underlying release of different pulsatile physiological signals. For example, it can be applied to thyroid hormone secretion or gonadal hormone secretion or growth hormone secretion. Since transcriptional program stimulated by pulses is very different from constant signaling, intermittent input design/control can be beneficial for treating some disorders related to pulsatile physiological signals optimally. Furthermore, inspired by the pulse controller proposed in this research, in BMI design, it is possible to design intermittent controllers instead of continuous controllers to improve the battery life of the brain implant. Moreover, this type of bio-inspired pulse controller can potentially be used to control psychiatric disorders such as post-traumatic stress disorder, major depression, and addiction. For example, in psychiatric disorders, in theory, one could potentially measure electrodermal activity and use the deconvolution algorithm presented in this chapter to detect the brain impulsive profile to eventually recover the emotional shocks experienced by the patient, and ideally utilize an intermittent controller to stimulate ventromedial prefrontal cortex to reverse the effect of the emotional shocks experienced by the patient (Faghih et al. 2015b). In conclusion, by using a sparse system identification approach for analyzing experimental pulsatile physiological signals presented here and then by designing bio-inspired intermittent controllers discussed in this chapter, it is potentially possible to design BMIs or wearable-machine interfaces for optimal treatment of disorders linked to pulsatile physiological signals that are generated due to impulsive brain signals. The potential applications of this type of architecture go beyond neuroendocrine and mental disorders presented here and can be applied to disorders that naturally arise in neuroscience.

Acknowledgements This chapter summarizes and generalizes the concepts that appeared in my PhD thesis and related publications. I would like to thank my PhD advisors and all coauthors who have contributed to previously published work. Especially, I am greatly indebted to my PhD advisors Professor Emery N. Brown and Professor Munther A. Dahleh; this work would not have been possible without their guidance, depth of knowledge, and invaluable advice. I would also like to express my gratitude to Professor George Verghese, Dr. Elizabeth Klerman, and Dr. Gail Adler for very helpful conversations, and insightful feedback. I am also grateful to the National Science Foundation (NSF) for supporting this research through the NSF Graduate Research Fellowship. In honor of Professor Emery N. Brown's 60th birthday, I would like to add that being Emery's student and having him as a role model has always been a great honor and privilege for me; I am very lucky that I could benefit from the light of Emery's wisdom and learn so much from him during my S.M. and Ph.D. studies as well as my postdoctoral training. He has been immensely encouraging and a great source of enthusiasm, inspiration, and positive energy.

References

Attouch, H., Bolte, J., Redont, P., & Soubeyran, A. (2010). Proximal alternating minimization and projection methods for nonconvex problems: An approach based on the kurdyka-łojasiewicz inequality. *Mathematics of Operations Research, 35*(2), 438–457.

Boufounos, P., Duarte, M. F., & Baraniuk, R. G. (2007). Sparse signal reconstruction from noisy compressive measurements using cross validation. In *Proceedings of 14th IEEE Workshop on Statistical Signal Processing (SSP'07)* (pp. 299–303). New York: IEEE.

Brown, E. N., Meehan, P. M., & Dempster, A. P. (2001). A stochastic differential equation model of diurnal cortisol patterns. *American Journal of Physiology-Endocrinology And Metabolism, 280*(3), E450–E461.

Candes, E. J., Wakin, M. B., & Boyd, S. P. (2008). Enhancing sparsity by reweighted ℓ_1 minimization. *Journal of Fourier Analysis and Applications, 14*(5), 877–905.

Conrad, M., Hubold, C., Fischer, B., & Peters, A. (2009). Modeling the hypothalamus–pituitary–adrenal system: homeostasis by interacting positive and negative feedback. *Journal of Biological Physics, 35*(2), 149–162.

Dallmant, M., & Yates, F. (1969). Dynamic asymmetries in the corticosteroid feedback path and distribution metabolism-binding elements of the adrenocortical system. *Annals of the New York Academy of Sciences, 156*(1), 696–721.

Faghih, R. T. (2010). *The FitzHugh-Nagumo model dynamics with an application to the hypothalamic pituitary adrenal axis.* Master's thesis, Massachusetts Institute of Technology.

Faghih, R. T. (2014). *System identification of cortisol secretion: Characterizing pulsatile dynamics.* Ph.D. thesis, Massachusetts Institute of Technology.

Faghih, R. T., Dahleh, M. A., Adler, G. K., Klerman, E. B., & Brown, E. N. (2014). Deconvolution of serum cortisol levels by using compressed sensing. *PLoS ONE, 9*(1), e85204.

Faghih, R. T., Dahleh, M. A., Adler, G. K., Klerman, E. B., & Brown, E. N. (2015a). Quantifying pituitary-adrenal dynamics and deconvolution of concurrent cortisol and adrenocorticotropic hormone data by compressed sensing. *IEEE Transactions on Biomedical Engineering, 62*(10), 2379–2388.

Faghih, R. T., Dahleh, M. A., & Brown, E. N. (2015b). An optimization formulation for characterization of pulsatile cortisol secretion. *Frontiers in Neuroscience, 9*, 228.

Faghih, R. T., Savla, K., Dahleh, M. A., & Brown, E. N. (2011). A feedback control model for cortisol secretion. In *Proceedings of IEEE Conference on Engineering in Medicine and Biology Society (EMBC)* (pp. 716–719). New York: IEEE.
Faghih, R. T., Stokes, P. A., Marin, M.-F., Zsido, R. G., Zorowitz, S., Rosenbaum, B. L., et al. (2015c). Characterization of fear conditioning and fear extinction by analysis of electrodermal activity. In *Proceedings of 37th IEEE Conference on Engineering in Medicine and Biology Society (EMBC)* (pp. 7814–7818). New York: IEEE.
Fazel, M. (2002). *Matrix rank minimization with applications*. Ph.D. thesis, Stanford University.
Golub, G. H., Heath, M., & Wahba, G. (1979). Generalized cross-validation as a method for choosing a good ridge parameter. *Technometrics, 21*(2), 215–223.
Gorodnitsky, I. F., & Rao, B. D. (1997). Sparse signal reconstruction from limited data using focuss: A re-weighted minimum norm algorithm. *IEEE Transactions on Signal Processing, 45*(3), 600–616.
Gupta, S., Aslakson, E., Gurbaxani, B. M., & Vernon, S. D. (2007). Inclusion of the glucocorticoid receptor in a hypothalamic pituitary adrenal axis model reveals bistability. *Theoretical Biology and Medical Modelling, 4*(1), 8.
Hansen, P. C. (1999). *The L-curve and its use in the numerical treatment of inverse problems*. IMM, Department of Mathematical Modelling, Technical University of Denmark.
He, Z., Xie, S., & Cichocki, A. (2012). On the convergence of focuss algorithm for sparse representation. arXiv preprint arXiv:1202.5470.
Johnson, T. D. (2003). Bayesian deconvolution analysis of pulsatile hormone concentration profiles. *Biometrics, 59*(3), 650–660.
Keenan, D. M., Chattopadhyay, S., & Veldhuis, J. D. (2005). Composite model of time-varying appearance and disappearance of neurohormone pulse signals in blood. *Journal of Theoretical Biology, 236*(3), 242–255.
Kettyle, W. M., & Arky, R. A. (1998). *Endocrine pathophysiology*. Philadelphia: Lippincott Williams & Wilkins.
Klerman, E. B., Goldenberg, D. L., Brown, E. N., Maliszewski, A. M., & Adler, G. K. (2001). Circadian rhythms of women with fibromyalgia. *The Journal of Clinical Endocrinology & Metabolism, 86*(3), 1034–1039.
Kyrylov, V., Severyanova, L. A., & Vieira, A. (2005). Modeling robust oscillatory behavior of the hypothalamic-pituitary-adrenal axis. *IEEE Transactions on Biomedical Engineering, 52*(12), 1977–1983.
Lidberg, L., & Wallin, B. G. (1981). Sympathetic skin nerve discharges in relation to amplitude of skin resistance responses. *Psychophysiology, 18*(3), 268–270.
Lightman, S. L., & Conway-Campbell, B. L. (2010). The crucial role of pulsatile activity of the hpa axis for continuous dynamic equilibration. *Nature Reviews Neuroscience, 11*(10), 710.
Linkowski, P., Mendlewicz, J., Leclercq, R., Brasseur, M., Hubain, P., Golstein, J., et al. (1985). The 24-hour profile of adrenocorticotropin and cortisol in major depressive illness. *Journal of Clinical Endocrinology & Metabolism, 61*(3), 429–438.
Lobo, M. S., Fazel, M., & Boyd, S. (2007). Portfolio optimization with linear and fixed transaction costs. *Annals of Operations Research, 152*(1), 341–365.
Lönnebo, A., Grahnén, A., & Karlsson, M. O. (2007). An integrated model for the effect of budesonide on acth and cortisol in healthy volunteers. *British Journal of Clinical Pharmacology, 64*(2), 125–132.
McMaster, A., Jangani, M., Sommer, P., Han, N., Brass, A., Beesley, S., et al. (2011). Ultradian cortisol pulsatility encodes a distinct, biologically important signal. *PLoS ONE, 6*(1), e15766.
Murray, J. F. (2005). *Visual recognition, inference and coding using learned sparse overcomplete representations*. Ph.D. thesis, University of California, San Diego.
Peters, A., Conrad, M., Hubold, C., Schweiger, U., Fischer, B., & Fehm, H. L. (2007). The principle of homeostasis in the hypothalamus-pituitary-adrenal system: new insight from positive feedback. *American Journal of Physiology-Regulatory, Integrative and Comparative Physiology, 293*(1), R83–R98.

Refetoff, S., Van Cauter, E., Fang, V., Laderman, C., Graybeal, M., & Landau, R. (1985). The effect of dexamethasone on the 24-hour profiles of adrenocorticotropin and cortisol in cushing's syndrome. *Journal of Clinical Endocrinology & Metabolism, 60*(3), 527–535.

Rhee, S. S., & Pearce, E. N. (2011). The endocrine system and the heart: a review. *Revista Española de Cardiología (English Edition), 64*(3), 220–231.

Sarabdjitsingh, R., Joëls, M., & De Kloet, E. (2012). Glucocorticoid pulsatility and rapid corticosteroid actions in the central stress response. *Physiology & Behavior, 106*(1), 73–80.

Savić, D., & Jelić, S. (2005). A mathematical model of the hypothalamo-pituitary-adrenocortical system and its stability analysis. *Chaos, Solitons & Fractals, 26*(2), 427–436.

Sethi, S. P., & Thompson, G. L. (2006). *Optimal control theory: Applications to management science and economics*. New York: Springer.

Spiga, F., Waite, E. J., Liu, Y., Kershaw, Y. M., Aguilera, G., & Lightman, S. L. (2011). Acth-dependent ultradian rhythm of corticosterone secretion. *Endocrinology, 152*(4), 1448–1457.

Stavreva, D. A., Wiench, M., John, S., Conway-Campbell, B. L., McKenna, M. A., Pooley, J. R., et al. (2009). Ultradian hormone stimulation induces glucocorticoid receptor-mediated pulses of gene transcription. *Nature Cell Biology, 11*(9), 1093.

Ten, S., New, M., & Maclaren, N. (2001). Addison's disease 2001. *Journal of Clinical Endocrinology & Metabolism, 86*(7), 2909–2922.

Van Cauter, E. (1981). Quantitative methods for the analysis of circadian and episodic hormone fluctuations. In *Human pituitary hormones* (pp. 1–28). Dordrecht: Springer.

Veldhuis, J. D., Iranmanesh, A., Lizarralde, G., & Johnson, M. L. (1989). Amplitude modulation of a burstlike mode of cortisol secretion subserves the circadian glucocorticoid rhythm. *American Journal of Physiology-Endocrinology And Metabolism, 257*(1), E6–E14.

Vidal, A., Zhang, Q., Médigue, C., Fabre, S., & Clément, F. (2012). Dynpeak: An algorithm for pulse detection and frequency analysis in hormonal time series. *PLoS ONE, 7*(7), e39001.

Vinther, F., Andersen, M., & Ottesen, J. T. (2011). The minimal model of the hypothalamic–pituitary–adrenal axis. *Journal of Mathematical Biology, 63*(4), 663–690.

Vis, D. J., Westerhuis, J. A., Hoefsloot, H. C., Pijl, H., Roelfsema, F., van der Greef, J., et al. (2010). Endocrine pulse identification using penalized methods and a minimum set of assumptions. *American Journal of Physiology-Endocrinology And Metabolism, 298*(2), E146–E155.

Vyzas, E., & Picard, R. W. (1999). Off-line and online recognition of emotion expression from physiological data. In *Proceedings of 3rd International Conference on Autonomous Agents–Workshop on Emotion-Based Agent Architectures* (pp. 135–142).

Walker, J. J., Spiga, F., Waite, E., Zhao, Z., Kershaw, Y., Terry, J. R., et al. (2012). The origin of glucocorticoid hormone oscillations. *PLoS Biology, 10*(6), e1001341.

Walker, J., Terry, J., Tsaneva-Atanasova, K., Armstrong, S., McArdle, C., & Lightman, S. (2010). Encoding and decoding mechanisms of pulsatile hormone secretion. *Journal of Neuroendocrinology, 22*(12), 1226–1238.

Wallin, B. G. (1981). Sympathetic nerve activity underlying electrodermal and cardiovascular reactions in man. *Psychophysiology, 18*(4), 470–476.

Wang, X., & Balakrishnan, S. N. (2008). Optimal neuro-controller synthesis for variable-time impulse driven systems. In *Proceedings of American Control Conference* (pp. 3817–3822).

Young, E. A., Carlson, N. E., & Brown, M. B. (2001). Twenty-four-hour ACTH and cortisol pulsatility in depressed women. *Neuropsychopharmacology, 25*(2), 267–276.

Zdunek, R., & Cichocki, A. (2008). Improved M-FOCUSS algorithm with overlapping blocks for locally smooth sparse signals. *IEEE Transactions on Signal Processing, 56*(10), 4752–4761.

Chapter 11
Neural Engine Hypothesis

Hideaki Shimazaki

11.1 Introduction

Humans and animals change sensitivity to sensory stimulus either adaptively to the stimulus conditions or following a behavioral context even if the stimulus does not change. A potential neurophysiological basis underlying these observations is gain modulation that changes responsiveness of neurons to stimulus; an example is contrast gain-control found in retina (Sakmann and Creutzfeldt 1969) and primary visual cortex under anesthesia (Ohzawa et al. 1985; Laughlin 1989), or in higher visual area caused by attention (Reynolds et al. 2000; Martínez-Trujillo and Treue 2002). Theoretical considerations suggested the gain modulation as a nonlinear operation that integrates information from different origins, offering ubiquitous computation performed in neural systems (see Salinas and Sejnowski (2001), Carandini and Heeger (2012) for reviews). Regulation of the level of background synaptic inputs (Chance et al. 2002; Burkitt et al. 2003), shunting inhibition (Doiron et al. 2001; Prescott and De Koninck 2003; Mitchell and Silver 2003), and synaptic depression (Abbott et al. 1997; Rothman et al. 2009) among others have been suggested as potential biophysical mechanisms of the gain modulation (see Silver (2010) for a review). While such modulation of the informative neural activity is a hallmark of computation performed internally in an organism, a principled view to quantify the internal computation has not been proposed yet.

Neurons convey information about the stimulus in their activity patterns. To describe probabilities of a combinatorially large number of activity patterns of the neurons with a smaller number of activity features, the maximum entropy principle has been successfully used (Schneidman et al. 2006; Shlens et al. 2006). This

H. Shimazaki (✉)
Kyoto University, Kyoto, Japan and Honda Research Institute Japan, Saitama, Japan
e-mail: h.shimazaki@i.kyoto-u.ac.jp

Z. Chen, S.V. Sarma (eds.), *Dynamic Neuroscience*,
https://doi.org/10.1007/978-3-319-71976-4_11

principle constructs the least structured probability distribution given the small set of specified constraints on the distribution, known as a maximum entropy model. It explains probabilities of activity patterns as a result of nonlinear operation on the specified features using a softmax function. Moreover, the model belongs to an exponential family distribution, or a Gibbs distribution. Equivalence of inference under the maximum entropy principle with aspects of the statistical mechanics and thermodynamics was explicated through the work by Jaynes (1957). Recently thermodynamic quantities were used to assess criticality of neural activity (Tkačik et al. 2014, 2015). However, analysis of neural populations under this framework only recently started to include "dynamics" of a neural population (Shimazaki et al. 2009, 2012; Shimazaki 2013; Kass et al. 2011; Kelly and Kass 2012; Granot-Atedgi et al. 2013; Nasser et al. 2013; Donner et al. 2017), and has not yet reached maturity to include computation performed internally in an organism.

Based on a neural population model obtained under the maximum entropy principle, this study investigates neural dynamics during which gain of neural response to a stimulus is modulated with a delay by an internal mechanism to enhance the stimulus information. The delayed gain modulation is observed at different stages of visual pathways (McAdams and Maunsell 1999; Reynolds et al. 2000; Lee et al. 2003). For example, effect of contrast gain-control by attention on response of V4 neurons to high contrast stimulus appears 200–300 ms after the stimulus presentation, but is absent during 100–200 ms time period during which the neural response is returning to a spontaneous rate (Reynolds et al. 2000). This process is expected for dynamics of neurons subject to a feedback gain-modulation mechanism, e.g., via recurrent networks (Salinas and Abbott 1996; Spratling and Johnson 2004; Sutherland et al. 2009). Similar modulation of the late activity component of neurons is discussed as underpinnings of working memory (Supèr et al. 2001), sensory perception (Cauller and Kulics 1991; Sachidhanandam et al. 2013; Manita et al. 2015), and reward value (Schultz 2016). We demonstrate that our hypothetical neural dynamics with delayed gain-modulation forms an information-theoretic cycle that generates entropy ascribed to the stimulus-related activity using entropy supplied by the internal gain-modulation mechanism. The process works analogously to a heat engine that produces work from heat supplied by reservoirs. We hypothesize that neurons in the brain act in this manner when it actively modulates the incoming sensory information to enhance perceptual capacity.

This chapter is organized as follows. In Sect. 11.2, we construct a maximum entropy model of a neural population by constraining two types of activities, one that is directly regulated by stimulus and the other that represents background activity of neurons, termed "internal activity." We point out that modulation of the internal activity realizes gain-modulation of stimulus response. In Sect. 11.3, we explain the conservation of entropy, equation of state for the neural population, and information on stimulus. In Sect. 11.4, we construct cycles of neural dynamics that model stimulus-evoked activity during which the stimulus information is enhanced by the internal gain-modulation mechanism. We define entropic efficiency of gain-modulation performed to retain the stimulus information. An ideal cycle introduced in this section achieves the highest efficiency. The chapter ends with discussion in which the state-space model of the neural population is argued as a potential approach to test the hypothesis. Thermodynamic formulation and derivations of free energies for a neural population are summarized in Appendix.

11.2 A Simple Model of Gain Modulation by a Maximum Entropy Model

11.2.1 Maximum Entropy Model of Spontaneous Neural Activity

We start by modeling spontaneous activity of N spiking neurons. We represent a state of the i-th neuron by a binary variable $x_i = (0, 1)$ $(i = 1 \cdots N)$. Here silence of the neuron is represented by "0" whereas activity, or a spike, of the neuron is denoted by "1." The simultaneous activity of the N neurons is represented by a vector of the binary variables, $\mathbf{x} = (x_1, \ldots, x_N)$. The joint probability mass function, $p(\mathbf{x})$, describes the probability of generating the pattern $\mathbf{x}$. There are 2^N different patterns. We characterize the combinatorial neural activity with a smaller number of characteristic features $F_i(\mathbf{x})$ $(i = 1, \ldots, d$, where $d < 2^N)$, based on the maximum entropy principle. Here $F_i(\mathbf{x})$ is the i-th feature that combines the activity of individual neurons. For example, these features can be the first and second order interactions, $F_i(\mathbf{x}) = x_i$ for $i = 1, \ldots, N$, and $F_{N+(N-i/2)(i-1)+j-i}(\mathbf{x}) = x_i x_j$ for $i < j$. The maximum entropy principle constructs the least structured probability distribution while expected values of these features are specified (Jaynes 1957). By representing expectation by $p(\mathbf{x})$ using a bracket $\langle \cdot \rangle$, these constraints are written as $\langle F_i(\mathbf{x}) \rangle = c_i$ $(i = 1, \ldots, d)$, where c_i is the specified constant.

Maximization of a function subject to the equality constraints is formulated by the method of Lagrange multipliers that alternatively maximizes the following Lagrange function

$$\mathcal{L}[p] = -\sum_{\mathbf{x}} p(\mathbf{x}) \log p(\mathbf{x}) - a \sum_{\mathbf{x}} p(\mathbf{x}) - \sum_{i} b_i \left\{ \sum_{\mathbf{x}} p(\mathbf{x}) F_i(\mathbf{x}) - c_i \right\}, \tag{11.1}$$

where a and b_i $(i = 1, \ldots, d)$ are the Lagrange multipliers. The Lagrange function is a functional of the probability mass function. By finding a zero point of its variational derivative, we obtain

$$p(\mathbf{x}) \sim \exp\left(-\sum_{i} b_i F_i(\mathbf{x}) \right). \tag{11.2}$$

The Lagrange parameters b_i are obtained by simultaneously solving $\frac{\partial \mathscr{L}}{\partial b_i} = \langle F_i(\mathbf{x}) \rangle - c_i = 0$ for $i = 1, \ldots, d$. Many gradient algorithms and approximation methods have been developed to search the parameters. Activities of retinal ganglion cells (Schneidman et al. 2006; Shlens et al. 2006; Tkačik et al. 2014, 2015), hippocampal (Shimazaki et al. 2015), and cortical neurons (Tang et al. 2008; Yu et al. 2008; Shimazaki et al. 2012) were successfully characterized using Eq. (11.2). In the following, we use a vector notation $\mathbf{b}_0 = (b_1, \ldots, b_d)^\top$ and $\mathbf{F}(\mathbf{x}) = (F_1(\mathbf{x}), \ldots, F_d(\mathbf{x}))^\top$. Here $\mathscr{H}_0 \equiv \mathbf{b}_0^\top \mathbf{F}(\mathbf{x})$ is a Hamiltonian of the spontaneously active neurons. In statistical mechanics, Eq. (11.2) is identified as the Boltzmann distribution with a unit thermodynamic *beta*. If the features contain only up to the second order interactions, the model is equivalent to the Ising or spin-glass model for ferromagnetism.

11.2.2 Maximum Entropy Model of Evoked Neural Activity

In this subsection, we model evoked activity of neurons caused by changes in extrinsic stimulus conditions. We define a feature of stimulus-related activity as $X(\mathbf{x}) = \mathbf{b}_1^\top \mathbf{F}(\mathbf{x})$, where elements of $\mathbf{b}_1$ dictate response properties of each feature in $\mathbf{F}(\mathbf{x})$ to a stimulus. For simplicity, we represent the stimulus-related activity by this single feature, and consider that the evoked activity is characterized by the two summarized features, $\mathscr{H}_0(\mathbf{x})$ and $X(\mathbf{x})$. To model it, we constrain expectation of the internal and stimulus features using U and X, respectively. Here we assume that $\mathbf{F}(\mathbf{x})$, $\mathbf{b}_0$, and $\mathbf{b}_1$ are known and fixed. For example, this would model responses of visual neurons when we change contrast of a stimulus while fixing the rest of the stimulus properties. The maximum entropy distribution subject to these constraints is again given by the method of Lagrange multipliers. The Lagrange function is given as

$$\mathscr{L}[p] = -\sum_{\mathbf{x}} p(\mathbf{x}) \log p(\mathbf{x}) \\ - a \sum_{\mathbf{x}} p(\mathbf{x}) - \beta \left\{ \sum_{\mathbf{x}} p(\mathbf{x}) \mathscr{H}_0(\mathbf{x}) - U \right\} + \alpha \left\{ \sum_{\mathbf{x}} p(\mathbf{x}) X(\mathbf{x}) - X \right\}. \tag{11.3}$$

Here a, β, and α are the Lagrange parameters. By maximizing the functional $\mathscr{L}$ with respect to p, we obtain the following maximum entropy model,

$$p(\mathbf{x}) = \exp[-\beta \mathscr{H}_0(\mathbf{x}) + \alpha X(\mathbf{x}) - \psi(\beta, \alpha)], \tag{11.4}$$

where $\psi(\beta, \alpha)(= 1 + a)$ is a logarithm of a normalization term. It is computed as

$$\psi(\beta, \alpha) = \log \sum_{\mathbf{x}} e^{-\beta \mathscr{H}_0(\mathbf{x}) + \alpha X(\mathbf{x})}. \tag{11.5}$$

We call $\psi(\beta, \alpha)$ a log-partition function. The Lagrange multipliers, β and α, are adjusted such that $\langle \mathscr{H}_0(\mathbf{x}) \rangle = U$ and $\langle X(\mathbf{x}) \rangle = X$. Equation (11.4) is a softmax function (generalization of a logistic function to multinomial outputs) that returns the population output from a linear sum of the features weighted by $-\beta$ and α. With this view, we may alternatively regard β or α as an input parameter that controls U and X. Hereafter we simply call U internal activity, and X stimulus-related activity. Similarly, we call β an internal component, and α a stimulus component. We consider that the stimulus component α can be controlled by changing extrinsic stimulus conditions that an experimenter can manipulate. The stimulus component is written as $\alpha(s)$ if it is a function of a scalar stimulus condition s, such as stimulus contrast for visual neurons. In contrast, the internal component β is not directly controllable by the stimulus conditions. The spontaneous activity is modeled at $\beta = 1$ and $\alpha = 0$.

11.2.3 Gain Modulation by Internal Activity

We give a simple example of the maximum entropy model to show how the internal activity modulates the stimulus-related activity. Figure 11.1a illustrates an exemplary model composed of 5 neurons. With these particular model parameters (see figure caption), the stimulus component α controls activity rates of the first three neurons and their correlations. The internal component β controls background activity rates of all neurons. In our settings, decreasing β increases the baseline activity level of all neurons. Figure 11.1b displays activity rates of the individual neurons ($\langle x_i \rangle$ for $i = 1, \ldots, 5$) as a function of the stimulus component α with a fixed internal component β. Increasing α under these conditions activates the first three neurons without changing the activity rates of Neuron 4 and 5.[1] Furthermore, the response functions of the three neurons shift toward left when the background activity rates of all neurons is increased by *decreasing* the internal component β (Fig. 11.1b dashed lines). Thus Neuron 1–3 increase sensitivity to stimulus component α. This type of modulation is called input-gain control. For example, if α is a logarithmic function of contrast s of visual stimulation presented to an animal while recording visual neurons ($\alpha(s) = \log s$), increasing the modulation (decreasing β) makes neurons respond to multiplicatively smaller stimulus contrast. This models the contrast gain-control observed in visual pathways (Sakmann and Creutzfeldt 1969; Ohzawa et al. 1985; Reynolds et al. 2000; Martínez-Trujillo and Treue 2002). Other types of nonlinearity in the input-output relation can be constructed, depending on the nonlinearity in $\alpha(s)$.

[1]The activity rates of Neuron 4, 5 do not depend on α because $\mathbf{b}_0$ does not contain interactions that relate Neuron 1–3 with Neuron 4, 5. If there are non-zero interactions between any pair from Neuron 1–3 and Neuron 4, 5 in $\mathbf{b}_0$, the activity rates of Neuron 4, 5 increase with the increased rates of Neuron 1–3.

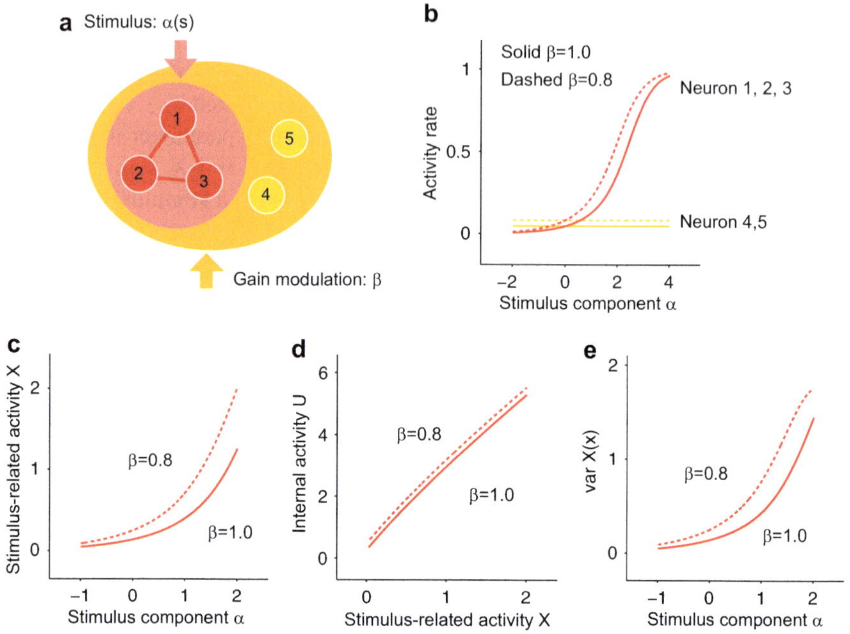

Fig. 11.1 A simple model of gain modulation by a maximum entropy model of 5 neurons. (**a**) An illustration of neurons that are activated by a stimulus (neurons in a pink area) and controlled by an internal mechanism (neurons in a yellow area). The model is constrained by features containing up to the second order statistics: $\mathbf{F}(\mathbf{x}) = (x_1, \ldots, x_5,\ x_1x_2, x_1x_3, x_2x_3, \ldots, x_4x_5)^\top$, where the first 5 elements are parameters for the individual activities x_i $(i = 1, \ldots, 5)$ and the rest of the elements is the joint activities of two neurons x_ix_j $(i < j)$. We assume that the stimulus-related activity is characterized by $\mathbf{b}_1 = (1, 1, 1, 0, 0,\ 0.3, 0.3, 0.3, 0, \ldots, 0)$. The first 3 elements are parameters for individual activity of the first three neurons x_i $(i = 1, 2, 3)$. The value 0.3 is assigned to the joint activities of the first three neurons, namely the features specified by x_1x_2, x_1x_3, and x_2x_3. The internal activity is characterized by $\mathbf{b}_0 = (2, 2, 2, 2, 2, 0, \ldots, 0)$, which regulates activity rates of individual neurons but does not change their interactions. (**b**) The activity rates of neurons as a function of the stimulus component α at fixed internal components, $\beta = 1.0$ (solid line) and $\beta = 0.8$ (dashed line). (**c**) The stimulus component X as a function of α at different internal components. (**d**) The relation between the stimulus-related activity X and internal activity U. (**e**) The Fisher information about the stimulus component α

Figure 11.1c displays a relation of the stimulus component α with the stimulus-related activity X at different internal component β. Similarly to the activity rates (Fig. 11.1b), the stimulus-related activity X is augmented if the internal component β is decreased. This nonlinear interaction between α and β is caused by the neurons that belong to both stimulus-related and internal activities. In this example, the stimulus component α also increases the internal activity U (Fig. 11.1d) because of increased activity rates of the shared neurons 1, 2, 3. Finally, Fig. 11.1e displays the variance of stimulus feature $X(\mathbf{x})$ as a function of α. It quantifies the information about the stimulus component α, which we will discuss in the next section.

11.3 The Conservation of Entropy, Equation of State, and Stimulus Information for a Neural Population

11.3.1 Conservation of Entropy for Neural Dynamics

The probability mass function, Eq. (11.4), belongs to the exponential family distribution. The Lagrange parameters are called natural or canonical parameters. The activity patterns of neurons are modeled as a linear combination of the two features $\mathscr{H}_0(\mathbf{x})$ and $X(\mathbf{x})$ using the canonical parameters $(-\beta, \alpha)$ in the exponent. Expectation of the features are called the expectation parameters U and X. Either natural or expectation parameters are sufficient to specify the probability distribution. We review dual structure of the two representations (Amari and Nagaoka 2000), and show that the relation provides the conservation law of entropy.

Negative entropy of the neural population is computed as

$$\begin{aligned} -S &= \langle \log p(\mathbf{x}) \rangle \\ &= -\beta \langle \mathscr{H}_0(\mathbf{x}) \rangle + \alpha \langle X(\mathbf{x}) \rangle - \psi(\beta, \alpha) \\ &= -U\beta + X\alpha - \psi(\beta, \alpha). \end{aligned} \tag{11.6}$$

Since the log-partition function of Eq. (11.4) is a cumulant generating function, U and X are related to the derivatives of $\psi(\beta, \alpha)$ as

$$\frac{\partial \psi(\beta, \alpha)}{\partial \beta} = -\langle \mathscr{H}_0(\mathbf{x}) \rangle = -U, \tag{11.7}$$

$$\frac{\partial \psi(\beta, \alpha)}{\partial \alpha} = \langle X(\mathbf{x}) \rangle = X. \tag{11.8}$$

Equations (11.6)–(11.8) form a Legendre transformation from $\psi(\beta, \alpha)$ to $-S(U, X)$. The inverse Legendre transformation is constructed using Eq. (11.6) as well: $\psi(\beta, \alpha) = -\beta U + \alpha X - (-S(U, X))$. Thus dually to Eqs. (11.7) and (11.8), the natural parameters are obtained as derivatives of the entropy with respect to the expectation parameters,

$$\left(\frac{\partial S}{\partial U} \right)_X = \beta, \tag{11.9}$$

$$\left(\frac{\partial S}{\partial X} \right)_U = -\alpha. \tag{11.10}$$

The natural parameters represent sensitivities of the entropy to the independent variables U and X. From these results, the total derivative of $S(U, X)$ is written as

$$\begin{aligned} dS &= \left(\frac{\partial S}{\partial U} \right)_X dU + \left(\frac{\partial S}{\partial X} \right)_U dX \\ &= \beta dU - \alpha dX. \end{aligned} \tag{11.11}$$

This explains a change of neurons' entropy by changes in the internal and stimulus-related activities. We denote an entropy change caused by the internal activity as $dS^{\text{int}} \equiv \beta dU$, and an entropy change caused by the extrinsic stimulus as $dS^{\text{ext}} \equiv \alpha dX$, respectively. Then Eq. (11.11) is written as

$$dS = dS^{\text{int}} - dS^{\text{ext}}. \tag{11.12}$$

We remark that dS is an infinitesimal difference of entropies at two close states, and its integral does not depend on a specific transition between the two states. In contrast, dS^{int} and dS^{ext} represent production of entropy separately by the internal and stimulus-related activities, and their integrals depend on the specific paths. Equation (11.12) constitutes the conservation of entropy for neural dynamics. We stress that although it is the first law of thermodynamics, the neurons considered here interact with an environment differently from conventional thermodynamic systems.[2] While internal energy of the conventional systems is indirectly controlled via work and heat, we consider that the internal activity of neurons is controlled directly by the organism's internal mechanism. Thus we use dS^{int} and dS^{ext}, rather than the work and heat, as quantities that neurons exchange with an environment.

11.3.2 Equation of State for a Neural Population

Equation (11.8) is an equation of the state for a neural population, which we rewrite here as

$$X(\beta, \alpha) = \frac{\partial \psi(\beta, \alpha)}{\partial \alpha}. \tag{11.13}$$

Through the log-partition function ψ, this equation relates state variables, β, α, and X, similarly to, e.g., the classical ideal gas law that relates temperature, pressure, and volume. Figure 11.1c displayed the equation of state. We note that ψ is related to the Gibbs free energy (see Appendix). Furthermore, without loss of generality, we can assume that the hamiltonian of the silent state is zero: $\mathscr{H}_0(\mathbf{0}) = X(\mathbf{0}) = 0$, where $\mathbf{x} = \mathbf{0}$ denotes the simultaneous silence of all neurons. We then obtain $p(\mathbf{0}) = e^{-\psi}$, namely

$$-\psi(\beta, \alpha) = \log p(\mathbf{0}). \tag{11.14}$$

[2] We obtain $dU = TdS - fdX$, using $\beta \equiv 1/T$ and $\alpha \equiv \beta f$ in Eq. (11.11). In this form, the expectation parameter U is a function of (S, X). According to the conventions of thermodynamics, we may call U internal energy, T temperature of the system, and f force applied to neurons by a stimulus. It is possible to describe the evoked activity of a neural population using these standard terms of thermodynamics. However, this introduces the concepts of work and heat, which may not be relevant quantities for neurons to exchange with environment.

Thus $-\psi(\beta,\alpha)$ is a logarithm of the simultaneous silence probability.[3] Since $d(\log p(\mathbf{0})) = dp(\mathbf{0})/p(\mathbf{0})$, $-d\psi$ gives a fractional increase of the simultaneous silence probability of the neurons. Accordingly, Eq. (11.13) states that the stimulus-related activity X equals to the fractional decrease of the simultaneous silence probability by a small change of α, given β.

The opposite representation of the equation of state, α as a function of X given β, is obtained as follows. From Eq. (11.13), we have $d\psi = X d\alpha$ given that β is fixed. Let ψ_0 and X_0 be ψ and X at $\alpha = 0$. Then, if the internal component β is fixed, the stimulus component α at X is given by

$$\alpha(\beta, X) = \int_{\psi_0}^{\psi} \left(\frac{1}{X}\right)_{\beta} d\psi' = \int_{X_0}^{X} \left(\frac{1}{X'}\frac{\partial\psi}{\partial X'}\right)_{\beta} dX'. \tag{11.15}$$

Here $\left(\frac{\partial\psi}{\partial X}\right)_{\beta}$ is a fractional decrease of the simultaneous silence probability when X shifts to $X + dX$ while β is fixed.

11.3.3 Information About Stimulus

The Fisher information $J(\alpha)$ provides the accuracy of estimating a small change in the stimulus component α by an optimal decoder. More specifically, the inverse of the Fisher information provides a lower bound of variance of an unbiased estimator for α from a sample. For the exponential family distribution, it is given as the second order derivative of the log-partition function with respect to α, which is also the variance of stimulus feature $X(\mathbf{x})$:

$$\begin{aligned} J(\alpha) \equiv \left\langle \left(\frac{\partial \log p(\mathbf{x})}{\partial\alpha}\right)^2 \right\rangle &= \frac{\partial^2\psi(\beta,\alpha)}{\partial\alpha^2} \\ &= \frac{\partial X}{\partial\alpha} = \langle X(\mathbf{x})^2\rangle - \langle X(\mathbf{x})\rangle^2. \end{aligned} \tag{11.16}$$

The first equality in the second line of Eq. (11.16) is obtained using the first order derivative of ψ, namely the equation of state (Eq. (11.13)). The second equality in Eq. (11.16) represents the fluctuation-dissipation relation of the stimulus feature. The equalities show that the Fisher information can be computed in three different manners given that the internal component β is fixed: (1) the second derivative of

[3]Importantly, $-\psi$ is a logarithm of the simultaneous silence probability predicted by the model, Eq. (11.4). The observed probability of the simultaneous silence could be different from the prediction if the model is inaccurate. For example, an Ising model may be inaccurate, and it was shown that neural higher-order interactions may significantly contribute to increasing the silence probability (Ohiorhenuan et al. 2010; Shimazaki et al. 2015).

ψ with respect to α using the simultaneous silence probability, (2) the derivative of X with respect to α using the equation of state, or (3) the variance of the stimulus feature.

The Fisher information computed at two fixed internal components was shown in Fig. 11.1e. The stimulus component α becomes relatively dominant in characterizing the neural activity if the internal component β decreases. This results in the larger Fisher information $J(\alpha)$ for the smaller internal component β at given α. If the stimulus condition s controls the stimulus component as $\alpha(s)$, and it is not related to β, the information about s is given as $\frac{\partial \alpha(s)}{\partial s} J(\alpha) \frac{\partial \alpha(s)}{\partial s}$.

11.4 Information-Theoretic Cycles by a Neural Population

We now introduce neural dynamics that models dynamical gain-modulation performed by an internal mechanism while neurons are processing stimulus. Since there are neurons that belong to both stimulus-related and internal activities, the internal mechanism changes not only the internal activity but also the stimulus-related activity, which realizes the modulation. From an information-theoretic point of view, this process converts entropy generated by the internal mechanism to entropy associated with stimulus-related activity after one cycle of the neural response is completed. To explain this in detail, we first provide an intuitive example of delayed gain-modulation using a dynamical model, and then provide an ideal cycle that efficiently enhances stimulus information. Using the latter model, we explain why the process works similarly to a heat engine, and show how to quantify efficiency of the gain-modulation performed by the internal mechanism.

11.4.1 An Example of Delayed Gain-Modulation

We first consider a simple dynamical model of delayed gain-modulation. We use the feature vector, $\mathbf{b}_0$ and $\mathbf{b}_1$ based on those used in Fig. 11.1. In this model, neurons are activated by a stimulus input, which subsequently increases modulation by an internal mechanism (Fig. 11.2a). Such a process can be modeled through dynamics of the controlling parameters given by

$$\tau_\alpha \dot{\alpha}(t) = -\alpha(t) + s\, e^{-t/\tau_\alpha} \tag{11.17}$$

$$\tau_\beta \dot{\beta}(t) = -\beta(t) + \beta_0 - \gamma \alpha(t) \tag{11.18}$$

for $t \geq 0$. Here s is intensity of an input stimulus. Neurons are initially at a spontaneous state: $\alpha(0) = 0$ and $\beta(0) = \beta_0 = 1$. The top panel of Fig. 11.2b displays the dynamics of $\alpha(t)$ and $\beta(t)$. The population activity is sampled from

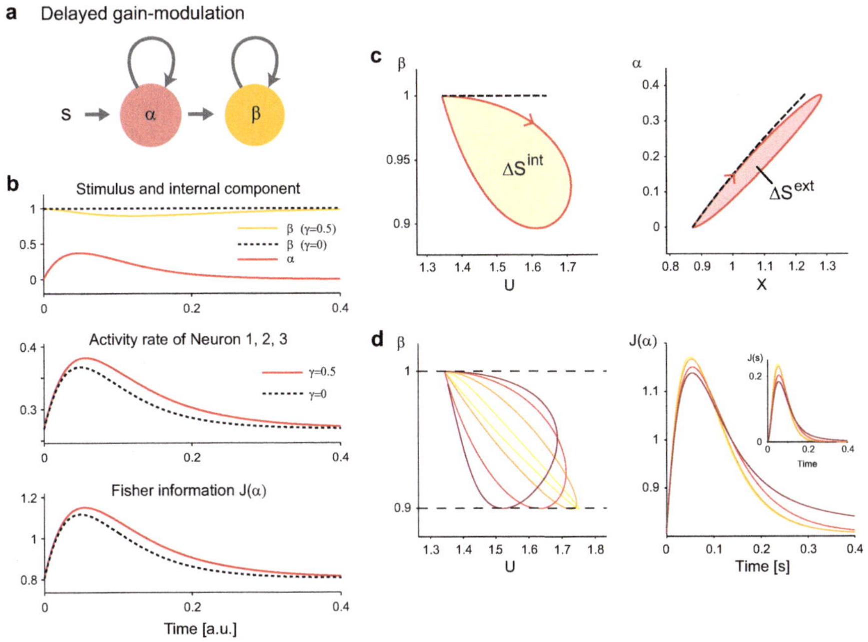

Fig. 11.2 The delayed gain-modulation by internal activity. The parameters of the maximum entropy model ($N = 5$) follow those in Fig. 11.1. (**a**) An illustration of delayed gain-modulation described in Eqs. (11.17) and (11.18). The stimulus increases the stimulus component α that activates Neuron 1, 2, and 3. Subsequently, the internal component β is increased, which increases the background activity of all 5 neurons. We assume a slower time constant for the gain-modulation than the stimulus activation ($\tau_\beta = 0.1$ and $\tau_\alpha = 0.05$). (**b**) *Top:* Dynamics of the stimulus and internal components (solid lines, $\gamma = 0.5$). The internal component β without the delayed gain-modulation ($\gamma = 0$) is shown by a dashed black line. *Middle:* Activity rates [a.u.] of Neuron 1–3 with (solid red) and without (dashed black) the delayed gain-modulation. *Bottom:* The Fisher information about stimulus component α (Eq. (11.16)). (**c**) The X-α (left) and U-β (right) phase diagrams. A red solid cycle represents dynamics when the delayed gain-modulation is applied ($\gamma = 0.5$). The dashed line is a trajectory when the delayed gain-modulation is not applied to the population ($\gamma = 0$). (**d**) *Left:* The U-β phase diagrams of neural dynamics with different combinations of τ_β and γ that achieve the same level of the maximum modulation (the minimum value of $\beta = 0.9$). *Right:* The Fisher information about the stimulus component α for different cycles. The color code is the same as in the left panel. The inset shows the Fisher information about the stimulus intensity s (Eq. (11.19))

the maximum entropy model with these dynamical parameters. Here we consider a continuous-time representation of the maximum entropy model[4] (Kass et al. 2011;

---

[4]Under the assumption that rates of synchronous spike events scale with $\mathcal{O}(\Delta^k)$, where Δ is a bin size of discretization and k is the number of synchronous neurons. It was proved (Kass et al. 2011) that it is possible to construct a continuous-time limit ($\Delta \to 0$) of the maximum entropy model that takes the synchronous events into account. Here we follow their result to consider the continuous-time representation.

Kelly and Kass 2012). The activity rates of neurons are increased by the delayed gain-modulation (solid lines in Fig. 11.2b, middle panel) from those obtained without the modulation ($\gamma = 0$; dashed lines). Accordingly, the information about the stimulus component α contained in the population activity as quantified by the Fisher information (Eq. (11.16)) increases and lasts longer by the delayed gain-modulation (Fig. 11.2b, bottom panel). Note that in this example, the information about the stimulus strength s is carried in both $\beta(t)$ and $\alpha(t)$ as time passes. The result obtained from the Fisher information about s using both $\beta(t)$ and $\alpha(t)$ is qualitatively the same as the result of the Fisher information about α (not shown).[5]

The U-β phase diagram (Fig. 11.2c, left panel) shows that dynamics without the gain-modulation is represented as a line because β is constant. In contrast, dynamics with the gain-modulation forms a cycle because weaker and then stronger modulation (larger and then smaller β) is applied to neurons when the internal activity U increases and then decreases, respectively. Similarly, the dynamics forms a cycle in the X-α plane (Fig. 11.2c, right panel) if the stimulus activity X is augmented by the delayed gain-modulation. By applying the conservation law for entropy (Eq. (11.12)) to the cycle, we obtain

$$0 = \oint \beta dU - \oint \alpha dX. \tag{11.20}$$

Here $\oint \beta dU \equiv \Delta S^{\text{int}}$ is entropy produced by the internal activity during the cycle due to the delayed gain-modulation, and $\oint \alpha dX \equiv \Delta S^{\text{ext}}$ is entropy produced by the activity related to extrinsic stimulus conditions. These are the areas within the circles in the phase diagrams. Equation (11.20) states that the two cycles have the same area ($\Delta S^{\text{int}} = \Delta S^{\text{ext}}$).

The left panel in Fig. 11.2d displays the U-β phase diagram for dynamics with given maximum strength of modulation (the minimum value of β). Among these cycles, larger cycles retain the information about the stimulus component α for a longer time period (Fig. 11.2d, right panel). The same conclusion is made from the Fisher information about s (Fig. 11.2d, an inset in right panel). The larger cycles were made because the modulation was only weakly applied to neurons when the internal activity U increased, then the strong modulation was applied when U decreased. Such modulation is considered to be efficient because it allows neurons to retain the stimulus information for a longer time period by using the slow time-scale of β without excessively increasing activity rates of neurons at its initial rise. In the

[5]When α and β are both dependent on the stimulus, the Fisher information about s is given as

$$J(s) = \frac{\partial \boldsymbol{\theta}(s)^{\top}}{\partial s} \mathbf{J} \frac{\partial \boldsymbol{\theta}(s)}{\partial s}, \tag{11.19}$$

where $\boldsymbol{\theta}(s) \equiv [-\beta, \alpha]^{\top}$ and $\mathbf{J}$ is a Fisher information matrix given by Eq. (11.24), which will be discussed in the later section. We computed Eq. (11.19) using analytical solutions of the dynamical equations given as $\alpha(t) = \frac{st}{\tau_\alpha} e^{-t/\tau_\alpha}$ and $\beta(t) = 1 - \frac{s\gamma}{\tau_\beta - \tau_\alpha} \left\{ \frac{\tau_\alpha \tau_\beta}{\tau_\beta - \tau_\alpha} (e^{-t/\tau_\beta} - e^{-t/\tau_\alpha}) - te^{-t/\tau_\alpha} \right\}$.

next section, we introduce the largest cycle that maximizes the entropy produced by the gain-modulation when the maximum strength of the modulation is given. Using this cycle, we explain how the cycle works analogously to a heat engine, and define efficiency of the cycle to retain the stimulus information.

11.4.2 The Efficient Cycle by a Neural Population

The largest cycle is made if the modulation is not applied when the internal activity U increases, then applied when U decreases. Figure 11.3 displays a cycle of hypothetical neural dynamics that maximizes the entropy production when the ranges of the internal component and activity are given. The model parameters follow those in Fig. 11.1. This cycle is composed of four steps. The process starts at the state A at which neurons exhibit spontaneous activity ($\beta = \beta_H = 1, \alpha = 0$). Figure 11.3a displays a sample response of the neural population to a stimulus change. Figure 11.3b and c display the X-α and U-β phase diagrams of the cycle. Heat capacity of the neural population and the Fisher information about α are shown in Fig. 11.3d. Details of the cycle steps are now described as follows.

A→B **Increased stimulus response** The stimulus-related activity X is increased by increasing the stimulus component α while the internal component is fixed at $\beta = \beta_H$. In this process the internal activity U also increases.

B→C **Internal computation** An internal mechanism decreases the internal component β while keeping the internal activity ($dU = 0$). In this process the stimulus-related activity X decreases. The process ends at $\beta = \beta_L$.

C→D **Decreased stimulus response** The stimulus-related activity X is decreased by decreasing the stimulus component α while the internal component is fixed at $\beta = \beta_L$. In this process the internal activity U also decreases.

D→A **Internal computation** An internal mechanism increases the internal component β while keeping the internal activity ($dU = 0$). In this process the stimulus-related activity X increases. The process ends at $\beta \equiv \beta_H$.

The processes B→C and D→A represent additional computation performed by an internal neural mechanism on the neurons' stimulus information processing. It is applied after the initial increase of stimulus-related activity during A→B, therefore manifests delayed modulation. Without these processes, the neural dynamics is represented as a line in the phase diagrams. The Fisher information about α also increases during the process between C and D (Fig. 11.3d, right panel). We reiterate that the Fisher information quantifies the accuracy of estimating a small change in α by an optimal decoder. Thus operating along the path between C and D is more advantageous than the path between A and B for downstream neurons if their goal is to detect a change in the stimulus-related activity of the upstream neurons that is not explained by the internal activity.

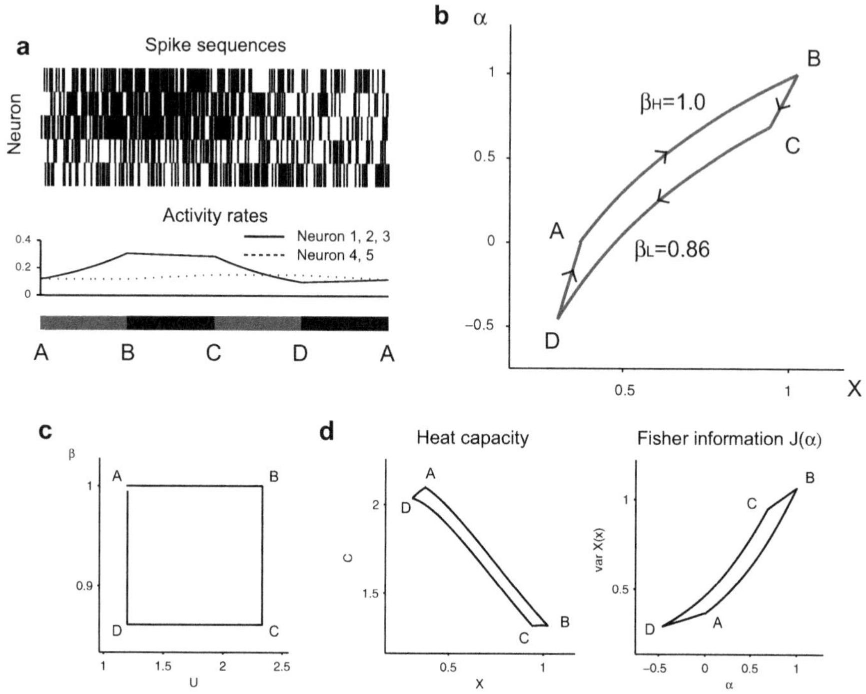

Fig. 11.3 The efficient circle by a neural population ($N = 5$). The parameters of the maximum entropy model follow those in Fig. 11.1. The cycle starts from the state A at which $\beta = \beta_H = 1$ and $\alpha = 0$. See the main text for details of the steps. The efficiency of this cycle is 0.14. (**a**) *Top:* Spike raster plots during the cycle. *Middle:* Activity rates of neurons. *Bottom:* The cycle steps. (**b**) The X-α phase diagram. (**c**) The U-β phase diagram. (**d**) *Left:* X v.s. heat capacity. The heat capacity is defined as $C = \langle h^2 \rangle - \langle h \rangle^2$, where $h = -\log p(\mathbf{x})$ is information content. *Right:* Fisher information about the stimulus component α

11.4.3 Interpretation as an Information-Theoretic Cycle

We start our analysis on the cycle by examining how much entropy is generated by the internal and stimulus-related activities at each step. First, we denote by $\Delta S_{\text{AB}}^{\text{int}}$ and $\Delta S_{\text{CD}}^{\text{int}}$ the entropy changes caused by the internal activity during the process A→B and C→D, respectively. Since the internal component β is fixed at β_H during the process A→B, we obtain $\Delta S_{\text{AB}}^{\text{int}} = \beta_H \Delta U$, where ΔU is a change of the internal activity (see Fig. 11.3c). This change in the internal activity is positive ($\Delta U > 0$). Since the internal activity does not change during B→C and D→A, a change of the internal activity during C→D is given by $-\Delta U$ (Note that the internal activity is a state variable). We obtain $\Delta S_{\text{CD}}^{\text{int}} = -\beta_L \Delta U$ for the process during C→D. The total entropy change caused by the internal activity during the cycle is given as $\Delta S_{\text{AB}}^{\text{int}} + \Delta S_{\text{CD}}^{\text{int}} = (\beta_H - \beta_L)\Delta U$, which is positive because $\beta_H > \beta_L$ and $\Delta U > 0$. Thus the internal activity contributes to increasing the entropy of neurons during the

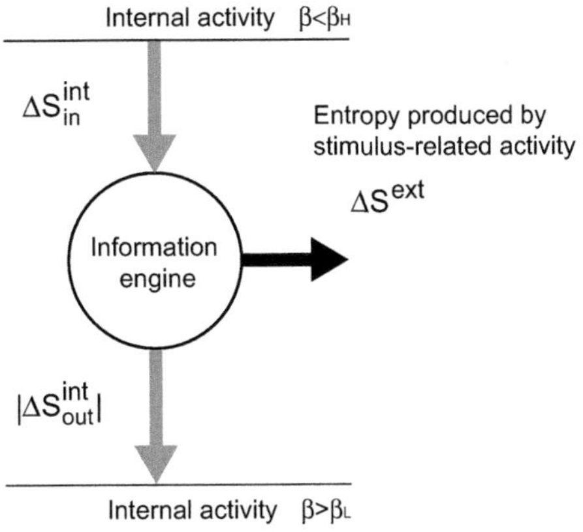

Fig. 11.4 An information-theoretic cycle by a neural population

cycle. Second, we denote by ΔS^{ext} the total entropy change caused by the stimulus-related activity during the cycle. According to the conservation law (Eq. (11.12)) applied to this cycle, we obtain

$$0 = \Delta S^{\text{int}}_{\text{AB}} + \Delta S^{\text{int}}_{\text{CD}} - \Delta S^{\text{ext}}. \tag{11.21}$$

Note that the sign of $\Delta S^{\text{ext}} = \Delta S^{\text{int}}_{\text{AB}} + \Delta S^{\text{int}}_{\text{CD}}$ is positive. Hence the stimulus-related activity contributes to decreasing the entropy of neurons during the cycle.

This cycle belongs to the following cycle that is analogous to a heat engine (Fig. 11.4). In this paragraph, we temporarily use *receive entropy* and *emit entropy* to express the positive and negative path-dependent entropy changes caused by the internal or stimulus-related activity in order to facilitate comparison with a heat engine.[6] In this cycle, neurons receive *entropy* as internal activity from an environment ($\Delta S^{\text{int}}_{\text{in}} > 0$) and emit *entropy* to the environment ($\Delta S^{\text{int}}_{\text{out}} < 0$). The received *entropy* as the internal activity is larger than the emitted *entropy* ($\Delta S^{\text{int}}_{\text{in}} + \Delta S^{\text{int}}_{\text{out}} > 0$). The surplus *entropy* is emitted to the environment in the form of the stimulus-related activity ($-\Delta S^{\text{ext}} < 0$). Thus we may regard the cycle as the process that produces stimulus-related entropy using entropy supplied by the internal mechanism. We hereafter denote this cycle as an information-theoretic cycle, or engine. The cycle in Fig. 11.2 is also regarded as an information-theoretic cycle by separating the process at which the internal activity is maximized. The conservation law prohibits a perpetual information-theoretic cycle that can indefinitely produce the stimulus-related entropy without entropy production by the internal mechanism.[7]

[6] Here we use *entropy* synonymously with heat in thermodynamics to facilitate the comparison with a heat engine. However this is not an accurate description because the entropy is a state variable.

[7] This is synonymous with the statement that the first law prohibits a perpetual motion machine of the first kind, a machine that can work indefinitely without receiving heat.

11.4.4 Efficiency of a Cycle

As we discussed for the example dynamics in Fig. 11.2, we may consider that the modulation is efficient if it helps neurons to retain stimulus information without excessively increasing the internal and stimulus-related activities during the initial response. Such a process was achieved when gain-modulation was only weakly applied to neurons when the internal activity U increased, then strong gain modulation was applied when U decreased. We can formally assess this type of efficiency by defining entropic efficiency, similarly to thermal efficiency of a heat engine. It is given by a ratio of the entropy change caused by the stimulus-related activity as opposed to the entropy change gained by the internal activity as:

$$\eta \equiv \frac{\Delta S^{\text{ext}}}{\Delta S^{\text{int}}_{\text{in}}} = 1 - \frac{|\Delta S^{\text{int}}_{\text{out}}|}{\Delta S^{\text{int}}_{\text{in}}}. \tag{11.22}$$

For the proposed information-theoretic cycle in Fig. 11.3, it is computed as

$$\eta_e = 1 - \frac{|\Delta S^{\text{int}}_{\text{CD}}|}{\Delta S^{\text{int}}_{\text{AB}}} = 1 - \frac{\beta_L}{\beta_H}, \tag{11.23}$$

which is a function of the internal components, β_H and β_L. This cycle is the most efficient in terms of the entropic efficiency defined by Eq. (11.22) when the highest and lowest internal components and activities are given. The square cycle in the U-β phase diagram (Fig. 11.3c) already suggests this claim, and we can formally prove this by comparing the information-theoretic cycle with an arbitrary cycle $\mathscr{C}$ whose internal component β satisfies $\beta_L \le \beta \le \beta_H$.[8] Thus the proposed cycle bounds efficiency of the additional computation made by the delayed gain-modulation mechanism. Here we now call the proposed cycle in Fig. 11.3, the ideal information-theoretic cycle. Note that this cycle is similar to, but different from the Carnot cycle (Carnot 1824) that can be realized by replacing the processes B→C and D→A with adiabatic processes. The Carnot cycle achieves the highest *thermal* efficiency.

[8]Let us consider the efficiency η achieved by an arbitrary cycle $\mathscr{C}$ during which the internal component β satisfies $\beta_L \le \beta \le \beta_H$. Let the minimum and maximum internal activity in the cycle be $U_{\min}$ and $U_{\max}$. We decompose $\mathscr{C}$ into the path $\mathscr{C}_1$ from $U_{\min}$ to $U_{\max}$ and the path $\mathscr{C}_2$ from $U_{\max}$ to $U_{\min}$ during which the internal component is given as $\beta_1(U)$ and $\beta_2(U)$, respectively. Because the cycle acts as an engine, we expect $\beta_1(U) > \beta_2(U)$. The entropy changes produced by the internal activity during the path C_i ($i = 1, 2$) is computed as $\Delta S^{\text{int}}_{\mathscr{C}_1} = \int_{U_{\min}}^{U_{\max}} \beta_1(U)\, dU \le \beta_H \int_{U_{\min}}^{U_{\max}} dU = \beta_H(U_{\max} - U_{\min})$ and $|\Delta S^{\text{int}}_{\mathscr{C}_2}| = |\int_{U_{\max}}^{U_{\min}} \beta_2(U)\, dU| \ge |\beta_L \int_{U_{\max}}^{U_{\min}} dU| = \beta_L(U_{\max} - U_{\min})$. Hence we obtain $|\Delta S^{\text{int}}_{\mathscr{C}_2}|/\Delta S^{\text{int}}_{\mathscr{C}_1} \ge \beta_L/\beta_H$, or $\eta \le \eta_e$.

11.4.5 Geometric Interpretation

Finally, we introduce geometric interpretation of the cycle, and consider conditions that realize the information-theoretic cycle. Let us denote the internal and stimulus components as $\boldsymbol{\theta} = [-\beta, \alpha]^{\top}$. In addition, we represent the expected internal and stimulus features by $\boldsymbol{\eta} = [U, X]^{\top}$. The parameters $\boldsymbol{\theta}$ and $\boldsymbol{\eta}$ form dually flat affine coordinates, and are called θ and η-coordinates in information geometry (Amari and Nagaoka 2000).

A small change in $\boldsymbol{\theta}$ is related to a change in $\boldsymbol{\eta}$ as $d\boldsymbol{\eta} = \mathbf{J}d\boldsymbol{\theta}$. Here $\mathbf{J}$ is the Fisher information matrix with respect to $\boldsymbol{\theta}$. It is given as

$$\mathbf{J} = \begin{bmatrix} \langle \mathbf{b}_0, \mathbf{b}_0 \rangle & \langle \mathbf{b}_0, \mathbf{b}_1 \rangle \\ \langle \mathbf{b}_1, \mathbf{b}_0 \rangle & \langle \mathbf{b}_1, \mathbf{b}_1 \rangle \end{bmatrix}, \tag{11.24}$$

where $\langle \mathbf{b}_i, \mathbf{b}_j \rangle \equiv \mathbf{b}_i^{\top} \mathbf{G} \mathbf{b}_j$ $(i, j = 0, 1)$ is an inner product of the vectors $\mathbf{b}_i$ and $\mathbf{b}_j$ with a metric given by $\mathbf{G} = \langle \mathbf{F}(\mathbf{x})\mathbf{F}(\mathbf{x})^{\top} \rangle - \langle \mathbf{F}(\mathbf{x}) \rangle \langle \mathbf{F}(\mathbf{x}) \rangle^{\top}$. Note that $\langle \mathbf{b}_0, \mathbf{b}_0 \rangle$ is equivalent to Eq. (11.16). In general, in order to make a change of the internal component β influence the stimulus-related activity X, therefore controls stimulus information, one requires $\langle \mathbf{b}_0, \mathbf{b}_1 \rangle \neq 0$ because $dX = -\langle \mathbf{b}_1, \mathbf{b}_0 \rangle d\beta + \langle \mathbf{b}_1, \mathbf{b}_1 \rangle d\alpha$ from $d\boldsymbol{\eta} = \mathbf{J}d\boldsymbol{\theta}$. This condition indicates that the modulation by an internal mechanism is achieved through the activity features shared by the two components. Accordingly, this condition is violated if neurons participate in the stimulus-related activity and neurons subject to the internal modulation do not overlap (namely if neurons that appear in the features corresponding to non-zero elements of $\mathbf{b}_0$ are separable from those of $\mathbf{b}_1$).

For the ideal information-theoretic cycle, we indicate the parameters at A, B, C, and D using a subscript of $\boldsymbol{\theta}$ or $\boldsymbol{\eta}$. For example, the parameters at A are $\boldsymbol{\theta}_A$ and $\boldsymbol{\eta}_A$. The first process A→B of the ideal information-theoretic cycle is a straight line (geodesic) between $\boldsymbol{\theta}_{\mathrm{A}}$ and $\boldsymbol{\theta}_{\mathrm{B}}$ in the curved space of θ-coordinates. It is called *e*-geodesic. In addition, the internal component β is fixed while the stimulus component decreases, therefore the *e*-geodesic is a vertical line in the θ-coordinates. The second process B→C is the shortest line between $\boldsymbol{\eta}_{\mathrm{B}}$ and $\boldsymbol{\eta}_{\mathrm{C}}$ in the curved space of η-coordinates. The path is called an *m*-geodesic. In addition, the internal activity U is fixed while the stimulus-related activity decreases, therefore the *m*-geodesic is a vertical line in the η-coordinates. Similarly, the process C→D is an *e*-geodesic, and the process D→A is an *m*-geodesic.

The change in the internal component β during the processes along *m*-geodesic manifested the internal computation in the ideal information-theoretic cycle. The small change in $\boldsymbol{\eta}$ is related to the change in $\boldsymbol{\theta}$ by $d\boldsymbol{\theta} = \mathbf{J}^{-1} d\boldsymbol{\eta}$. Since the *m*-geodesic processes B→C and D→A are characterized by $d\boldsymbol{\eta} = [0, dX]^{\top}$, the small change in θ-coordinates is given as

$$d\boldsymbol{\theta} = \begin{bmatrix} -\langle \mathbf{b}_0, \mathbf{b}_1 \rangle \\ \langle \mathbf{b}_0, \mathbf{b}_0 \rangle \end{bmatrix} |\mathbf{J}|^{-1} dX, \tag{11.25}$$

Conversely, the internal mechanism needs to change the internal and stimulus component according to the above gradient in order to accomplish the most efficient cycle. Thus the internal mechanism need to access the stimulus component α in order to realize the ideal information-theoretic cycle. Again, if $\langle \mathbf{b}_0, \mathbf{b}_1 \rangle = 0$, the internal component β is not allowed to change, which however means that the entire process does not form a cycle. Therefore we impose $\langle \mathbf{b}_0, \mathbf{b}_1 \rangle \neq 0$.

11.5 Discussion

In this study, we provided hypothetical neural dynamics that efficiently encodes stimulus information with the aid of delayed gain-modulation by an internal mechanism, and demonstrated that the dynamics forms an information-theoretic cycle that acts similarly to a heat engine. This view provided us to quantify the efficiency of the gain-modulation in retaining the stimulus information. The ideal information-theoretic cycle introduced here bounded the entropic efficiency.

As an extension of a logistic activation function of a single neuron to multinomial outputs, the maximum entropy model explains probabilities of activity patterns by a softmax function of the features, therefore allows nonlinear interaction of the inputs (here β and α) in producing the stimulus-related activity X (Fig. 11.1). This interaction was caused by shared activity features in $\mathbf{b}_1$ and $\mathbf{b}_0$. The gain modulation more effectively changes the stimulus-related activity if the features of the stimulus-related and internal activities resemble (i.e., $\langle \mathbf{b}_1, \mathbf{b}_0 \rangle$ is close to 1), which may have implications in similarity between evoked and spontaneous activities (Kenet et al. 2003) that can be acquired during development (Berkes et al. 2011).

The model's statistical structure common to thermodynamics (the Legendre transformation; see Appendix) allowed us to construct the first law for neural dynamics (Eq. (11.12)), the equation of state (Eq. (11.13)), fluctuation-dissipation relation (Eq. (11.16)), and neural dynamics similar to a thermodynamic cycle (Figs. 11.2 and 11.3) although we emphasized the differences from conventional thermodynamics in terms of the controllable quantities. The dynamics forms a cycle if the gain modulation is applied after the initial increase of the stimulus-related activity. This scenario is expected when the stimulus response is modulated by a feedback mechanism of recurrent networks (Salinas and Abbott 1996; Spratling and Johnson 2004; Sutherland et al. 2009), and is associated with short-term memory of the stimulus (Salinas and Abbott 1996; Salinas and Sejnowski 2001; Supèr et al. 2001). Consistently with the idea of efficient stimulus-encoding by a cycle, effect of attentional modulation on neural response typically appears several hundred milliseconds after stimulus onset (later than the onset of the stimulus response) (Motter 1993; Luck et al. 1997; McAdams and Maunsell 1999; Seidemann and Newsome 1999; Reynolds et al. 2000; Ghose and Maunsell 2002) although the temporal profile can be altered by task design (Luck et al. 1997; Ghose and Maunsell 2002). Further, the modulation of late activity components is ubiquitously observed in different neural systems (Cauller and Kulics 1991; Supèr et al. 2001; Sachidhanandam et al. 2013; Manita et al. 2015; Schultz 2016).

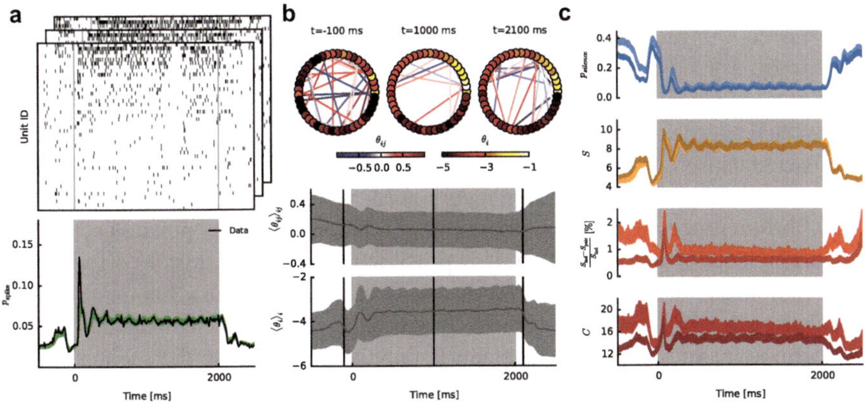

Fig. 11.5 The state-space method for estimating time-varying Ising model for monkey V4 data. (**a**) *Top:* Simultaneously recorded spiking data from 45 neurons while grating stimulus is presented to a monkey. *Bottom:* spiking probability (black, data; green, model fit). Gray area indicates the period of stimulus presentation. (**b**) *Top:* Time-varying parameters of an Ising model (i.e., individual and pairwise interaction parameters) are estimated by fitting the state-space model using an expectation-maximization (EM) algorithm. *Bottom:* the means and standard deviations of the Ising parameters. (**c**) Estimated dynamics of thermodynamic quantities (from top to bottom: silence probability, entropy, fractional entropy for correlations, heat capacity). The figure is modified from (Donner et al. 2017)

To test the hypothesis that neurons act as an information-theoretic engine using empirical data, the internal and stimulus feature need to be specified. Since even spontaneous neural activity is known to exhibit ongoing dynamics (Kenet et al. 2003), estimation of these features is nontrivial. The optimal sequential Bayesian algorithms have been proposed to smoothly estimate the parameters of the neural population model when they vary in time (Shimazaki et al. 2009, 2012; Shimazaki 2013; Donner et al. 2017), based on the paradigm developed by Brown and colleagues (Brown et al. 1998; Smith and Brown 2003) for joint estimation of the state-space and parameter estimation for point process observations. With the recent advances in applying various approximation methods to this model, it was demonstrated that the method is applicable to simultaneously analyzing a large number of neurons, and trace dynamics of thermodynamic quantities of the network such as the free energy, entropy, and heat capacity (Donner et al. 2017) (see Fig. 11.5). Hence this and similar approaches can be used to select dominant features of spontaneous and evoked activities, and then to estimate the time-varying internal and stimulus-related components. Efficiency of the cycles computed from the data can be used to test the hypothesis that the neurons are working as an information-theoretic engine. Further, by including multiple stimulus features in the model, the theory is expected to make quantitative predictions on competitive mechanisms of selective attention (Moran and Desimone 1985; Motter 1993; Luck et al. 1997; Reynolds et al. 1999). The conservation law of entropy imposes competition among the stimuli given a limited entropic resource generated by the internal mechanism.

The current theory assumes a quasi-static process for a neural response as we use an equilibrium model of the neural population at each point of time. For this to be a good approximation of neural dynamics, network activity caused by stimulus presentation may need to change more slowly than the time-scale of individual neurons under the examination, which may be expected as several tens of milliseconds for cortical neurons based on synaptic and membrane time constants and axonal delays. Otherwise, the theory needs to be extended to account for non-equilibrium processes by considering causal relations of past population activity on a current state of the population. It is possible to include the history effect on the population activity in the model (Shimazaki et al. 2012) or by using non-equilibrium models such as a kinetic Ising model. It will be an important challenge to consider a thermodynamic paradigm for a neural population including the second law for such non-equilibrium processes based on the recent advances in the field, where the second law of thermodynamics was generalized for a causal system with feedback (Sagawa and Ueda 2010, 2012; Ito and Sagawa 2013, 2015).

In summary, a neural population that works as an information-theoretic engine produces entropy ascribed to stimulus-related activity out of entropy supplied by an internal mechanism. This process is expected to appear during stimulus response of neurons subject to feedback gain-modulation. It is thus hoped that quantitative assessment of the neural dynamics as an information-theoretic engine contributes to understanding neural computation performed internally in an organism.

Acknowledgements This chapter is an extended edition of the manuscript submitted to the arXiv (Shimazaki H., Neurons as an information-theoretic engine. arXiv:1512.07855, 2015). The author thanks J. Gaudreault, C. Donner, D. Hirashima, S. Koyama, S. Amari, and S. Ito for critically reading the original manuscript.

Appendix: Free Energies of Neurons

In this appendix, we introduce thermodynamic formulation and free energies of a neural population. Let us first discuss the relation of state variables and free energies that appear in our analysis of the neural population with those found in conventional thermodynamics. Assume that the small change in internal activity of neurons has the following linear relations to entropy S, expected feature X, and the number of neurons N:

$$dU = TdS + fdX + \mu dN. \tag{11.26}$$

Equation (11.26) is the first law of thermodynamics, and the parameters are temperature T, force f, and chemical potential μ. The first law describes the internal activity as a function of (S, X, N). In thermodynamics, the Helmholtz free energy $F = U - TS$, Gibbs free energy $G = F - fX$, or enthalpy $H = U - fX$ is introduced to change the independent variables to (T, X, N), (T, f, N), and (S, f, N), respectively.

These free energies are useful to analyze isothermal or other processes in which only one of the independent variables is changed. For example, the Helmholtz free energy can be used to compute the work done by force f under the isothermal condition. However, the concepts of the force and work may not be directly relevant to information-theoretic analysis of a neural population. Here we introduce the free energies that are more consistent with the framework based on entropy changes.

The first law is alternatively written as

$$dS = \beta dU - \alpha dX - \gamma dN, \tag{11.27}$$

Here we used $\beta = 1/T$, $\alpha = f/T$, and $\gamma = \mu/T$. This first law describes a small entropy change as a function of (U, X, N). The parameters are defined as

$$\beta(U,X,N) = \left(\frac{\partial S}{\partial U}\right)_{X,N}, \tag{11.28}$$

$$\alpha(U,X,N) = -\left(\frac{\partial S}{\partial X}\right)_{N,U}, \tag{11.29}$$

$$\gamma(U,X,N) = -\left(\frac{\partial S}{\partial N}\right)_{U,X}. \tag{11.30}$$

We change the independent variable U to β. For this goal, here we define the *scaled* Helmholtz free energy $\mathscr{F}$ as

$$\mathscr{F} = S - \beta U. \tag{11.31}$$

Note that $\mathscr{F} = -\beta F$. It is a function that changes the independent variables from (S, X, N) to (β, X, N). This can be confirmed from the total derivative of $\mathscr{F}$: $d\mathscr{F} = dS - d(\beta U) = -Ud\beta - \alpha dX - \gamma dN$. From this equation, we have

$$U(\beta,X,N) = -\left(\frac{\partial \mathscr{F}}{\partial \beta}\right)_{X,N}, \tag{11.32}$$

$$\alpha(\beta,X,N) = -\left(\frac{\partial \mathscr{F}}{\partial X}\right)_{N,\beta}, \tag{11.33}$$

$$\gamma(\beta,X,N) = -\left(\frac{\partial \mathscr{F}}{\partial N}\right)_{\beta,X}. \tag{11.34}$$

The entropy change caused by the stimulus-related activity when X changes from X_1 to X_2 is given by the area under the curve of $\alpha(\beta, X, N)$ in the X-α phase plane. From Eq. (11.33), if the process satisfies $d\beta = dN = 0$, the entropy change is computed as reduction of the scaled Helmholtz free energy as

$$\Delta S^{\text{ext}} = \int_{X_1}^{X_2} \alpha(\beta, X, N)\, dX = \mathscr{F}(\beta, X_2, N) - \mathscr{F}(\beta, X_1, N). \tag{11.35}$$

Further change of the independent variables from (β, X, N) to (β, α, N) is done by introducing the *scaled* Gibbs free energy:

$$\mathscr{G} = \mathscr{F} + \alpha X = S - \beta U + \alpha X. \quad (11.36)$$

Note that $\mathscr{G} = -\beta G$. The independent variables of the Gibbs free energy are (β, α, N) since $d\mathscr{G} = d\mathscr{F} + (d\alpha X + X d\alpha) = -U d\beta + X d\alpha - \gamma dN$. From this equation, we find

$$\left(\frac{\partial \mathscr{G}}{\partial \beta}\right)_{\alpha,N} = -U(\beta, \alpha, N), \quad (11.37)$$

$$\left(\frac{\partial \mathscr{G}}{\partial \alpha}\right)_{\beta,N} = X(\beta, \alpha, N). \quad (11.38)$$

Note that the definition of the Gibbs free energy by Eq. (11.36) is obtained from Eq. (11.6) if we identify $\mathscr{G} = \psi$. Accordingly, Eqs. (11.37) and (11.38) coincide with Eqs. (11.7) and (11.8).

The Legendre transformation that changes the state variable N to μ is given by

$$\mathscr{G} + \gamma N = S - \beta U + \alpha X + \gamma N. \quad (11.39)$$

Since $d(\mathscr{G} + \mu N) = d\mathscr{G} + (d\gamma N + \gamma dN) = -U d\beta + X d\alpha + N d\gamma$, the natural independent variables is now (β, α, γ). From the extensive property of S, X, and N, we have the Gibbs-Duhem relation,

$$-U d\beta + X d\alpha + N d\gamma = 0. \quad (11.40)$$

Thus this free energy is identical to zero, and we obtain $\mathscr{G} = -\gamma N$.

References

Abbott, L. F., Varela, J. A., Sen, K., & Nelson, S. B. (1997). Synaptic depression and cortical gain control. *Science, 275*(5297), 220–224.

Amari, S.-I., & Nagaoka, H. (2000). *Methods of information geometry*. Providence: The American Mathematical Society.

Berkes, P., Orbán, G., Lengyel, M., & Fiser, J. (2011). Spontaneous cortical activity reveals hallmarks of an optimal internal model of the environment. *Science, 331*(6013), 83–87.

Brown, E. N., Frank, L. M., Tang, D., Quirk, M. C., & Wilson, M. A. (1998). A statistical paradigm for neural spike train decoding applied to position prediction from ensemble firing patterns of rat hippocampal place cells. *Journal of Neuroscience, 18*(18), 7411–7425.

Burkitt, A. N., Meffin, H., & Grayden, D. B. (2003). Study of neuronal gain in a conductance-based leaky integrate-and-fire neuron model with balanced excitatory and inhibitory synaptic input. *Biological Cybernetics, 89*(2), 119–125.

Carandini, M., & Heeger, D. J. (2012). Normalization as a canonical neural computation. *Nature Review Neuroscience, 13*(1), 51–62.

Carnot, S. (1824). *Réflexions sur la puissance motrice du feu et sur les machines propres à développer cette puissance*, Bachelier, Paris.

Cauller, L. J., & Kulics, A. T. (1991). The neural basis of the behaviorally relevant N1 component of the somatosensory-evoked potential in SI cortex of awake monkeys: Evidence that backward cortical projections signal conscious touch sensation. *Experimental Brain Research, 84*(3), 607–619.

Chance, F. S., Abbott, L. F., & Reyes, A. D. (2002). Gain modulation from background synaptic input. *Neuron, 35*(4), 773–782.

Doiron, B., Longtin, A., Berman, N., & Maler, L. (2001). Subtractive and divisive inhibition: Effect of voltage-dependent inhibitory conductances and noise. *Neural Computation, 13*(1), 227–248.

Donner, C., Obermayer, K., & Shimazaki, H. (2017). Approximate inference for time-varying interactions and macroscopic dynamics of neural populations. *PLoS Computational Biology, 13*(1), e1005309.

Ghose, G. M., & Maunsell, J. H. R. (2002). Attentional modulation in visual cortex depends on task timing. *Nature, 419*(6907), 616–620.

Granot-Atedgi, E., Tkačik, G., Segev, R., & Schneidman, E. (2013). Stimulus-dependent maximum entropy models of neural population codes. *PLoS Computational Biology, 9*(3), e1002922.

Ito, S., & Sagawa, T. (2013). Information thermodynamics on causal networks. *Physics Review Letter, 111*(18), 180603.

Ito, S., & Sagawa, T. (2015). Maxwell's demon in biochemical signal transduction with feedback loop. *Nature Communication, 6*, Article number: 7498.

Jaynes, E. T. (1957). Information theory and statistical mechanics. *Physical Review, 106*(4), 620–630.

Kass, R. E., Kelly, R. C., & Loh, W.-L. (2011). Assessment of synchrony in multiple neural spike trains using loglinear point process models. *Annals of Applied Statistics, 5*, 1262–1292.

Kelly, R. C., & Kass, R. E. (2012). A framework for evaluating pairwise and multiway synchrony among stimulus-driven neurons. *Neural Computation, 24*(8), 2007–2032.

Kenet, T., Bibitchkov, D., Tsodyks, M., Grinvald, A., & Arieli, A. (2003). Spontaneously emerging cortical representations of visual attributes. *Nature, 425*(6961), 954–956.

Laughlin, S. B. (1989). The role of sensory adaptation in the retina. *Journal of Experimental Biology, 146*, 39–62.

Lee, B. B., Dacey, D. M., Smith, V. C., & Pokorny, J. (2003). Dynamics of sensitivity regulation in primate outer retina: The horizontal cell network. *Journal of Vision, 3*(7), 513–526.

Luck, S. J., Chelazzi, L., Hillyard, S. A., & Desimone, R. (1997). Neural mechanisms of spatial selective attention in areas V1, V2, and V4 of macaque visual cortex. *Journal of Neurophysiology, 77*(1), 24–42.

Manita, S., Suzuki, T., Homma, C., Matsumoto, T., Odagawa, M., Yamada, K., et al. (2015). A top-down cortical circuit for accurate sensory perception. *Neuron, 86*(5), 1304–1316.

Martínez-Trujillo, J., & Treue, S. (2002). Attentional modulation strength in cortical area MT depends on stimulus contrast. *Neuron, 35*(2), 365–370.

McAdams, C. J., & Maunsell, J. H. (1999). Effects of attention on orientation-tuning functions of single neurons in macaque cortical area V4. *Journal of Neuroscience, 19*(1), 431–441.

Mitchell, S. J., & Silver, R. A. (2003). Shunting inhibition modulates neuronal gain during synaptic excitation. *Neuron, 38*(3), 433–445.

Moran, J., & Desimone, R. (1985). Selective attention gates visual processing in the extrastriate cortex. *Science, 229*(4715), 782–784.

Motter, B. C. (1993). Focal attention produces spatially selective processing in visual cortical areas V1, V2, and V4 in the presence of competing stimuli. *Journal of Neurophysiology, 70*(3), 909–919.

Nasser, H., Marre, O., & Cessac, B. (2013). Spatio-temporal spike train analysis for large scale networks using the maximum entropy principle and monte carlo method. *Journal of Statistical Mechanics, 2013*(03), P03006.

Ohiorhenuan, I. E., Mechler, F., Purpura, K. P., Schmid, A. M., Hu, Q., & Victor, J. D. (2010). Sparse coding and high-order correlations in fine-scale cortical networks. *Nature, 466*(7306), 617–621.

Ohzawa, I., Sclar, G., & Freeman, R. D. (1985). Contrast gain control in the cat's visual system. *Journal Neurophysiology, 54*(3), 651–667.

Prescott, S. A., & De Koninck, Y. (2003). Gain control of firing rate by shunting inhibition: roles of synaptic noise and dendritic saturation. *Proceedings of National Academy of Science USA, 100*(4), 2076–2081.

Reynolds, J. H., Chelazzi, L., & Desimone, R. (1999). Competitive mechanisms subserve attention in macaque areas V2 and V4. *Journal of Neuroscience, 19*(5), 1736–1753.

Reynolds, J. H., Pasternak, T., & Desimone, R. (2000). Attention increases sensitivity of V4 neurons. *Neuron, 26*(3), 703–714.

Rothman, J. S., Cathala, L., Steuber, V., & Silver, R. A. (2009). Synaptic depression enables neuronal gain control. *Nature, 457*(7232), 1015–1018.

Sachidhanandam, S., Sreenivasan, V., Kyriakatos, A., Kremer, Y., & Petersen, C. C. (2013). Membrane potential correlates of sensory perception in mouse barrel cortex. *Nature Neuroscience, 16*(11), 1671–1677.

Sagawa, T., & Ueda, M. (2010). Generalized Jarzynski equality under nonequilibrium feedback control. *Physics Review Letter, 104*(9), 090602.

Sagawa, T., & Ueda, M. (2012). Fluctuation theorem with information exchange: Role of correlations in stochastic thermodynamics. *Physics Review Letter, 109*(18), 180602.

Sakmann, B., & Creutzfeldt, O. D. (1969). Scotopic and mesopic light adaptation in the cat's retina. *Pflügers Archiv: European Journal of Physiology, 313*(2), 168–185.

Salinas, E., & Abbott, L. F. (1996). A model of multiplicative neural responses in parietal cortex. *Proceedings of National Academy of Sciences USA, 93*(21), 11956–11961.

Salinas, E., & Sejnowski, T. J. (2001). Gain modulation in the central nervous system: Where behavior, neurophysiology, and computation meet. *Neuroscientist, 7*(5), 430–440.

Schneidman, E., Berry, M. J., Segev, R., & Bialek, W. (2006). Weak pairwise correlations imply strongly correlated network states in a neural population. *Nature, 440*(7087), 1007–1012.

Schultz, W. (2016). Dopamine reward prediction-error signalling: A two-component response. *Nature Review Neuroscience, 17*(3), 183–195.

Seidemann, E., & Newsome, W. T. (1999). Effect of spatial attention on the responses of area MT neurons. *Journal of Neurophysiology, 81*(4), 1783–1794.

Shimazaki, H. (2013). Single-trial estimation of stimulus and spike-history effects on time-varying ensemble spiking activity of multiple neurons: a simulation study. *Journal of Physics: Conference Series, 473*, 012009.

Shimazaki, H., Amari, S.-I., Brown, E. N., & Grün, S. (2009). State-space analysis on time-varying correlations in parallel spike sequences. In *Proceedings of IEEE ICASSP*, pp. 3501–3504.

Shimazaki, H., Amari, S.-i., Brown, E. N., & Grün, S. (2012). State-space analysis of time-varying higher-order spike correlation for multiple neural spike train data. *PLoS Computational Biology, 8*(3), e1002385.

Shimazaki, H., Sadeghi, K., Ishikawa, T., Ikegaya, Y., & Toyoizumi, T. (2015). Simultaneous silence organizes structured higher-order interactions in neural populations. *Scientific Reports, 5*, 9821.

Shlens, J., Field, G. D., Gauthier, J. L., Grivich, M. I., Petrusca, D., Sher, A., et al. (2006). The structure of multi-neuron firing patterns in primate retina. *Journal of Neuroscience, 26*(32), 8254–8266.

Silver, R. A. (2010). Neuronal arithmetic. *Nature Review Neuroscience, 11*(7), 474–489.

Smith, A. C., & Brown, E. N. (2003). Estimating a state-space model from point process observations. *Neural Computation, 15*(5), 965–991.

Spratling, M. W., & Johnson, M. H. (2004). A feedback model of visual attention. *Journal of Cognitive Neuroscience, 16*(2), 219–237.

Supèr, H., Spekreijse, H., & Lamme, V. A. (2001). A neural correlate of working memory in the monkey primary visual cortex. *Science, 293*(5527), 120–124.

Sutherland, C., Doiron, B., & Longtin, A. (2009). Feedback-induced gain control in stochastic spiking networks. *Biological Cybernetics, 100*(6), 475–489.

Tang, A., Jackson, D., Hobbs, J., Chen, W., Smith, J. L., Patel, H., et al. (2008). A maximum entropy model applied to spatial and temporal correlations from cortical networks in vitro. *Journal of Neuroscience, 28*(2), 505–518.

Tkačik, G., Marre, O., Amodei, D., Schneidman, E., Bialek, W., & Berry, M. J. (2014). Searching for collective behavior in a large network of sensory neurons. *PLoS Computational Biology, 10*(1), e1003408.

Tkačik, G., Mora, T., Marre, O., Amodei, D., Palmer, S. E., Berry, M. J., et al. (2015). Thermodynamics and signatures of criticality in a network of neurons. *Proceedings of National Academy of Sciences USA, 112*(37), 11508–11513.

Yu, S., Huang, D., Singer, W., & Nikolic, D. (2008). A small world of neuronal synchrony. *Cerebral Cortex, 18*(12), 2891–2901.

Chapter 12
Inferring Neuronal Network Mechanisms Underlying Anesthesia-Induced Oscillations Using Mathematical Models

Sujith Vijayan and Michelle McCarthy

12.1 Introduction

General anesthesia is administered to over 100,000 people daily, yet we are just beginning to understand the mechanisms through which anesthetics act. During the administration of a general anesthetic, as the depth of anesthesia increases, patients transition through a series of stereotypical brain states, each marked by prominent oscillatory activity in the EEG. Examples of oscillatory activity with anesthesia include increases in beta (13–30 Hz) frequency oscillations with low doses of propofol (McCarthy et al. 2008) and increases in alpha (8–13 Hz) oscillations at anesthetic doses of $GABA_a$-potentiating drugs such as propofol and sevoflurane (Akeju et al. 2014). In contrast, the EEG during ketamine displays prominent low gamma ~30–40 Hz oscillations (Akeju et al. 2016). These dynamical states also correlate with different behavioral states (for example, drowsy, unable to respond, etc.). Therefore, gaining an understanding of the systems and circuit level mechanisms through which general anesthetics act may provide insight into how these dynamics bring about the concomitant behavioral states, allowing for better design of anesthetics, as well as providing a deeper understanding of the neural mechanisms underlying brain dynamics more generally.

One approach to understanding the mechanisms underlying the neural dynamics seen during general anesthesia is via mathematical modeling, which allows the investigator to examine and perturb the neural activity of a system, especially at large scale levels, in a very precise manner that is not often easily achieved using

S. Vijayan (✉)
School of Neuroscience, Virginia Tech, Blacksburg, VA, USA
e-mail: neuron99@vt.edu

M. McCarthy
Boston University, Boston, MA, USA
e-mail: mmccart@math.bu.edu

Z. Chen, S.V. Sarma (eds.), *Dynamic Neuroscience*,
https://doi.org/10.1007/978-3-319-71976-4_12

traditional physiological techniques. Mathematical modeling investigations can identify novel mechanisms that may be responsible for the observed brain dynamics and provide experimentally testable hypotheses. Examples of mathematical models providing insight into brain dynamics include theoretical work on "interneuronal network gamma" in which, as the name suggests, interneurons form network oscillations at gamma frequency (Cannon et al. 2014). These mathematical models have been experimentally tested in recent experiments in which cortical fast spiking interneurons were optogenetically driven in vivo resulting in a selective increase in gamma oscillations (Cardin et al. 2009).

In this chapter, we will discuss modeling work that has been used to understand the systems and circuit level mechanisms underlying the various brain states during the administration of the general anesthetic propofol. Along the way we will discuss the common mathematical modeling techniques used to identify candidate generative circuit and systems level mechanisms responsible for given brain dynamics.

12.2 Biophysical Modeling

Various types of models are employed to investigate brain dynamics. Some commonly employed models are mean field models, leaky integrate and fire models, and Hodgkin-Huxley models. Each has its strengths and weaknesses. Mean field models, such as the Wilson-Cowan model (Wilson and Cowan 1972), easily allow for the investigation of large scale population activity, but at the cost of biophysical detail. Hodgkin-Huxley models are more biophysically detailed than mean field models, often incorporating many of the ionic currents known to exist in a given class of neurons, and ranging in anatomical detail from a single compartment to models that incorporate the fine details of dendritic arborizations. However, Hodgkin-Huxley models are not as conducive to examining large scale population activity as mean field models, since Hodgkin-Huxley models require more computational resources due to their greater biophysical detail (Hodgkin and Huxley 1952). In the propofol work presented below, single compartment Hodgkin-Huxley models are employed so we provide more details about such models here. The membrane potentials of such single compartment Hodgkin-Huxley models are governed by the ordinary differential equation:

$$C_M \frac{dV}{dt} = -\sum I_M - \sum I_{\mathrm{Syn}} \tag{12.1}$$

where the product of the change in the membrane potential with respect to time (dV/dt) and the membrane capacitance (C_M) is equal to the product of negative one and the sum of the membrane currents (I_M) and the synaptic currents (I_{Syn}). The membrane currents that are incorporated into a model of a particular cell type are usually dictated by experimental evidence. In the original equations developed by Hodgkin and Huxley (1952), the membrane currents consisted only of the spiking currents, which include the fast sodium current (I_{Na}), the fast potassium current (I_K), and a leak current (I_L) as well as an applied current (I_{app}). The spiking currents were Ohmic and governed by the following equation:

$$I_{\text{ion}} = g_{\text{ion}}(V - E_{\text{ion}}) \tag{12.2}$$

In this equation, $I_{\text{ion}} = I_{Na}$, I_K or I_L. The conductance of each ionic current is represented by g_{ion} and E_{ion} is the equilibrium potential for each respective ionic current. In Hodgkin and Huxley's original experiments, they found that the conductance of the ionic channels was best described by incorporating activation and inactivation gating variables to the ionic conductance:

$$g_{\text{ion}} = \bar{g}_{\text{ion}} m^a h^b. \tag{12.3}$$

where $\bar{g}_{\text{ion}}$ is the maximal conductance of the ionic current, m is the fraction of open activation gates, and h is the fraction of open inactivation gates. Accordingly, $m, h \in [0, 1]$. The constants a and b are the number of activation or inactivation gates, respectively, and take integer values ≥ 0. The gating variables (m,h) are first-order ordinary differential equations written as:

$$\frac{dx}{dt} = \frac{x_\infty(V) - x}{\tau_x(V)} \qquad \text{for } x = m, h \tag{12.4}$$

The kinetics of the gating variables are determined by the time constant of decay (τ_x) and the steady state activation curve (x_∞). Both these functions are often dependent on the voltage (V) of the neuron and both can be measured experimentally.

The synaptic currents are also Ohmic and follow the equation:

$$I_{\text{syn}} = g_{\text{syn}}(V - E_{\text{syn}}) \tag{12.5}$$

Here syn stands for the post-synaptic receptor type, which could be GABA (gamma-aminobutyric acid, including GABA_a and GABA_b), AMPA, or NMDA among others. The reversal potential of the synaptic current (E_{syn}) depends on the ionic species which flows across the membrane after the synaptic receptor is activated. For example, chloride ions flow across the GABA_a channel, so E_{GABAa} is the reversal potential for chloride ion. The synaptic conductance also has a gating variable that depends on the pre-synaptic voltage:

$$g_{\text{syn}} = \bar{g}_{\text{syn}} s. \tag{12.6}$$

Since the gating variable s represents the fraction of open channels, $s \in [0, 1]$. Like the gating variables for the membrane currents, the dynamics of the variable s follow a first order differential equation:

$$\frac{ds}{dt} = \frac{g(V_i)}{\tau_r}(1 - s) - \frac{s}{\tau_{\text{syn}}} \tag{12.7}$$

Here, τ_{syn} is the time-constant of decay of the synaptic gating variable and τ_r is the synaptic rise time-constant. The rate function for the open state of the synaptic receptor, $g(V_i)$, is dependent on the pre-synaptic voltage (V_i) and can be functionally defined as:

$$g(V_i) = \frac{1 + \tanh(V_i/10)}{2} \tag{12.8}$$

Here we briefly described basic Hodgkin-Huxley-type equations. For the interested reader, there are many good textbooks that cover more details about modeling neurons using Hodgkin-Huxley dynamics (Koch and Segev 1998; Izhikevich 2007; Börgers 2017).

12.3 State Changes During the Administration of Propofol

As the patient is administered propofol and the depth of anesthesia is increased, there is a characteristic sequence of spatio-temporal changes in brain dynamics. Prior to the administration of propofol, as the patient is lying down with their eyes closed, alpha activity (8–13 Hz) is often observed over the occipital cortex. At low concentrations of propofol, patients become excited rather than sedated. This excitation, aptly referred to as paradoxical excitation, is marked by the disinhibition of motor activity and the emergence in the EEG of relatively fast oscillatory activity in the beta band (12.5–25 Hz). As the concentration of propofol increases and patients enter a deeper anesthetic state, there is a spatial shift in the alpha power from the back of the brain to the front of the brain. This spatial shift in power is referred to as anteriorization. At a behavioral level anteriorization co-occurs with behaviorally defined loss of consciousness (LOC), when patients stop responding (e.g., making a button press) in the context of a behavioral task. As the concentration of propofol increases further, patients enter an even deeper state of anesthesia referred to as burst suppression, which is marked at the level of the EEG by alternations between periods of quiescence and high amplitude activity at a wide spatial scale. If the concentration is increased even further, the EEG activity becomes relatively flat.

12.3.1 Modeling Cortical Networks During Propofol-Induced Paradoxical Excitation

Many anesthetics, including propofol, work by potentiating $GABA_a$ receptors, which are the main receptors providing fast inhibition in CNS circuits. However, low doses of propofol paradoxically can increase behavioral excitation along with increased EEG beta oscillations (Gugino et al. 2001).

Since the EEG can record electrical activity from up to a billion neurons (Nunez and Srinivasan 2006), rhythmic activity in the EEG is thought to represent coordinated activity of networks of neurons. Using biophysical models with Hodgkin-Huxley-type conductances introduces computational constraints on the number of neurons that can be simulated. However, even networks of 240 biophysical neurons can be sufficient to give insight into the underlying physiological and network mechanisms that may be producing rhythmic activity in the EEG. Moreover, critical insight into dynamics can be obtained with even a two- or three-neuron network. For example, the main insight into the dynamics potentially underlying propofol-induced paradoxical excitation arose from two- and three-cell models.

With a simple network of two neurons, (a pyramidal cell receiving $GABA_a$ inhibition from an inhibitory interneuron) one can reproduce a commonly known phenomenon: post-inhibitory rebound spiking (Fig. 12.1). Post-inhibitory rebound spiking can occur when certain membrane currents are present that cause the membrane to respond to hyperpolarization with excitation. Examples of such membrane currents needed to produce post-inhibitory rebound spiking include the hyperpolarization-activated H-current (Ascoli et al. 2010) or the T-type calcium current (Alvina et al. 2010). Thus, a simple two-cell model can explain at least one means of producing excitation from inhibition.

A similar two-cell model with an intrinsic membrane M-current in the post-synaptic neuron can also produce post-inhibitory rebound spiking (Fig. 12.1). This occurs because the M-current is a non-inactivating potassium current with a slow time constant of decay. Thus, the M-current decreases in response to inhibition, which in turn gives excitation to the neuron. Since the time constant of the M-current is generally slower than the time constant of the $GABA_a$ response, inhibition due to $GABA_a$ can be followed by excitation due to lowered M-current and this can result in rebound spiking. However, we note that the cell containing an M-current does

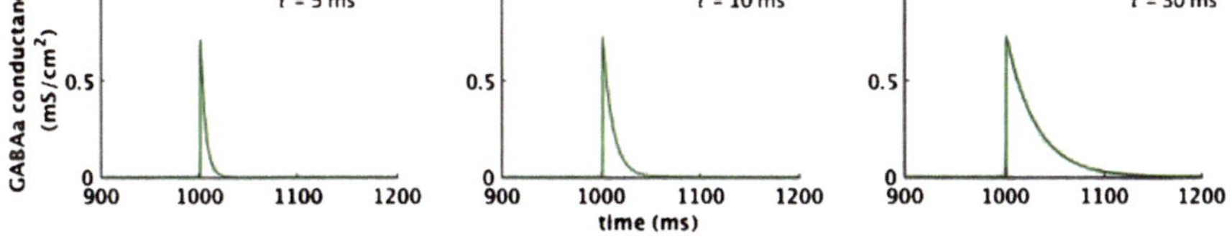

Fig. 12.1 Post-inhibitory rebound spiking of a neuron with an M-current as a function of the time-constant of inhibition (τ) of the $GABA_a$ current

not always respond to GABAa-mediated inhibition with rebound excitation. If the time course of the $GABA_a$ conductance is too brief or too long, then no rebound excitation occurs (Fig. 12.1).

This simple two-cell model thus gives a critical insight into dynamics that might be at work in propofol-induced paradoxical excitation. This is because propofol dose-dependently increases the conductance and time-constant of the $GABA_a$ synapse. Thus, at baseline, the $GABA_a$ kinetics are fast enough that rebound excitation does not occur by this mechanism, while with low-doses of propofol, post-inhibitory rebound can occur and with even greater $GABA_a$ potentiation, rebound excitation again ceases. The two-cell model shows us that the kinetics of the M-current can create a window of excitation dependent on the level of potentiation of the inhibitory $GABA_a$ synapse.

We can observe a related though different form of excitation with potentiated inhibition in our two-neuron model if we add an excitatory AMPA connection from the pyramidal neuron to the inhibitory interneuron (Fig. 12.2a). In this system, the spiking frequency of the two cell network is set by the level of background

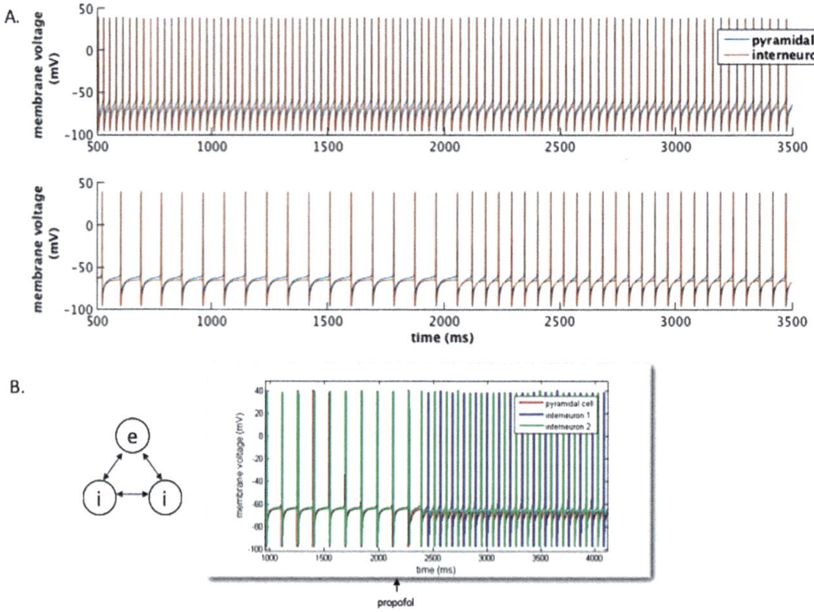

Fig. 12.2 Two and three-cell networks show paradoxical excitation with propofol. (**a**) Two-cell network of one pyramidal cell with and M-current and one inhibitory interneuron reciprocally connected. Baseline conditions exist up to 2000 ms, then low-dose propofol is simulated after 2000 ms. Note that the spiking frequency decreases with propofol if the baseline spiking rate is gamma, whereas the spiking rate increases with propofol if the baseline spiking rate is at alpha range (~11 Hz). (**b**) Three-neuron network of one pyramidal cell and two interneurons (all cells with M-current) connected all-to-all. Propofol is simulated after 2200 ms. Note the interneurons form an anti-synchronous beta rhythm with propofol and the pyramidal cell is suppressed

excitation (applied current) to the pyramidal cell. When the pyramidal cell spikes, the interneuron is excited by the AMPA current and also spikes a few milliseconds after the pyramidal cell. If the background excitation to the pyramidal cell is high such that the network has a low-gamma (30–45 Hz) frequency rhythm, then potentiating $GABA_a$, consistent with low doses of propofol, slows the rhythm to a beta. This is the expected consequence of extra inhibition to a network. However, unexpectedly, if the network has a low baseline spiking rate in the alpha range ($\sim$10 Hz), potentiating the $GABA_a$ synapse has the effect of speeding up the rhythm to the beta frequency range ($\sim$14 Hz). As a consequence of the expected slowing down of higher than beta frequency rhythms and the unexpected speeding up of lower than beta frequency rhythms, independent two-cell oscillators over a range of frequencies from alpha to gamma will converge into the beta frequency range (McCarthy et al. 2008).

The speeding up of alpha frequency rhythms into the beta frequency range following $GABA_a$ potentiation comes about from the interaction of the $GABA_a$ kinetics with the M-current kinetics. As we observed in the case of post-inhibitory rebound excitation, the slow return of the M-current to its baseline following hyperpolarization can create a temporal window of excitation after the GABAa inhibition has mostly worn off. If the baseline rhythm is slow enough (e.g., alpha frequency range), then the pyramidal neuron is given enough time between spikes for this temporal window of excitation to be expressed resulting in advancement of the pyramidal cell spiking before its usual spike time based purely on its applied current level. Thus, the rhythm is sped up, in this case from alpha to beta frequency. If the network rhythm is faster and in the gamma frequency range, then this window of excitation provided by the M-current is never reached before the next pyramidal cell spike. Thus, for the higher frequency rhythms, the dominant effect of increasing $GABA_a$ potentiation is slowing of the network rhythm. These phenomena, that of lower frequency rhythms speeding up and higher frequency rhythms slowing down can also be observed by using phase response curves (McCarthy et al. 2008).

Another paradoxically exciting phenomenon occurs in three-cell networks. In the three-cell network, two inhibitory interneurons and one pyramidal cell neuron are connected all-to-all (Fig. 12.2b). As in the reciprocally connected two-cell network, the spiking rate of the pyramidal cell determines the network spiking rate, with the two interneurons spiking in response to AMPA input from the pyramidal cells. Potentiating $GABA_a$ consistent with low doses of propofol speeds up a baseline theta (6 Hz) rhythm to mid-beta frequency (19 Hz). Although this phenomenon is similar to what was observed before (i.e., speeding up of alpha frequency rhythms in the two-cell network), the underlying mechanism is distinct. Indeed, the two-cell network can only speed up the network rate by several hertz at most, whereas in the three-cell network, we observe more than a tripling of the baseline frequency.

The underlying mechanism for beta frequency generation in the three-cell model is a switch from synchrony of the interneurons at baseline to the formation of an anti-synchronous interneuron rhythm after potentiation of the $GABA_a$ kinetics. This phenomenon is only observed if the interneurons have an M-current and thus they are a type of interneuron we label as a low-threshold spiking cell, or LTS cell.

Since the LTS cells have an M-current and are reciprocally connected by $GABA_a$ synapses, the same $GABA_a$-M-current interactions that cause rebound spiking in our two-cell models are present also in the LTS cell interactions. However, at baseline each LTS cell only spikes in response to the pyramidal cell input and does not rebound spike in response to inhibition from the other LTS cell. This is due to a phase-dependence of the effect of the $GABA_a$ current on the M-current. Specifically, when inhibition arrives during the repolarization phase of neuronal spiking, the M-current is only minimally suppressed and the neuron does not achieve the level of excitation necessary to rebound spike. However, this phase-dependence is lost when the $GABA_a$ current is potentiated by propofol. Since the LTS cells can respond to inhibition with a rebound spike and since the rebound spikes are on a faster time scale than the baseline network rhythm, the LTS cells form a stable anti-synchronous beta frequency oscillation. This has the effect of suppressing the one pyramidal cell in the three-cell network, since the pyramidal cell's natural spiking rate (6 Hz) is slower than the beta inhibition it is receiving. However, in slightly larger networks consisting of 12 pyramidal cells and 2 reciprocally connected LTS cells, the pyramidal cells can participate sparsely in the beta rhythm (McCarthy et al. 2008).

It is interesting that the $GABA_a$-M-current interaction gives the correct timescale for the production of beta frequency rhythms. It is thought that the time-constants of the $GABA_a$ current and the M-current interact to create this time scale. However, the exact mechanism is not clear as the M-current has a theta time scale without inhibition and the $GABA_a$ current is relatively fast with a time constant of 10 ms when simulating low-dose propofol. The beta-determining factors in the $GABA_a$-M-current interaction, to the best of our knowledge, are yet to be fully explained.

The small two- and three-cell networks gave the underlying intuition into the dynamical interactions on the neuronal and network level that form the basis of two forms of paradoxically exciting phenomena that might be observed as $GABA_a$ potentiation is increased by propofol. Larger scale models of up to 240 neurons confirm that the dynamics that appear in the two- and three-neuron models are also appreciated in larger scale models. Indeed, networks consisting of 240 neurons of three commonly found cell types in cortex (pyramidal cells, LTS cells and fast spiking interneurons) show that the network produces a prominent beta frequency rhythm upon potentiation of $GABA_a$ consistent with low doses of propofol (McCarthy et al. 2008). The LTS neurons form anti-synchronous clusters of cells, which help to pattern the pyramidal cells into a beta frequency rhythm. Model EEG rhythms also increase in the beta frequency range with simulations of low doses of propofol largely due to the patterning of the pyramidal cells by the interneurons.

Thus, we use this example to demonstrate how simple biophysical models can be used to infer critical membrane interactions within neurons and synaptic interactions between neurons that lead to oscillatory network dynamics in larger scale systems. Identifying potential underlying mechanisms producing beta frequency rhythmicity with propofol not only suggests network mechanisms at work during propofol-induced paradoxical excitation, but also gives insight into other neuronal systems which may be using the same underlying mechanism to produce beta frequency oscillations. In particular, the M-current/GABA$_a$ interaction has been used to suggest striatal circuits as a potential source of the exaggerated beta-frequency rhythms that emerge in basal ganglia circuits in Parkinson's disease (McCarthy et al. 2011). Thus, unraveling the circuit mechanism underlying anesthetic states provides additional insight into normal, baseline network activity as well as how neuronal circuits may become altered due to neuromodulatory changes that occur with neurologic disease.

12.3.2 Dynamical System Analysis of Propofol Paradoxical Excitation Model

By potentiating GABA$_a$ synapses, propofol produces a window of excitation in post-synaptic cells that have an M-current such that post-inhibitory rebound spiking is observed only if the GABA$_a$ potentiation is not too small or too large. Unlike the classic version of post-inhibitory rebound spiking in which the voltage of the cell is held hyperpolarized for a length of time and then depolarized in a step-like manner, the voltage of the post-synaptic cells depends on progressively changing interaction between the GABA$_a$ inhibition and the M-current. This dynamic creates a window of excitation (e.g., during low doses of propofol) as opposed to a GABA$_a$-potentiation-threshold resulting in a rebound spike (Fig. 12.1).

Mathematical methods exist that can further disentangle the interaction of time-scales when the time-scales are separated "enough." Specifically, the methods of *geometric singular perturbation theory* (GSPT) are widely used to lend insight into the dynamics of fast-slow systems, in which some variables in an interacting system of equations operate on a time scale substantially faster than other variables. Such is the case in the cortical network producing propofol-induced paradoxical excitation beta: the time scale of the M-current and the GABA$_a$ current are substantially slower than the time scales of the other variables in the model equations.

Visualization of the dynamics is best if the system is under 3 dimensions. Our original 200+ neuron model of propofol-induced paradoxical excitation has several hundreds of dimensions and is thus intractable in this form. However, we recall that the beta oscillations emerge from post-inhibitory rebound spiking of the LTS cells: a phenomenon we have shown by simulations to depend on the interaction of the GABA$_a$ and the M-current in these cells. Thus, we can reduce the complexity of our system by looking at the response of one LTS cell to inhibition, with inhibition introduced as a step function with exponential decay. This approximation of GABA$_a$ inhibition does not significantly change the GABA$_a$ dynamics and moreover releases the system of equations from explicit dependence on a pre-synaptic LTS cell. Such a reduction reduces the dimensions of our system from 1000+ to 6 dimensions, which can be written as:

$$c_m \frac{dv}{dt} = \overbrace{-\bar{g}_{Na} m^3 h(v - E_{Na}) - \bar{g}_k n^4 (v - E_k) - \bar{g}_L (v - E_L)}^{\text{spiking currents}}$$
$$\overbrace{-\bar{g}_m w(v - E_k)}^{\text{M-current}} \quad \overbrace{-\bar{g}_i s(v - E_i)}^{\text{inhibitory (GABAa) current}} \quad +I_{\text{app}} \tag{12.9}$$
$$\frac{dx}{dt} = \frac{x_\infty(v) - x}{\tau_x(v)} \qquad \text{for } x = m, h, n, w$$
$$\frac{ds}{dt} = -\frac{s}{\tau_s}.$$

In these equations, the time constant of the M-current (τ_w) depends on the voltage (v), which makes it difficult to directly compare with the time constant of the GABA$_a$ synapse, τ_s, which is constant. Thus, in order to determine the interaction of these two time-constants in further analysis, τ_w is allowed to be constant. With τ_w as a parameter, we still observe a τ_s-dependent window of rebound spiking. Thus the dynamics we wish to characterize using GSPT are still present. Within the physiological range, τ_w functions to determine the range of the τ_s-dependent window, with the width of the window increasing as τ_w increases (see Fig. 6 in McCarthy et al. 2012).

These simplifications and the reduction to 6-dimensions aids our ability to analytically examine the dynamical interaction between the M-current and the GABA$_a$ current, the w and s variables in the system of equations, respectively. To begin, an analysis of the time scales of the system is necessary. By a method called "non-dimensionalization," we can make each of our 6 variables unit-less, and thus directly compare the time scales of each (Segel and Edelstein-Keshet 2013). By non-dimensionalization, we find that all the spiking variables (m, h, n, and v) are fast variables. Interestingly, the number of slow variables depends on the decay time constant of the GABA$_a$ inhibition, τ_s. This is important in light of the finding that the decay time of GABA$_a$-mediated inhibition in CNS circuits can vary widely from at least 1–70 ms (Traub et al. 2005). Additionally, it is important to the problem of propofol-induced rebound excitation since as the dose of propofol is increased, the time constant of GABA$_a$ inhibition can increase (Kitamura et al. 2003). Specifically, in this set of equations, when τ_s is at its baseline value of $\sim$5 ms then w (the M-current variable) is the only slow variable and the entire system of equations has three time scales (fast, medium and slow), with τ_s being the medium time-scale variable. However, when τ_s is slightly higher ($\geq$10 ms) as observed with the addition of propofol, then both w and s are considered slow variables and the system of equations has two time scales (fast and slow).

As τ_s increases and transitions from spiking to non-spiking solutions, the values of τ_s around this transition make the problem a two time-scale problem. As such, we can use the methods of geometric singular perturbation theory to understand the underlying mathematical nature of the transition from spiking to non-spiking solutions as the time-constant of GABAa increases. In order to do this, we first define our singular perturbation parameters as $\varepsilon = \tau_{\max}/\tau_s$ and $\bar{\varepsilon} = \tau_s/\tau_w$ and rescale time by letting $\tau = \varepsilon t$. In these equations, $\tau_{\max} = 3$, which is an approximation of the maximum value of all τ_x for all x in Eq. (12.9) in the subthreshold range. Thus, we obtain the slow (singularly perturbed) system of equations:

$$\varepsilon c_m \frac{dv}{d\tau} = -\bar{g}_{Na} m^3 h(v - E_{Na}) - \bar{g}_k n^4 (v - E) - \bar{g}_L(v - E_L) - \bar{g}_m w(v - E) - \bar{g}_i s(v - E) + I_{\text{app}}$$

$$\varepsilon \frac{dx}{d\tau} = \frac{x_\infty(v) - x}{\tau_x(v)} \quad \text{for } x = m, h, n \tag{12.10}$$

$$\frac{dw}{d\tau} = \frac{\bar{\varepsilon}(w_\infty(v) - w)}{\tau_{\max}}$$

$$\frac{ds}{d\tau} = -\frac{s}{\tau_{\max}}$$

In the singular limit of Eq. (12.10), when $\varepsilon = 0$, we have a 2D system of differential equations evolving on a 2D surface (the critical manifold) in 3D (v, s, w)-space.

$$0 = F(v) - \bar{g}_m w(v - E_k) - \bar{g}_i s(v - E_i) \equiv f(v, w, s) \tag{12.11}$$

$$\dot{w} = \frac{\bar{\varepsilon}(w_\infty(v) - w)}{\tau_{\max}} \tag{12.12}$$

$$\dot{s} = -\frac{s}{\tau_{\max}}. \tag{12.13}$$

In these equations, $F(v) = -\bar{g}_{Na} m_\infty^3 h_\infty(v - E_{Na}) - \bar{g}_k n_\infty^4 (v - E_k) - \bar{g}_L(v - E_L) + I_{\text{app}}$. We note that $f(v, s, w)$ is an important geometric structure called the critical manifold. It is the set of equilibrium points of the fast (singularly perturbed) system (i.e., the system of Eq. (12.10) with the singular perturbation parameters ε and $\bar{\varepsilon}$ but without the rescaling of time). The critical manifold is also the geometric structure on which the dynamics of the slow system evolve in the singular limit. In the subthreshold range of voltage (less than approximately 50 mV), the critical manifold forms two sheets separated by a fold curve in (v, s, w)-space (Fig. 12.3, the projection of the fold curve onto the (v, s)-plane is represented by the red line). Interestingly, in this problem the value of v on the fold curve is a constant (i.e., $v_c = -63.29$ mV). This need not be the case. The fold curve is of importance as it represents the spiking threshold in the singular limit. It is a spiking threshold because for values of $v < v_c$, the critical manifold is an attracting surface, along which dynamics of the slow system flow in the singular limit. However for values of $v > v_c$ the critical manifold is a repelling surface. Thus the fold curve is the

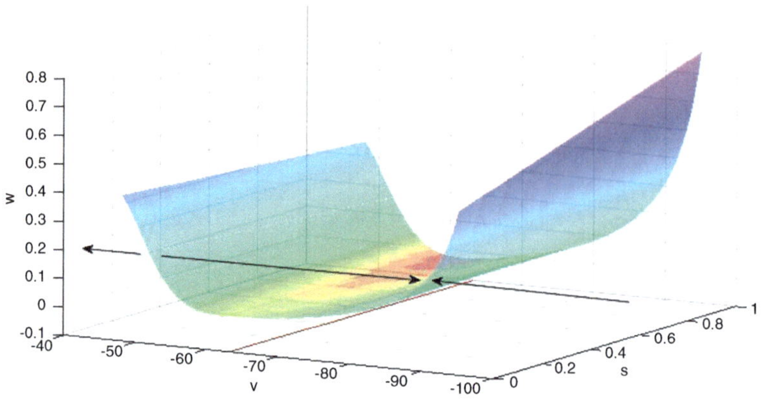

Fig. 12.3 The critical manifold in (v, s, w)-space. The black arrows indicate fast attracting and fast repelling dynamics towards and away from the critical manifold, respectively. The red line represents the projection of the fold curve onto the (v, s)-plane

curve in the singular limit at which the slow dynamics cease and the fast dynamics again become relevant. Thus, most solution that can reach the fold curve in the singular limit is a "spiking solution." (Canard solutions are exceptions and will not be discussed here). Those solutions that fail to reach the fold curve are "non-spiking solutions ." Special structures on the critical manifold determine what separates spiking solutions from non-spiking solutions, as we now demonstrate with this example.

Projecting the critical manifold onto the (v, s)-plane allows for better visualization of the dynamics around the fold curve (Szmolyan and Wechselberger 2001). Since the projection of the system is singular at the fold curve, the structure that separates "spiking" from "non-spiking" solutions, we desingularize the projected system by rescaling time by $\tau = -f_v \bar{\tau}$. This desingularization allows us to characterize the dynamics near the fold curve. In this rescaling, f_v is the first derivative of f as defined in Eq. (12.11), with respect to v. Note that points on the critical manifold for which $f_v < 0$ are attracting with respect to the fast (singularly perturbed) system (Eq. (12.9)) and the points for which $f_v > 0$ are repelling with respect to the fast (singularly perturbed) system. The points $f_v = 0$ are the fold curve. Note that by rescaling time by $-f_v$, we have reversed the direction field on the repelling surface (the surface where $f_v > 0$) of the critical manifold. After desingularization, our projected system follows the equations:

$$v' = -\frac{1}{\tau_{\max}} \bar{\varepsilon} \bar{g}_m (v - E)(w_\infty - w(v, s)) + \frac{1}{\tau_{\max}} \bar{g}_i (v - E) s \equiv H(v, s) \tag{12.14}$$

$$s' = \frac{1}{\tau_{\max}} \left(F'(v) - \frac{F(v)}{(v - E)} \right) s \equiv G(v, s). \tag{12.15}$$

In this system, the derivative is taken with respect to $\bar{\tau}$. We find folded singularities (singularities of this system of equations that lie on the fold curve) by setting $H(v,s) = 0$. We note that $G(v,s) = 0$ on the fold curve because $F'(v) - \frac{F(v)}{(v-E)} = f_v$ and $f_v = 0$ on the fold curve. Solving for the folded singularities, we find they are described by:

$$s_c = \left[w_\infty(v_c) - \frac{F(v_c)}{\bar{g}_m(v_c - E)} \right] \left(\frac{\bar{g}_m}{\bar{g}_i} \right) \left(\frac{\bar{\varepsilon}}{1 - \bar{\varepsilon}} \right). \tag{12.16}$$

Equation (12.16) shows us how the position of the folded singularity depends on the parameters of the original system (Eq. (12.9)), especially $\bar{g}_m$ and $\bar{g}_i$, the maximal conductances of the M-current and the GABA$_a$-current, respectively. Furthermore, after linearization of Eqs. (12.14) and (12.15), we can determine the stability of the folded equilibrium by looking at the determinate of the Jacobian matrix. We find that for all values of τ_w and τ_s within the physiological range, the folded equilibrium is a folded saddle. Figure 12.4 shows a representative phase diagram for specific values of τ_s and τ_w. The stable eigenspace of the folded saddle forms a separatrix separating "spiking" and "non-spiking" solutions in the singular limit. The spiking solutions are the solutions that can reach the fold curve and thus are the solutions to the right of the stable eigenspace of the folded saddle in Fig. 12.4. The non-spiking solutions

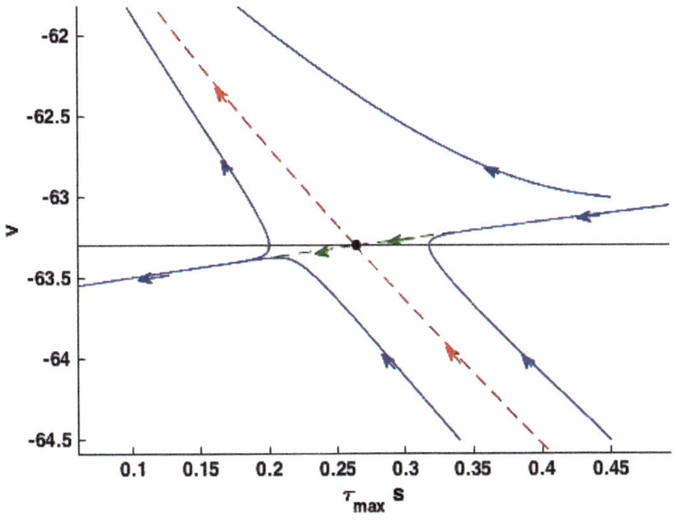

Fig. 12.4 The (v,s)-phase plane for $\tau_s = 100$ ms and $\tau_w = 117$ ms. The fold curve is represented by the black horizontal line. The folded saddle equilibrium is denoted with a black dot. The stable and unstable eigenspaces of the folded saddle are the red-dashed and green-dashed lines, respectively. Solution trajectories for initial conditions starting off the stable and unstable eigenspaces are shown in blue

fall to the left of the stable eigenspace of the folded saddle and are prevented from reaching the fold curve as the unstable eigenspace of the folded saddle draws these solutions away from the fold curve.

Our analysis of the slow system now allows us to understand the transition from spiking to non-spiking solutions as τ_s increases. On the critical manifold, initial conditions to the right of the stable manifold of the folded saddle are "spiking" solutions and those that are on the left are "non-spiking" solutions. Since the variable s moves slowly compared to v, we can approximate the initial condition of s on the critical manifold as its initial condition in the full system. In particular, $s = 0.727$ at the start of inhibition. Furthermore, for a fixed value of s, there exists one equilibrium point of the fast subsystem and it is to this point that v approaches on the critical manifold (McCarthy et al. 2012). Numerically, this equilibrium point takes the value -67.71 for the variable v. Thus, we approximate the initial condition on the critical manifold as $(v, s) = (-67.71, 0.727)$. By numerically finding the value of s on the critical manifold where the stable manifold of the folded saddle intersects with $v = -67.71$, we can determine if the initial condition for s (i.e., $s = 0.727$) is on the spiking or non-spiking side of the stable manifold of the folded saddle. In the particular case when $\tau_w = 117$ ms, we find that when τ_s is greater than approximately 100 ms we have non-spiking solutions and when τ_s is less than 100 ms, we have spiking solutions. This confirms what we observed in simulations (as τ_s is increased the dynamics of the neuron transition from spiking to non-spiking in response to inhibition (Fig. 12.1)) and also gives us geometric insight into why this transition occurs.

One benefit of characterizing the system using GSPT is that it allows us to visualize how rebound spiking depends on the specific parameters in our system. For example, as $\bar{g}_m$ increases the value of the folded saddle equilibrium (s_c) increases (Eq. (12.16)), allowing for a potentially larger region of the critical manifold that contains non-spiking solutions (note that the region of the critical manifold that allows for non-spiking solutions also depends on the position of the stable manifold of the folded saddle). This supports our intuition, since $\bar{g}_m$ represents the maximal conductance of the M-current, which is a non-inactivating potassium current that stabilizes the neuronal voltage against depolarization. Thus, increasing the maximal conductance of this current makes spiking less likely.

In contrast, increasing $\bar{g}_i$, decreases the value of s_c and potentially allows for more spiking solutions in the singular limit. This is an unintuitive result since we expect increasing inhibition in our system will result in less spiking. However, simulations confirm this prediction as we find solutions to the full system of equations (Eq. (12.9) with s dependent on the pre-synaptic voltage of a spiking GABAergic neuron) with $\tau_s = 10$ ms do not rebound spike after inhibition when $\bar{g}_i$ is low but will rebound spike if $\bar{g}_i$ is higher (Fig. 12.5). Thus, GSPT can not only lead to non-intuitive insight into the range of dynamics of spiking neuron solutions but also allow us to determine the underlying mathematical structure that forms the basis of these dynamics.

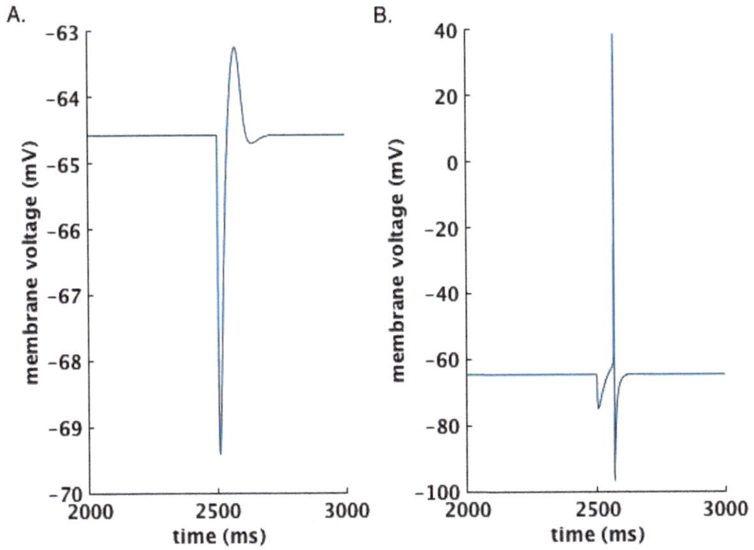

Fig. 12.5 Computer simulations of a neuron with M-current receiving inhibition. Simulations with (**a**) a low GABAa maximal conductance ($g_i = 0.165$ mS/cm^2) and (**b**) a maximal conductance that is 4 times higher. Both simulations use $\tau_i = 10$ ms

12.3.3 Anteriorization

As mentioned earlier, during the administration of propofol, there is a stage called anteriorization during which alpha power undergoes a spatial shift from the back to the front of the brain. Behaviorally defined LOC occurs concomitant with anteriorization. Furthermore, if propofol administration is stopped and the patient allowed to recover, as the frontal alpha power dissipates the patient regains consciousness (e.g., begins to respond again within the context of a behavioral task). Therefore, gaining a mechanistic and circuits-level understanding of anteriorization may provide insight into what leads to loss of consciousness and allow the development of better anesthetics. In order to gain a mechanistic understanding of anteriorization, Vijayan et al. (2013) constructed a computational model, combining a frontally projecting thalamocortical model (Fig. 12.6) that is capable of generating frontal alpha (Ching et al. 2010) with a thalamic model capable of generating occipital alpha (Vijayan and Kopell 2012). In their model, the thalamic part of each component consists of reticular nucleus (RE) cells and thalamocortical (TC) cells (see Destexhe et al. (1996) for details of the currents incorporated into each cell type). The TC cells excite the RE cells via AMPA and the RE cells in turn inhibit the TC cells via GABA. The thalamic part of the occipital component also contains a specialized subset of TC cells called high-threshold thalamocortical (HTC) cells.

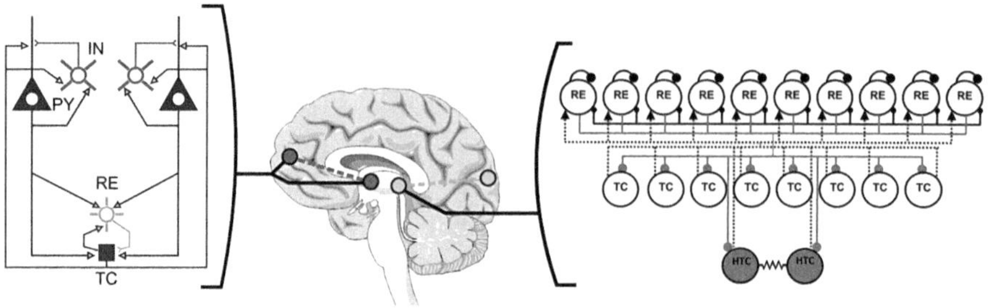

Fig. 12.6 Cartoon of network architecture. The frontal component consists of a thalamic component (TC and RE cells) and a cortical component (PY and IN cells), while the occipital component consists of only a thalamic component (TC cells, RE cells, and HTC cells, the putative generators of occipital alpha). Modified from Vijayan et al. (2013)

HTC cells have the same connectivity as TC cells, but are gap junction-coupled and burst at the alpha frequency at relatively depolarized membrane potentials (Hughes et al. 2004; Lorincz et al. 2009; Vijayan and Kopell 2012). HTC cells are thought to be the generators of occipital alpha in vivo (Lorincz et al. 2009). A variant of I_T channels, a calcium current, I_{THT}, which is active at more depolarized membrane potentials, is thought to mediate the alpha frequency bursting in HTC cells (Hughes et al. 2004; Lorincz et al. 2009; Vijayan and Kopell 2012); I_{THT} currents are incorporated into HTC cells in the model of Vijayan and Kopell (2012). The TC cells in the frontal component of the model have excitatory projections to cortical pyramidal cells (PY) and interneurons (IN), and the PY neurons in turn send excitatory projections (via AMPA) back to the TC cells and RE cells. The PY cells also send excitatory projections to the IN, which inhibits the PY cells via GABA (see Ching et al. (2010), Vijayan and Kopell (2012) for additional details about the cortical component). When the model parameters are set to quiet awake conditions, the cells in the frontal component fire irregularly and the frontal LFP is relatively flat (Fig. 12.7a), while the HTC cells in the occipital component burst at the alpha frequency and alpha power is seen in the occipital LFP (Fig. 12.7b). The anesthetic propofol is known to increase $GABA_a$ conductances and reduce I_h, a hyperpolarization activated current. To mimic the effects of propofol, Vijayan et al. reduced I_h conductances and increased $GABA_a$ conductances in their model (Vijayan et al. 2013). In the occipital component, these changes resulted in the silencing of the HTC cells, causing a relatively flat occipital LFP (Fig. 12.7d). The HTC cells are silenced because the reduction in I_h conductances causes them to become more hyperpolarized (I_h has a relatively high reversal potential), moving the I_{THT} currents out of their operating range and thus preventing HTC cells from bursting at the alpha frequency (Fig. 12.7d). In contrast, in the frontal component, the actions of propofol result in all frontal cell types bursting at the alpha frequency, producing alpha power in the LFP (Fig. 12.7c).

The frontal alpha emerges due to several actions of propofol. In the cortical portion of the frontal component, increased $GABA_a$ conductances result in an

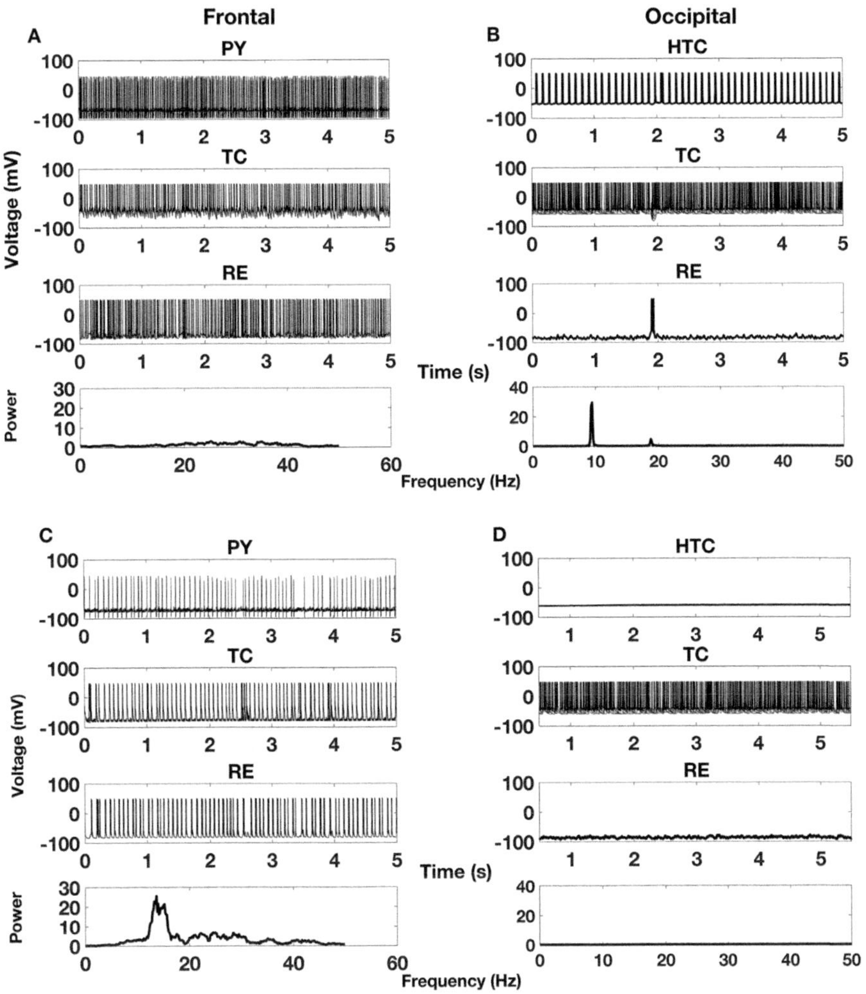

Fig. 12.7 Simulation traces during anteriorization. (**a**) Activity of the frontal component during quiet awake activity. Spiking activity of PY cells (first row), TC cells (second row), and RE cells (third row), and power spectrum of LFP (bottom row). (**b**) Activity of the occipital component during quiet awake activity. Spiking activity of HTC cells (first row), TC cells (second row), and RE cells (third row), and LFP power spectrum (bottom row). (**c**) Frontal component after administration of propofol. Rows same as in (**a**). (**d**) Occipital component after administration of propofol. Rows same as in (**b**). Modified from Vijayan et al. (2013)

increase in the inhibition of PY cells by the IN, reducing the firing rate of PY cells into the alpha range (Fig. 12.8a-b). In the thalamic portion of the frontal component, increased $GABA_a$ conductances cause the RE cells to increase their inhibition of TC cells, causing the TC cells to become hyperpolarized. This hyperpolarization allows the currents I_T and I_h, which operate at alpha time constants, to become engaged.

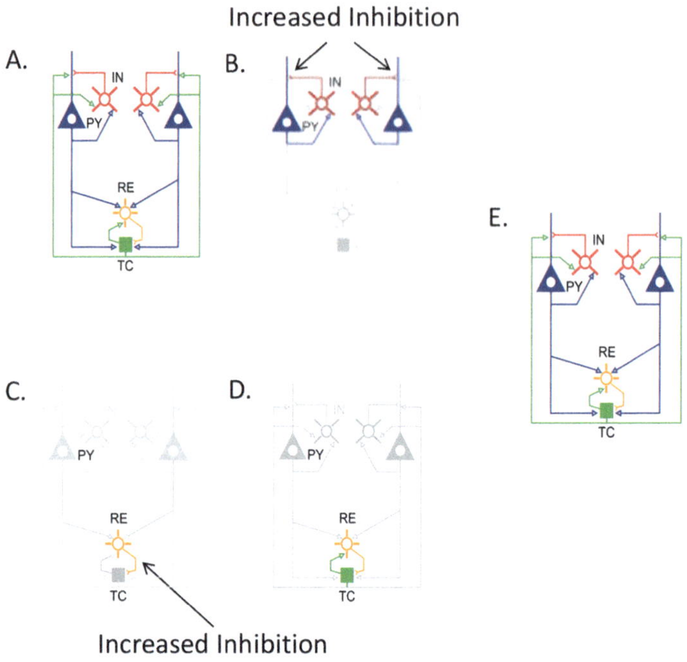

Fig. 12.8 Cartoon of mechanism responsible for the emergence of frontal alpha during the administration of propofol. (**a**, **b**) Increased inhibition from IN to PY cells, due to the increase in GABA conductances, brings frontal spiking activity into the alpha range. (**c**, **d**). Increased inhibition from RE cells to TC cells, due to the increase in GABA conductances, results in hyperpolarization of TC cells. Hyperpolarization-activated currents in TC cells cause them to rebound spike and excite RE cells, which in turn inhibit TC cells, starting the cycle again. This ping-pong-like interaction between TC and RE cells happens at an alpha time scale because the currents engaged in TC cells at relatively hyperpolarized membrane potentials have an alpha time scale. (**e**) The alpha activity in the thalamic and cortical components reinforces each other, resulting in hypersynchronous frontal alpha activity. Modified from Ching and Brown (2014)

The activation of these currents causes TC cells to rebound spike and excite the RE cells, which in turn inhibit TC cells again (Fig. 12.8c-d). The back-and-forth activity between TC cells and RE cells occurs at the alpha frequency due to the alpha time scales of the intrinsic currents that are involved. Furthermore, the cortical and thalamic alpha activity reinforce each other, resulting in a hypersynchronous alpha activity (Fig. 12.8e). The underlying mechanism of the emergent frontal alpha activity provides us with clues and predictions as to why anteriorization may lead to loss of consciousness. For example, the hypersynchronous alpha activity, which in our model is both thalamic and cortical in nature, may prevent external stimuli from being processed properly or routed from the thalamus to the cortex. Thus external stimuli may not be perceived, resulting in LOC that coincides with anteriorization.

12.4 Discussion and Conclusion

Models based on known biophysical properties of neural systems allow us to investigate neural dynamics and to gain insight into underlying circuit-level mechanisms in a fashion that is often difficult to do in an experimental setting. The neural models employed in such investigations vary in their level of detail. There is a trade-off between the level of biophysical detail used in a particular model type and the computational power needed to simulate networks of the same size. Somewhat related is that specific neural model types may be more conducive to looking at specific types of phenomena. For example, mean field models, which typically lack the biophysical detail of Hodgkin-Huxley models, may be particularly well suited to investigating large scale phenomena, where such detail may not be critical to capturing and explaining the observed neural dynamics.

In this chapter we describe how modeling has been employed to investigate the neural dynamics observed during the administration of the general anesthetic propofol and to decipher how concomitant behavioral changes may relate to the observed neural dynamics. In particular, we discuss our use of single compartment Hodgkin-Huxley models, both large and reduced models, and the tools of dynamical systems theory, such as geometric singular perturbation theory, to elucidate the neural mechanisms underlying the dynamics observed during propofol-induced anesthesia. These investigations provided not only possible explanations for the concomitant behavioral phenomena, but also experimentally testable hypotheses about how propofol-induced neural dynamics emerge. For example, the modeling work suggests that the frontal alpha that emerges during anteriorization is the result of a hypersynchronous loop in which the thalamus and the frontal cortex reinforce each other's firing in the alpha band. The reason behaviorally defined loss of consciousness coincides with anteriorization may be because this hypersynchronous thalamocortical alpha activity may prevent or disrupt external stimuli from being routed via the thalamus to the cortex. These modeling results point to an experimentally testable prediction, that during the emergence of frontal alpha there is highly coherent alpha activity between the frontal cortex and frontally projecting thalamic nuclei.

Much still remains to be understood about how general anesthetics act, that is, how circuit-level mechanisms bring about the neural activity that arises during their administration. In particular, different anesthetics, when administered, produce diverse sets of changes in the spatio-temporal neural dynamics and the circuit-level mechanisms responsible for these neural activity patterns remains to a large extent a mystery. We hope we have imparted how mathematical modeling can serve as an invaluable tool in unraveling these remaining mysteries of general anesthesia and more generally in deciphering the circuit-level mechanisms of brain dynamics.

Appendix

For the model of propofol-induced paradoxical excitation, all cells had the spiking currents (I_{Na}, I_K, I_L) formulated using Hodgkin-Huxley-type conductances, which were used in previous modeling work (Olufsen et al. 2003). The gating functions for the sodium current activation (m) and inactivation (h) variables have rate functions defined by

$$\alpha_m = \frac{0.32(V+54)}{1-\exp[-(V+54)/4]} \quad (12.17)$$

$$\beta_m = \frac{0.28(V+27)}{\exp[(V+27)/5]-1} \quad (12.18)$$

$$\alpha_h = 0.128\exp[-(V+50)/18] \quad (12.19)$$

$$\beta_h = \frac{4}{1+\exp[-(V+27)/5]} \quad (12.20)$$

The fast potassium channel had four activation gates described by the rate functions

$$\alpha_m = \frac{0.032(V+52)}{1-\exp[-(V+52)/5]} \quad (12.21)$$

$$\beta_m = 0.5\exp[-(V+57)/40] \quad (12.22)$$

The leak current (I_L) had no gating variables. The M-current as formulated in Mainen and Sejnowski (1996) has one activation gate and no inactivation gates. The gating function for activation gate is described by the rate functions

$$\alpha_m = \frac{Q_s 10^{-4}(V+30)}{1-\exp[-(V+30)/9]} \quad (12.23)$$

$$\beta_m = -\frac{Q_s 10^{-4}(V+30)}{1-\exp[(V+30)/9]} \quad (12.24)$$

$$Q_s = Q_{10}^{(37\,^\circ\mathrm{C}-23\,^\circ\mathrm{C})/10} = 3.209 \quad (12.25)$$

where $Q_{10} = 2.3$.

The two synaptic currents (I_{AMPA} and $I_{\mathrm{GABA}a}$) have an activation gate dependent on the pre-synaptic voltage. The AMPA current is

$$I_{\mathrm{AMPA}} = \bar{g}_e s_e (V - E_e) \quad (12.26)$$

where s_e is the sum of the synaptic activation variables from all pre-synaptic pyramidal cells in the network. Thus, for the j-th interneuron in the network,

$$s_e = \frac{1}{N}\sum_{k=1}^{N} S_{e_k i_j} \tag{12.27}$$

where N denotes the number of pyramidal cells pre-synaptic to the j-th interneuron. The kinetics of the AMPA activation variable from pyramidal cell k to interneuron j are described by:

$$\frac{dS_{e_k i_j}}{dt} = g_{ee}(V_k)(1 - S_{e_k i_j}) - \frac{S_{e_k i_j}}{\tau_e} \tag{12.28}$$

The decay time constant for the AMPA current is $\tau_e = 2\,\text{ms}$ and the rate for the open state of the AMPA current, $g_{ee}(V_k)$ is dependent on the membrane voltage of the k-th e-cell:

$$g_{ee}(V_k) = 5\left(1 + \tanh\left(\frac{V_k}{4}\right)\right) \tag{12.29}$$

The GABA$_a$ current is formulated as:

$$I_{\text{GABAa}} = \bar{g}_i s_i (V - E_i) \tag{12.30}$$

where s_i is the gating variable for inhibitory GABA$_a$ current:

$$s_i = \frac{1}{N}\sum_{k=1}^{N} S_{i_k i_j} \quad \text{for interneuron to interneuron connections} \tag{12.31}$$

$$s_i = \frac{1}{N}\sum_{k=1}^{N} S_{i_k e_m} \quad \text{for interneuron to pyramidal cell connections} \tag{12.32}$$

where N represents the number of pre-synaptic interneurons.

Furthermore, the GABA$_a$ synaptic kinetics of the gating variable from the k-th interneuron to the j-th interneuron ($S_{i_k i_j}$) follows:

$$\frac{dS_{i_k i_j}}{dt} = g_{ii}(V_k)(1 - S_{i_k i_j}) - \frac{S_{i_k i_j}}{\tau_i} \tag{12.33}$$

When propofol is not present, $\tau_i = 5\,\text{ms}$; for low-dose propofol $\tau_i = 10$ and the rate function for the open state of the GABA$_a$ receptor is:

$$g_{ix}(V_k) = 2 + 2\tanh\left(\frac{V_k}{4}\right), \quad \text{for } x = i \text{ or } e \tag{12.34}$$

More details about this model can be found in McCarthy et al. (2008).

References

Akeju, O., Westover, M. B., Pavone, K. J., Sampson, A. L., Hartnack, K. E., Brown, E. N., et al. (2014). Effects of sevoflurane and propofol on frontal electroencephalogram power and coherence. *Anesthesiology, 121*(5), 990–998.

Akeju, O., Song, A. H., Hamilos, A. E., Pavone, K. J., Flores, F. J., Brown, E. N., et al. (2016). Electroencephalogram signatures of ketamine anesthesia-induced unconsciousness. *Clinical Neurophysiology, 127*(6), 2414–2422.

Alvina, K., Ellis-Davies, G., & Khodakhah, K. (2010). T-type calcium channels mediate rebound firing in intact deep cerebellar neurons. *Neuroscience, 158*(2), 635–641.

Ascoli, G. A., Gasparini, S., Medinilla, V., & Migliore, M. (2010). Local control of postinhibitory rebound spiking in CA1 pyramidal neuron dendrites. *Journal of Neuroscience, 30*(18), 6434–6442.

Börgers, C., (Ed.) (2017). *An introduction to modeling neuronal dynamics*. Cham: Springer.

Cannon, J., McCarthy, M. M., Lee, S., Lee, J., Borgers, C., Whittington, M. A., et al. (2014). Neurosystems: Brain rhythms and cognitive processing. *European Journal of Neuroscience, 39*(5), 705–719.

Cardin, J. A., Carlén, M., Meletis, K., Knoblich, U., Zhang, F., Deisseroth, K., et al. (2009). Driving fast-spiking cells induces gamma rhythm and controls sensory responses. *Nature, 459*, 663–667.

Ching, S., & Brown, E. N. (2014). Modeling the dynamical effects of anesthesia on brain circuits. *Current Opinion in Neurobiology, 25*, 116–122.

Ching, S., Cimenser, A., Purdon, P. L., Brown, E. N., & Kopell, N. J. (2010). Thalamocortical model for a propofol-induced alpha-rhythm associated with loss of consciousness. *Proceedings of National Academy of Sciences USA, 107*, 22665–22670.

Destexhe, A., Bal, T., McCormick, D. A., & Sejnowski, T. J. (1996). Ionic mechanisms underlying synchronized oscillations and propagating waves in a model of ferret thalamic slices. *Journal of Neurophysiology, 76*(3), 2049–2070.

Gugino, L. D., Chabot, R. J., Prichep, L. S., John, E. R., Formanek, V., & Aglio, L. S. (2001). Quantitative EEG changes associated with loss and return of consciousness in healthy adult volunteers anaesthetized with propofol or sevoflurane. *British Journal of Anaesthesia, 87*(3), 421–428.

Hodgkin, A. L., & Huxley, A. F. (1952). A quantitative description of membrane current and its application to conduction and excitation in nerve. *Journal of Physiology, 117*(4), 500–544.

Hughes, S. W., Lorincz, M. L., Cope, D. W., Blethyn, K. L., Kékesi, K. A., Parri, H. R., et al. (2004). Synchronized oscillations at alpha and theta frequencies in the lateral geniculate nucleus. *Neuron, 42*(2), 253–268.

Izhikevich, E. M., ed. (2007). *Dynamical systems in neuroscience: The geometry of excitability and bursting*. Cambridge: MIT Press.

Kitamura, A., Marszalec, W., Yeh, J. Z., & Narahashi, T. (2003). Effects of halothane and propofol on excitatory and inhibitory synaptic transmission in rat cortical neurons. *Journal of Pharmacology Experimental Theory, 304*(1), 162–171.

Koch, C., & Segev, I., (Eds.) (1998). *Methods in neuronal modeling: From Ions to networks*. Cambridge: MIT Press.

Lorincz, M. L., Kékesi, K. A., Juhász, G., Crunelli, V., & Hughes, S. W. (2009). Temporal framing of thalamic relay-mode firing by phasic inhibition during the alpha rhythm. *Neuron, 63*(5), 683–696.

Mainen, Z. F., & Sejnowski, T. J. (1996). Influence of dendritic structure on firing pattern in model neocortical neurons. *Nature, 382*, 363–366.

McCarthy, M. M., & Kopell, N. (2012). The effect of propofol anesthesia on rebound spiking. *SIAM Journal on Applied Dynamical Systems, 11*(4), 1674–1697.

McCarthy, M. M., Brown, E. N., & Kopell, N. (2008). Potential network mechanisms mediating electroencephalographic beta rhythm changes during propofol-induced paradoxical excitation. *Journal of Neuroscience, 28*, 13488–13504.

McCarthy, M. M., Moore-Kochlacs, C., Gu, X., Boyden, E. S., Han, X., et al. (2011). Striatal origin of the pathologic beta oscillations in Parkinson's disease. *Proceedings of National Academy of Sciences USA, 108*, 11620–11625.

Nunez, P., & Srinivasan, R. (2006). *Electric fields of the brain, the neurophysics of EEG* (2nd ed.). Oxford: Oxford University Press.

Olufsen, M. S., Whittington, M. A., Camperi, M., & Kopell, N. (2003). New roles for the gamma rhythm: Population tuning and preprocessing for the beta rhythm. *Journal of Computational Neuroscience, 14*, 33–54.

Segel, L. A., & Edelstein-Keshet, L. (2013). *A primer on mathematical models in biology*. Philadelphia, PA: Society for Industrial and Applied Mathematics.

Szmolyan, P., & Wechselberger, M. (2001). Canards in R3. *Journal of Differential Equations, 177*(2), 419–453.

Traub, R. D., Contreras, D., Cunningham, M. O., Murray, H., LeBeau, F. E., Roopun, A., et al. (2005). Single-column thalamocortical network model exhibiting gamma oscillations, sleep spindles, and epileptogenic bursts. *Journal of Neurophysiology, 93*, 2194–2232.

Vijayan, S., & Kopell, N. J. (2012). Thalamic model of awake alpha oscillations and implications for stimulus processing. *Proceedings of National Academy of Sciences USA, 109*, 18553–18558.

Vijayan, S., Ching, S., Purdon, P. L., Brown, E. N., & Kopell, N. J. (2013). Thalamocortical mechanisms for the anteriorization of alpha rhythms during propofol-induced unconsciousness. *Journal of Neuroscience, 33*, 11070–11075.

Wilson, H. R., & Cowan, J. D. (1972). Excitatory and inhibitory interactions in localized populations. *Biophysical Journal, 12*(1), 1–24.

Retraction Note to: Artifact Rejection for Concurrent TMS-EEG Data

Wei Wu, Corey Keller, and Amit Etkin

Retraction Note to: Chapter 6 "Artifact Rejection for Concurrent TMS-EEG Data" in: Z. Chen, S.V. Sarma (eds.), *Dynamic Neuroscience*, https://doi.org/10.1007/978-3-319-71976-4_6

Chapter 6 (Artifact Rejection for Concurrent TMS-EEG Data) by Wei Wu, Corey Keller, and Amit Etkin, that was originally published in the volume has been removed by the request of the authors, as this work will be appearing in another publication.

The retracted online version of this chapter can be found at
https://doi.org/10.1007/978-3-319-71976-4_6

Z. Chen, S.V. Sarma (eds.), *Dynamic Neuroscience*,
https://doi.org/10.1007/978-3-319-71976-4_13

Epilogue

I do not know where to begin. The news of this monograph, written as part of my recent 60th birthday celebration, was a complete surprise. I am immensely grateful to Sridevi V. Sarma and Zhe (Sage) Chen for their brilliant efforts organizing and editing the final work. Sri and Sage are truly a pair extraordinaire: great scientists with boundless energy and the ultimate can-do attitudes.

I want to thank Rob Kass and Ralph Lydic for the kind sentiments expressed about me and about my work in their forewords. I was honored when nearly 25 years ago, Rob like me was bitten by the neuroscience bug and suggested that we begin regular conversations about the challenging statistical problems in neural data analysis. It was great to know that a statistician of Rob's stature thought these problems were important/interesting and that moreover, he was willing to think hard about how to solve them. I have known Ralph since the 1980s and my days as an M.D. Ph.D. student at Harvard. At that time, Ralph was embarking on his path-breaking work with Helen Baghdoyan characterizing the neural circuitry of sleep regulation. Later, he turned his research to understanding the links between general anesthesia and sleep. Therefore, Ralph was the ideal partner, along with our neurologist colleague Nicholas Schiff, to work out in 2010, the relationships between general anesthesia, sleep and coma (Brown, Lydic & Schiff, New England Journal of Medicine, 2010).

I am especially grateful to each of the authors most of whom I had the privilege of working with either as one of my graduate students, one of my postdocs or both. They are: Behtash Babadi, Zhe Chen, ShiNung Ching, Uri T. Eden, Amit Etkin, Rose T. Faghih, Loren M. Frank, Corey Keller, Michelle McCarthy, Michael J. Prerau, Pierre Sacré, Srideva V. Sarma, Maryam M. Shanechi, Hideaki Shimazaki, Long Tao, Sujith Vijayan, and Wei Wu. It was kind and generous of them to contribute to this monograph work highlighting some of their newest ideas on neural signal processing, neural control and anesthesiology. I read each chapter with great interest, knowing that here, as in the past, I would learn a lot from the careful, insightful thinking of my trainees.

Z. Chen, S.V. Sarma (eds.), *Dynamic Neuroscience*,
https://doi.org/10.1007/978-3-319-71976-4

I admit frankly that I am biased and that I believe in two things. Neuroscience has the coolest signal processing and data analysis questions around. Anesthesiology is the coolest clinical neuroscience discipline and the coolest medical specialty around. Working with my colleagues, I have learned to savor the good fortunes of conducting research guided by this biased perspective. Once again, I thank you all!

Index

Z. Chen, S.V. Sarma (eds.), *Dynamic Neuroscience*,
https://doi.org/10.1007/978-3-319-71976-4

Q

R

S

T

GPSR Compliance

The European Union's (EU) General Product Safety Regulation (GPSR) is a set of rules that requires consumer products to be safe and our obligations to ensure this.

If you have any concerns about our products, you can contact us on ProductSafety@springernature.com

In case Publisher is established outside the EU, the EU authorized representative is:

Springer Nature Customer Service Center GmbH
Europaplatz 3
69115 Heidelberg, Germany

Batch number: 10739583

Printed by Printforce, the Netherlands